DOUGLAS MACARTHUR

Douglas MacArthur
The Far Eastern General

Michael Schaller

New York   Oxford
OXFORD UNIVERSITY PRESS

Oxford University Press

Oxford   New York   Toronto
Delhi   Bombay   Calcutta   Madras   Karachi
Petalng Jaya   Singapore   Hong Kong   Tokyo
Nairobi   Dar es Salaam   Cape Town
Melbourne   Auckland

and associated companies in
Berlin   Ibadan

For Sue, Nicholas,
and Gabriel

Preface

Born into an army family and shaped by the values of that institution, Douglas MacArthur searched his entire life for the personal and professional recognition that the military never seemed able to provide. Unsatisfied with any army command, he yearned for the power to shape, not merely implement, national policy. The higher he rose in rank, the more frustrated he became with the constraints imposed by the traditions of civilian supremacy and the complexities of modern war and diplomacy. Each promotion left him under the control of some higher military or civilian authority whom he resented and against whom he felt compelled to rebel. Even when he became Army Chief of Staff, the Great Depression and President Franklin D. Roosevelt's New Deal mocked the Republican orthodoxy and military tradition he championed.

In an army that eschewed flamboyance, MacArthur possessed a unique flair. With his chiseled face and a voice that, contemporaries recalled, "could trumpet and drum," he looked the part of a "heaven born general." Although hs admirers celebrated MacArthur's intelligence, he had little intellectual depth. An exceptionally good memory, a repertoire of quotes and anecdotes, a love of convoluted or even archaic language and a wealth of historical trivia formed the core of the general's wisdom. The geopolitical sweep and absolute certainty of his assertions left audiences gasping. Even if he said little original or profound, the vivid style, cadence, and metaphor convinced many they had heard an oracle.

Although he never said so, MacArthur found only two effective outlets for his repressed ambitions. Service abroad, especially in Asia, provided release from the constraints and supervision of obscure stateside posts. Also, even if he denied it, by the 1930s, he dreamed of a political career leading to the White House. As president, he would control the levers of power that had impeded the fortunes of his father and himself. Here again, service in the Orient provided the exceptional freedom to prove his abilities. Commands in the Philippines, the Southwest Pacific, Japan, and Korea served as MacArthur's platforms in a long campaign for influence in the United States.

In Asia, no other great American generals shared the limelight, allies were too weak to complicate planning, and Washington provided relatively little direct oversight. Far away from the War Department and Joint Chiefs of Staff, he surrounded himself with sycophants whom George C. Marshall described as a court befitting an oriental satrap. MacArthur often complained that American officials ignored or slighted Asia. But he thrived in an environment where few policy makers knew enough to second-guess his decisions.

A ceaseless promoter of the idea that he possessed a unique knowledge of "oriental psychology," the general functioned as a sort of American proconsul for almost two decades. Through his service in Asian posts, he personified the American experience in twentieth century Asia to much of the public. In fact, he understood very little about the great political upheavals sweeping East Asia during his lifetime. Still, as an early biography observed, "he was an American prophet honored at home largely because of the lavish ways in which he was honored abroad."

Along with the army, Douglas MacArthur's family shaped his career. He lived his entire life tied to his parents by unusually tight psychological bonds. As one of the army's senior officers, his father, Arthur MacArthur, Jr., guided Douglas's early career. His mother, Pinkie, surpassed even her husband's efforts on behalf of her son. Only death removed her as a direct influence on the general's career.

Douglas seemed driven to replay ritually many of his father's notable accomplishments and conflicts, on and off the battlefield. Like Arthur, Douglas sought recognition for conspicuous, if premeditated, acts of bravery. Arthur's removal from his command during the Philippine–American Revolution and subsequent failure to be named army chief of staff convinced Douglas that civilians had no business telling generals how to make war. The youth blamed political "enemies" in Washington for sabotaging his father's career and later credited such enemies for undermining his own. Like Arthur, Douglas believed that his personal destiny and America's was linked to the control of Asia and the Pacific.

In a penetrating psychological study of MacArthur, historian Carol Petillo has observed the powerful, unresolved psychological drives that buffeted him. Throughout his life, Douglas tried to win parental approval for accomplishments that mirrored or surpassed those of his father. However, his inability to separate himself fully from his parents called his own achievements constantly into question.

Even the physical restiveness of the general's life, Petillo shows, reflected this inner turmoil. Raised on a series of dusty army posts, he spent fifty years moving from one temporary billet to another. During most of the 1930s and 1940s he lived in hotels and guesthouses. In retirement, he spent the last thirteen years of his life in yet another hotel, waiting, in vain, for a call to higher duty. The Philippines were as close to a home as anything the general knew. In America, MacArthur focused his ambition on the one home whose attainment might finally secure his parents' symbolic approval, the White House.

Douglas MacArthur's unique career provides a lens both for focusing on the man and examining the tensions in American foreign policy in his time. Although trained as a soldier, the commands he held in the Philippines, the Pacific, Japan, and Korea all involved great political and diplomatic responsibilities. Asia served as the general's springboard to success as well as the site of his ultimate failure; his passage through the region reflected a personal destiny as well as a national policy.

MacArthur has been the subject of dozens of popular biographies, but of only three or four serious studies. Scholarly and popular accounts of MacArthur's service to 1950 are remarkably positive. In their memoirs, even such visceral rivals as Harry S. Truman and Dean Acheson praise the general's pre-Korean war achievements. Many biographies contrast the egotistical MacArthur of 1950–51 with the military "genius" and "statesman" who defeated and remade Japan. I dispute this interpretation.

D. Clayton James has written the most compelling and authoritative work on the general. His three-volume *The Years of MacArthur* traces the general from birth to death, analyzing his personal life and military career. Comprehensive and elegant, it is the standard reference for all subsequent biographers.

Carol Petillo's psychologically informed study, *Douglas MacArthur: The Philippine Years,* focuses on the interplay between the general's personality and his years of service in the Philippines. Along the way, she provides fascinating detail about MacArthur's family, military service, and Philippine society and politics. It stands as a model for research blending psychological analysis with political history.

The enormously popular *American Caesar,* by William Manchester, is both a marvelous read and a disappointment to scholars. Drawing much of its information from James's work and color from the popular press, Manchester has written a "heroic" biography, which seldom probes beneath the surface of the general's self-proclaimed accomplishments. Aside from some mild criticism of MacArthur's ego, the tone is worshipful. Treatment of the years after 1945, surely the most important part of MacArthur's public life, is thin.

Richard Rovere and Arthur Schlesinger, Jr., produced a critical portrait of MacArthur amid the drama of his recall from Korea. Although not based on archival research, *The General and the President* is rich with insights relating to the general's life before and during the Korean War. Its style and nuance also evokes for a later audience the electric atmosphere of 1951.

This study builds on the strengths of earlier work by James, Petillo, Rovere, and Schlesinger. Unlike them, I have focused almost entirely on evaluating MacArthur's post-1935 activities in Asia as a means for understanding both the man and American foreign policy. The general's policies in the Philippines, during World War II, in occupied Japan and Korea are assayed in the context of Asian nationalism, the cold war, and American politics.

The drama and nostalgia still evoked by MacArthur's name obscure many of the ironies of his life. Six months before the Pearl Harbor attack he lived as an expatriate in Manila, a famous but now-retired officer. Recalled to service in hope of rallying the defense of the Philippines, his forces collapsed only hours after the war began. With the help of sympathetic journalists, he parlayed defeat into a major role in the Pacific war. His hatred for Franklin Roosevelt prompted MacArthur to encourage a presidential draft in 1944, although it fizzled in a minor scandal. The general's genius for publicity convinced many Americans that his command accounted almost exclusively for Japan's defeat, despite sabotage by enemies in Washington.

Although celebrated as the founder of postwar democratic Japan, his selection as occupation commander resulted from a series of bureaucratic compromises and a desire by civilian and military leaders to keep him out of Europe and the United States. State and War Department officials had drafted most of his celebrated reform program months before he reached Tokyo—and spent much of the next three years trying to force him to carry it out. He undertook serious economic reform (in 1947–1948) only after Washington abandoned the goal. His motivation stemmed as much from a desire to win political support in America than from any belief in social justice.

After MacArthur's second bid for the White House collapsed, the Truman administration gradually reduced the general's authority in Tokyo and seldom followed his advice. Yet the outbreak of war in Korea in June 1950 thrust him back into the limelight. He believed a military victory there would turn the tide of the cold war and thereby resurrect his political crusade because of his position as the senior commander in Asia. When China's intervention in Korea dashed the general's plans, he lashed out at his superiors' refusal to expand the war against the People's Republic. As in his previous disputes with civilian and military authorities, MacArthur sought to rally the Republican Party and patriotic fervor against Democratic rivals. Although intimidation had worked on several previous occasions, in 1951 the general's luck finally ran out.

Driven by measures of talent, opportunism, vision, egotism, and jealousy, MacArthur's life exemplifies an era in which the United States government as well as individual Americans tried to shape the destiny of Asia.

I am especially grateful to Professors D. Clayton James and Carol Petillo for scrutinizing an earlier draft of this study. Their probing questions and skepticism compelled me to reexamine many of my assertions. Professors Leonard Dinnerstein and Robert Schulzinger read the manuscript with equal care and contributed immeasurably to its improvement. Jonathan Pollack and Professors Bruce Cumings and Allen Whiting read and suggested many improvements to the chapters covering the Korean War. The conclusions offered are, of course, my own.

Many groups provided research support for this study. I wish to thank the University of Arizona Social and Behavioral Sciences Research Institute and its Small Grants Program, the Franklin D. Roosevelt Foundation, the Harry S. Truman Library Institute, the Herbert Hoover Library Foundation, the Lyndon Baines Johnson Foundation, and the National Endowment for the Humanities Summer Stipend and Travel to Research Collections programs. Archivists at the Hoover, Roosevelt, Truman, and Johnson presidential libraries and at the Library of Congress and National Archives were especially helpful in locating material. My editor at Oxford, Nancy Lane, provided encouragement during all stages of this project.

Tucson   M. S.
August 1988

Contents

DOUGLAS MACARTHUR

1
The Making of a General

The chosen names of men in the MacArthur family reveals the intense concern it cultivated with paternity and tradition: Douglas MacArthur's grandfather (Arthur), father (Arthur, Jr.), eldest brother (Arthur III), and son (Arthur IV) shared the same name; his nephew (Douglas II) bore his own. Arthur, Jr., came of age during the Civil War, in which he fought as an officer in the Wisconsin volunteer infantry. The son of a successful lawyer-politician in Milwaukee with strong ties to the new Republican Party, Arthur Jr., achieved fame by leading his unit in a bloody charge against Confederate positions in the battle of Missionary Ridge in November 1863. Seriously wounded later in the war, the young volunteer left military service to study law under his father.

Soon bored with his apprenticeship, Arthur, Jr., enlisted in the Regular Army as a second lieutenant, quickly gaining promotion to the rank of captain. Despite native ability and lobbying by his well-connected father, the young captain stayed at this rank for the next twenty-three years. After the Civil War, the army had reverted to its traditional marginal place in American society. Aside from Reconstruction duties in the South, it acted as a mobile constabulary policing the displaced Indian tribes in the western territories. The army had few ties with urban, industrial America or with the major social changes sweeping the country in the postwar decades. Salaries remained low and promotions rare. Although the elder MacArthur became a prominent federal judge in Washington, D.C., he found few strings to pull on his son's behalf among his Republican friends.

In 1875, while on Reconstruction duty in Louisiana, Arthur, Jr., met and married Mary Pinckney ("Pinkie"), a woman originally from Virginia. She bore three sons, Arthur III, Malcolm, and Douglas in 1876, 1879, and 1880, respectively. (Malcolm died shortly after Douglas's birth. Arthur attended Annapolis and became a high ranking naval officer before he died of an illness in the 1920s.) Through most of the 1880s, the MacArthur family shifted between army posts in the West and Southwest. Young

Douglas's earliest recorded memories were of long, hot marches between frontier forts, tales he heard from troopers about nearby fighting with Apache tribes, and of learning to shoot and ride.[1]

In 1883, hoping to escape from the drudgery of frontier duty, Arthur MacArthur sought the unusual post of military attaché in China. He prepared a long report for former President Ulysses S. Grant, hoping he would speak up for the captain. This "Chinese Memorandum and Notes" echoed the current fascination with the fabled "China Market" and contained a heavy dose of Social Darwinism and other rationales for imperialism. MacArthur predicted that Czarist Russia and the Western nations would soon go to war to divide Asia, the "region of true romance." The outcome would affect "the commerce of the world" and American security since the United States needed to safeguard the China Market, the "real stepping stone to supremacy throughout the commercial world." The captain warned that mounting industrial surpluses would soon cause a social and political crisis unless America found new markets, "principally in the Far East, and there, naturally and inevitably, we must go." Although the Peking post went to another officer, fifteen years later fate, war, and clashing imperialism finally took Arthur MacArthur, Jr., east.[2]

Captain MacArthur's luck began to improve in 1889 when he gained promotion to major and the post of Assistant Adjutant General in Washington, D.C. Although his official duties were quite routine, his father, Judge MacArthur, introduced him to many politically influential persons. The major cultivated interest in the nearly forgotten heroic episode of the Civil War. Belatedly, in 1890, he received the Congressional Medal of Honor for his bravery at Missionary Ridge, twenty-seven years before.

From the time of his childhood, Douglas recounted proudly the dramatic details of his father's Civil War exploits. The highly stylized retelling echoed the published accounts popularized by his father's friends around 1900, by which time Arthur, Jr., was a senior officer commanding forces in the Philippines. Like most tales of heroism, this one grew over time in the mind of young Douglas as well as in public mythology.[3]

Major MacArthur's four-year tour in Washington ended in 1893. Like other officers, he was still expected to resume service in the West—this time at Fort Sam Houston in San Antonio, Texas. The thirteen-year old Douglas attended a local military academy, graduating in 1897. A year before, his father received a promotion to the rank of lieutenant colonel and, shortly afterward, a transfer to St. Paul, Minnesota.

Rather than accompanying her husband, Pinkie took her son to Milwaukee, where the MacArthur family had roots, and spent the next eighteen months tutoring Douglas and writing letters to important political figures requesting a nomination for him to West Point. Success came in 1898 when she secured a nomination and he passed the entrance examination. The next year he entered the academy as a cadet, while Pinkie took up residence at Craney's hotel on the edge of the campus.

Meanwhile, political events had transformed Colonel MacArthur's life.

The war with Spain, beginning in April 1898 bought a promotion to the rank of brigadier general of volunteers. Arthur expected to train citizen-soldiers for the impending invasion of Cuba. Instead, he was transferred to San Francisco and assigned to command troops being sent to the Philippines. During the next three years, while fighting Spanish troops and Filipino rebels, Arthur achieved national fame. At the same time, Douglas prepared for a military career of his own.

Military historians describe turn-of-the century West Point as an anachronistic shrine most notable for outdated military instruction, brutal hazing, and an archaic curriculum. Even though Cadet MacArthur stood at the top of his class in three out of four years, he experienced some especially severe hazing as a result of his father's prominence. Graduating in June 1903, a newly commissioned second lieutenant, he selected the Philippines for his first post.

Even though the Filipino revolt against American annexation had been crushed by this time, the islands were still unsettled and an exciting opportunity for an ambitious officer. The assignment also permitted him to follow in his father's footsteps, who had left the Philippines in 1901. In letters to his son at West Point, visits with him, in congressional testimony, and in public speeches, General Arthur MacArthur had stressed the importance of American control of the Philippines as an element in shaping the future of East and Southeast Asia. A strong colony, he argued, would block Japan, secure vital economic interests, and enhance America's stature as a world power. All these arguments resonated in his son's later actions.[4]

Before Arthur MacArthur arrived in the Philippines the Spanish had been struggling to suppress a protracted rebellion in its colony composed of thousands of islands. In the nineteenth century a mestizo class had developed which occupied a role midway between the vast majority of Malay natives and the small Spanish ruling class. These mestizo priests, landlords, intellectuals, and merchants chafed under Iberian rule and initiated a movement for greater autonomy. Relying on repression rather than compromise, colonial governors made martyrs out of reformers who initially hoped to share power with the Spanish.

The revolutionaries of the late-nineteenth century were divided between those concerned with the peasant masses and others interested in power sharing with the colonial rulers. Emilio Aguinaldo, one of the most prominent rebels with a foot in both camps, joined several comrades in a temporary truce in 1897. The Spanish offered him a pot of cash and sent him off to Hong Kong. Aguinaldo insisted the money and safe passage would allow him to rearm his forces for future struggle.

When Commodore George Dewey sailed from Hong Kong to Manila in the spring of 1898, he arranged to bring Aguinaldo back. Dewey may have considered him a useful ally or merely a device to frighten the Spanish. The rebels believed they were being cultivated as a nascent government even though Washington disavowed any political promises.

By the time American ground forces reached Manila early in August 1898, the navy and Aguinaldo's rebels had put the Spanish garrison under siege. In a carefully plotted drama, the American commander permitted his Spanish counterpart to make a symbolic show of resistance which would assuage Iberian honor. With Aguinaldo on the sidelines, Dewey's ships lobbed a few shells at prearranged targets after which the enemy surrendered.

One significant hitch occurred when (now) Major General Arthur MacArthur's brigade clashed with a Spanish unit that had probably misunderstood the script. After hasty consultation, the American army took Manila—cutting Aguinaldo's insurgents out. Ironically, the American and Spanish governments had agreed to a cease fire one day before the "battle" of Manila.

For the next six months General MacArthur oversaw the occupation of Manila. He kept disgruntled Filipino rebels from entering the city, babysat the Spanish until their repatriation, and provided a variety of basic urban services. However, in February 1899 the calm broke when the U.S. Senate ratified the Treaty of Paris, which ended the war and transferred the Philippines to American control. The insurgents charged betrayal and launched attacks against the new colonial masters. During the remainder of the year MacArthur conducted a successful, conventional war against the rebels whom, in May 1900, he declared crushed. At that time Washington named him as the colony's military governor.

In reality, the bloodiest phase of the insurrection had only begun. Filipino rebels initiated a guerrilla war against the American army, causing high casualties and frequent atrocities on both sides. President William McKinley appointed a new, civilian Philippine commission at this time, headed by William Howard Taft, to oversee the pacification program. The group arrived in June 1900 and clashed with General MacArthur for the next thirteen months.

The military governor objected bitterly to the presence of civilian overseers in the midst of a guerrilla war. He also disliked the commission's informal administrative style, which contrasted to his own military discipline. Taft and his colleagues badgered the general to win over the mestizo elite through promises of power sharing. MacArthur believed such compromise should await the decisive military defeat of the rebels.

In July 1901 the War Department reassigned MacArthur to the United States, allowing Taft to implement his version of counterinsurgency. During the next two years the policy of co-opting the native elite (supplemented by continued military operations) achieved the desired result. Many rebel leaders defected while rampant factionalism among insurgent bands sapped the power of the rebellion. Gradually, the fighting dissipated.

Growing reports of army mistreatment of Filipinos caused a scandal in America and eventually a congressional inquiry. The investigation largely exonerated Arthur MacArthur but could not alleviate the implication that his reassignment represented a punishment. Naturally, all of this publicity, as well as visits his father had made to West Point, made a vivid impression on Douglas.

Douglas MacArthur Discovers the Philippines and Asia

Anticipating adventure and his first extended separation from both parents, Lieutenant Douglas MacArthur arrived in the Philippines in October 1903. His engineering duty in the outlying provinces immersed him fully in the tropical beauty and violence of the islands. Here he had his first combat experience, shooting dead two bandits who attacked his small survey team in the jungle. Several months later Douglas discovered more of the islands' allure when he was transferred to Manila.

News of his father's political difficulties may have tempered his own excitement a bit. Arthur had given a speech about the mounting German threat to America, bringing down a formal reprimand from the War Department. Shortly afterward, his old nemesis, William Howard Taft, became Secretary of War. Possibly because of Taft's antipathy, a more junior officer than he was named Army Chief of Staff. Both Arthur and Douglas believed that political enemies had blocked a deserved promotion.

While in Manila, Douglas had the good fortune to meet a pair of rising Filipino politicians, Maneul Quezon and Sergio Osmeña, who would influence his later career. Less fortunately that summer, he contracted malaria and by October 1904 its lingering affects forced his transfer back to the United States. After his recuperation, he found little pleasure in military engineering assignments in California. Happily, his family intervened and made it possible for him to return to Asia.

Arthur MacArthur chafed under a series of stateside posts after his return from the Philippines in 1901. Taft's likely role in denying him the job of Chief of Staff added to the frustration. The out break of the Russo-Japanese War in 1904 whetted his appetite for a return to the excitement of the Far East. Early in 1905 he convinced the War Department to name him military attaché in Tokyo with responsibility for observing the fighting in Manchuria. Although he arrived too late to observe many battles, he secured permission from the War Department to conduct a military survey of China, Southeast Asia, and India. The general also persuaded Secretary of War Taft (who visited Japan in September 1905) to assign Douglas as an aide on the upcoming tour. The young officer joined his parents in Japan at the end of October.

The MacArthur family embarked on a Kiplingesque grand tour through much of East and Southest Asia. From November 1905 until the following July, the three MacArthurs visited China, Malaya, Singapore, Burma, Indochina, India, Ceylon, the East Indies, and Thailand. They toured shrines, met colonial administrators, had an audience with the king of Siam, and undoubtedly came away with an appreciation for the richness and diversity of Asia. Although the two officers spent very little time surveying military affairs, both considered the tour one of the most influential events in their lives.[5]

During the trip Arthur received a promotion to the rank of lieutenant general. In spite of the honor and significantly higher salary the promotion carried, the general fretted about his future. The War Department had twice passed him over for the army's top position. With Taft in the War Department, his future looked bleak. In order to avoid eventual reassignment to an obscure command, in 1907 General MacArthur proposed that he be "detached from command" and assigned to the midwestern city of Milwaukee, where he might conduct "special work" for the army. In effect, he desired informal retirement at current rank and promised to devote his time to writing a report (never completed) on his Asian tour.

Finding no other useful tasks for him, the War Department permitted the general to spend his last two years of service in this sinecure. Following Taft's election as President in 1909, he retired. Arthur died three years later, still so bitter over inadequate recognition by the army that his will forbade both a military funeral or burial at Arlington National Cemetery.

Shortly before his own death, Douglas MacArthur described the family's grand tour as the occasion when he first fused his personal destiny with that of Asia and the Pacific. The event, he wrote, would "color and influence all the days of my life." He had discovered "western civilization's last earth frontier." It seemed "crystal clear" that the future, "indeed the very existence of America, were irrevocably entwined with Asia and its island outposts." The "mystic hold" of the Far East was "upon me."[6]

In spite of this enthusiasm, sixteen years passed before Douglas could return to this new frontier. After he and his parents reached America in 1906, the War Department assigned him to engineering school in Washington, D.C. Although most of his peers considered this a plum, he found the course mundane and felt relieved when named a part-time military aide to President Theodore Roosevelt. This job provided him an occasional opportunity to discuss the Far East with the energetic president, but his work consisted mainly of announcing visitors to the White House and dressing up for ceremonial occasions. In 1907, he eagerly accepted assignment to a post in the army engineering district in Milwaukee, again close to his parents.

Soon his family ties superceded his devotion to work. MacArthur's commanding officer complained that the lieutenant lacked a "zeal to learn," and spent inordinate amounts of time with his family. The current army chief, a friend of Arthur's, intervened by reassigning Douglas as a company commander at Fort Leavenworth, one of the army's training centers. There, between 1908 and 1911, he blossomed and won promotion to the rank of captain.[7]

Family ties still exercised their hold, however. Following Arthur's death in 1912, Douglas assumed responsibility for his mother. Worried about Pinkie's declining health (she actually lived until 1935), he requested a transfer to Milwaukee or to any large city that would meet her needs for care. The plea impressed Chief of Staff Leonard Wood, another old family friend, who recalled Arthur's "distinguished service" in a letter to Secretary of War Henry Stimson. Stimson responded in December 1912, approving the transfer of Captain MacArthur to Washington. There he would serve directly under Wood. Four years before he faced a bleak future in the army; now, partly because of his father's network, he was apprenticed to one of the military's highest commanders. In September 1913, he was selected to join the army's General Staff, confirming his entry into influential circles.[8]

When American troops occupied Vera Cruz amidst the turmoil of the Mexican Revolution in 1914, Wood selected Captain MacArthur as a personal emissary to lead a reconnaissance mission behind Mexican lines. In defiance of the local occupation commander, he took a small party through enemy territory in search of steam engines that might be used for transportation. En route, the group fought with Mexican army units and bandits.

Several fellow officers urged that he be granted a Congressional Medal of Honor for the action. However, a War Department board dismissed Captain MacArthur's bravery as an "error of judgment." The board considered his performance routine, not a case of extraordinary valor. The incident is important only because MacArthur vigorously protested the ruling, probably seeing it as a parallel to the long deferred recognition of his father's Civil War feats.

The outbreak of war in Europe in August 1914 quickly overshadowed these problems. Although a neutral for nearly three years, the United States began a "preparedness" campaign under the National Defense Act of 1916. Promoted to major, MacArthur worked on the expansion of the Regular Army and cooperated with such rising stars as Assistant Secretary of the Navy Franklin D. Roosevelt in defense planning. The major also assumed responsibilities in the army's Bureau of Information, a post in which he acted as a liaison with the press. At this time he developed a friendship with Steve Early, a journalist destined to become Franklin Roosevelt's press secretary.

When the nation entered the Great War in April 1917, MacArthur, like his peers, clamored for a combat assignment. Ironically, his growing importance as a member of the General Staff made this problematical. His mediation of a bureaucratic dispute provided a way out. Secretary of War Newton Baker favored expanding the army rapidly by integrating existing National Guard units. This could be done more quickly than waiting for new draftees to complete training. MacArthur sided with Baker over the objections of many officers. In a conference, he helped convince President Woodrow Wilson to support the plan and attracted Wilson's attention with his idea of organizing a special army division drawn from guard units across the nation. Shortly afterwards, the president approved MacArthur's promotion to colonel and his selection as chief of staff of the new 42nd "Rainbow" Division, based on this idea.

Combat provided ample scope for MacArthur's dramatic flair and military skills. In Europe from October 1917 until April 1919, he fought in eight major battles, gained numerous American and Allied decorations, honed his talents as a battlefield commander, and rose to the rank of brigadier general.

As previous biographers have noted, sometime during World War I MacArthur adopted a conspicuous style, which he maintained throughout his long career. His acts of premeditated bravery included intentionally exposing himself to danger, shunning protective gear (such as a helmet or gas mask), and disputing orders from headquarters he thought slighted the honor of the Rainbow Division. His attention-getting behavior included a penchant for nonregulation dress that distinguished him from fellow officers. He sported a distinctive cap, riding crop, sweater, and long muffler, which made him appear something of a dandy. All these prefigured the cap, sunglasses, and corncob pipe of the next war.

Following the armistice of November 11, 1918, MacArthur assumed command of a small occupation zone in a bucolic corner of Germany. During the next six months, this rural zone remained calm even while urban Germany heaved with revolutionary turmoil. Occupation duty provided MacArthur with two valuable benefits. At a time when most recently promoted officers reverted to their prewar rank, his extended service abroad permitted him to remain a brigadier general longer than his cohorts. Also, as one of the senior army officers still in Europe, American journalists found him an attractive source of information. Building on his previous experience as a military liaison with the press, MacArthur grew increasingly adept at cultivating journalists' attention.

MacArthur returned to America in April 1919, uncertain about his future in a peacetime army. Once again, however, family connections helped him along. Chief of Staff Peyton March, yet another old comrade of Arthur MacArthur, selected Douglas as superintendent of West Point. A prestigious and highly visible post, the assignment ensured he would retain permanently his wartime rank, thereby securing permanent seniority over many of his peers. In the rigid seniority system of the pre-World War II army, this provided a vital edge for promotion.

The general relished the opportunity to reform his alma mater. Unfortunately, his two and a half years as superintendent proved exceptionally frustrating. Older faculty resented his efforts to modernize the archaic curriculum. The War Department provided only minimal encouragement, weakening his bargaining power.

The situation deteriorated in 1921 when General John J. Pershing became army chief. A series of minor disputes had marred their relations when Pershing commanded American forces in Europe. Now a personal animosity developed between them. In January 1922 MacArthur announced his engagement to Louise Cromwell Brooks, a wealthy, divorced mother of two children he had met shortly before. Brooks had recently been close to Pershing and many friends had expected them to marry. Two weeks after the engagement, Pershing relieved MacArthur of his position at West Point and ordered him to the Philippines. (The marriage pleased Pinkie MacArthur no more than it did Pershing. She declined to attend the ceremony or to accompany the newlyweds to Manila.)

Whatever Pershing's motives, the transfer was not wholly undesirable to MacArthur. Foreign service provided a gracious exit from the stalemate at West Point. It also placed him under the supervision of two important friends, Philippine Governor-General Leonard Wood and General George Read, commander of the American army's Philippine Department. He looked forward to commanding both American and Filipino units and to renewing his ties with the islands' leading politician, Manuel Quezon, now head of the Philippine Senate.[9]

Although Douglas complained to friends that Pershing's "little clique" hoped to wreck his career in the army, the chief of staff showed no vindictiveness once MacArthur had left America. In 1925, unknown to MacArthur, Pinkie resumed her lobbying on her son's behalf. She wrote to Pershing, pleading that he approve Douglas's promotion to major general. The wheels were set in motion and MacArthur received his second star ten days after Pershing retired.[10]

Shortly after this, Major General MacArthur returned to America. From 1925–27 he commanded the III Corps area near Washington and Baltimore. This permitted him and Louise to live in her Baltimore area estate while MacArthur enjoyed the advantages of service in the capital. His visibility increased when he sat as a judge in the controversial court-martial of Brigadier General William (Billy) Mitchell. Although he claimed to have voted for acquittal, MacArthur shared the army's disdain for Mitchell's "fanatical" views of air power. In 1927, he served as president of the American Olympic committee, a platform from which he extolled the connection between sport, character, and patriotism.

Despite these professional achievements, MacArthur's personal life began to unravel. He and Louise separated in mid-1927. She found life as a general's wife dull, objected to her husband's devotion to work, and chafed under the scrutiny of a meddling mother-in-law. Louise also attributed the estrangement to sexual problems. In contrast to her inexperienced husband, Louise had led a sexually active life before she met MacArthur. After their first night of marriage, she complained to her brother that despite Douglas's high rank, "he's a buck private in the boudoir."[11]

MacArthur's reputed sexual problems had political implications in later years. Partly because of Louise's revelations, gossip circulated thereafter concerning the general's alleged sexual inadequacies. His political enemies argued that lack of manliness accounted for either his excessive bravado or, alternatively, failure of leadership. During the Second World War, Interior Secretary Harold Ickes and columnist Drew Pearson (both of whom hated the general) linked their criticism of MacArthur's political ambitions and military strategy with reports of his failure as a lover.

Pearson relished telling friends that when the former Mrs. MacArthur "has had a few drinks," she explained that "Doug didn't think that his penis was for anything except to pee with." At parties, Louise reportedly bent "her little forefinger so as to make it look limp and held it out to illustrate MacArthur's virility." Regardless of their validity, these stories must have stung terribly.[12]

As twice before in his life and twice thereafter, MacArthur returned to the Philippines after his divorce as a sort of safe harbor. At the end of 1928 he was assigned to Manila as commander of the army's Philippine Department. The small military budgets of the interwar years, as well as the comparative stability of the world, afforded him few opportunities to implement numerous ideas for upgrading the islands' defenses. He took great pride, however, in improving the training, pay, and status of the Philippine Scouts, a Filipino unit in the American army. MacArthur looked to these forces as the key to future defense. His struggle against a two-tiered military racial division won him much local popularity.

Compared to the racism so prevalent among his colleagues and the overseas American community, MacArthur had an enlightened attitude toward Asians. He cultivated and treated as social equals many elite Filipinos, including Manuel Quezon. MacArthur also endorsed the concept of gradual independence so popular among Filipinos. Quezon hoped his sympathetic American friend might succeed to the post of governor-general after Henry Stimson stepped down in 1929. That April, the *New York Times* speculated that MacArthur desired the governorship as the first step in a political career leading to the presidency; civilian service in the Philippines could be parlayed into a cabinet post and, "eight or twelve years hence" a campaign for the White House.[13]

Such former governors-general as William H. Taft and Henry Stimson did have illustrious political careers before and after serving in Manila. During the 1930s, three governors-general, Paul McNutt, Francis Sayre, and Frank Murphy, also had national political aspirations. MacArthur probably saw the challenge of combining civilian and military administration as an exciting test of his talent. As the top civilian in the colony, he would occupy the position from which Taft had driven out Arthur MacArthur in 1901. If, as hoped, he were permitted to serve both as civilian administrator and chief military official, he would have vindicated and surpassed his father.

Historian Carol Petillo reveals that MacArthur desired this appointment so much that he ghosted numerous letters Quezon sent to Washington in 1928 recommending the general's candidacy. Over Quezon's signature, MacArthur informed Secretary of State Henry Stimson that a canvass of "political, industrial, and labor groups" revealed "almost unanimous agreement on General MacArthur." Naming him governor-general would prove a "master stroke of statesmanship and diplomacy," surpassing "anything these islands have ever known." Although the job went to former Secretary of War Dwight Davis, MacArthur mounted campaigns for the office three more times before World War II. On those occasions he recycled the same effusive praise he penned for Quezon.[14]

Despite the setback, President Herbert Hoover and Secretary of State Stimson maintained a high regard for the young general. When the time came in 1930 to select a new army chief, he placed high on their list. Secretary of War Patrick J. Hurley (who later befriended the general) complained that any "man who couldn't hold his women shouldn't be Chief of Staff." MacArthur sought to charm the easily flattered secretary of war. When Hurley issued a routine statement on the Philippines, MacArthur notified him that he considered the remarks "the most comprehensive and statesmanlike paper that has ever been presented with reference to this complex and perplexing problem." Even if nothing else were ever written, "your treatise would be complete and absolute." It left "nothing to be said" and "brought confidence and hope out of the morass of chaos and confusion which has existed in the minds of millions of people." Hurley apparently dropped his opposition and in August Hoover named Mac-Arthur army chief, effective that November.[15]

Even after Hoover tapped him, the general experienced a wave of self-doubt. Initially, MacArthur admitted, his spirit "shrank" from the "dreadful ordeal" of the job. He also worried that given his relatively young age (fifty in 1930), his army career would be over at age fifty-four when he completed a four-year tour. The army had few uses for former chiefs.

Another, more private, anxiety clouded his return to Washington. In 1929, MacArthur had begun an affair with a young woman in Manila. Of mixed Philippine and Scottish ancestry, Elizabeth "Isabel" Cooper had no place in the job description for a chief of staff. The nearly thirty year difference in their ages and the issue of race, not to mention Pinkie's likely reaction, made it impossible for the general either to marry or acknowledge her as a lover in the United States. Although MacArthur arranged to bring Isabel to America, subsequent problems justified his concern.[16]

When Pinkie got wind of Douglas's inner turmoil (although she did not connect it to the problem with Isabel) she cabled from Washington that he must accept. "She said my father would be ashamed if I showed timidity." This admonition carried him back to America, to a reunion with Pinkie, and into the job denied Arthur MacArthur, Jr., more than two decades before.[17]

The general made arrangements for Isabel to follow him after a discreet interval. During their several month separation, he wrote to her almost daily. The ship taking him from Manila to California stopped at several Chinese and Japanese ports. At each he bought her an elaborate gift and posted letters to "my own darling baby girl." MacArthur, who signed his love letters as "Daddy," was clearly mad for Isabel. In an ornate, nineteenth century style he declared his eternal and desperate love, which sent him "groping, blindly in the dark" for "my own true little fellah." Until Isabel reached Washington, he complained that he was "dying by inches" for want of her. Following her arrival, the general set up an apartment for her and they continued the clandestine affair for some time. Whatever its physical dimension, emotionally the affair with Isabel filled a deep need for MacArthur.[18]

The Chief of Staff

The havoc wrought by the Great Depression, in addition to its other horrors, made MacArthur's term as head of the army miserable. From November 1930, when he assumed the post, until his departure, late in 1935, the general fought a rearguard action against a political climate of pacifism and an economic catastrophe that compelled severe cuts in military spending. MacArthur tried desperately to preserve a viable officer corps for an army whose size already ranked as low as seventeenth in the world. Forced to distribute inadequate funds among jealous branches, the high command all but abandoned technical innovation. As a result, both the air and mechanized forces slipped further behind those of America's potential enemies. Even though MacArthur probably did as good a job as any Chief of Staff could have, his constant budget battles with Congress, squabbles with other agencies, and disputes with dissatisfied officers won him few accolades.

Growing friendships with President Herbert Hoover and Secretary of War Patrick J. Hurley provided some solace. The dour Hoover and mercurial Hurley (suddenly a great admirer of MacArthur) respected the general's professionalism and shared his traditional, conservative values. As self-made men, Hoover and Hurley made a fetish of their belief in rugged individualism. They opposed calls for federal programs to alleviate hunger, unemployment, and homelessness. MacArthur embraced Hoover's faith that private enterprise, not collective action, would solve individual and national economic problems. Hoover, Hurley, and MacArthur all shared a distrust of labor unions, immigrants, and socialism. The president and his army chief remained lifelong friends and political allies.

As the financial crisis forced deeper cuts in the army budget, MacArthur vented publicly his frustration with Congress, pacifists, and exponents of "radical" ideas by sending a letter in 1931 to a religious journal, *The World Tomorrow*. In a poll that had earlier appeared in the magazine, a large majority of nearly 20,000 clergymen had decried military training in high school and college and urged larger defense cuts even if it weakened the nation. Just over half opposed participation in *any* war, while a third condemned service by military chaplains.

The Chief of Staff responded by condemning "all those who would refuse to take up arms in defense of their country." Pacifists, he charged, exercised privilege but shunned responsibility. The opposition of the clergy to the "laws of the land" would "hearten every criminal and malefactor who either has or contemplates breaking some other law," he wrote. "Religion and patriotism" went "hand in hand" while "radicalism, communism, bolshevism and other enemies of free government" were the hand-maids of "atheism." Subsequently, MacArthur denounced "pacifism and its bedfellow communism" as responsible for all protests against established order.[19]

Mounting protests by unemployed veterans during the summer of 1932 raised a new revolutionary specter before the general's eyes. In June, 22,000 veterans of World War I tramped to Washington, set up makeshift quarters, and demanded immediate payment of a $1000 benefit, or bonus, promised to them in earlier legislation. An eclectic, grass-roots movement, the so-called Bonus Expeditionary Force (B.E.F.) drew men from around the nation. Despite the unfounded claims of MacArthur and other critics, Communists held few if any leadership roles among the law-abiding protesters. Yet, for the rest of his life, MacArthur insisted that "the Communists hoped to incite revolutionary action" against the federal government by means of the protest.[20]

The impoverished veterans assembled peacefully to lobby Congress to pass a bill introduced by Wright Patman granting them immediately the promised pension bonus. The House passed the Patman bill in mid-June but President Hoover, who favored deferring the payment until 1945, promised a veto should it pass the Senate. On June 17, the Senate defeated the measure. As a sop, the legislative body offered to buy tickets home for the protesters. About 6,000 accepted the offer. The district police chief urged patience while waiting for the rest to drift away. By July, however, the city fathers, President Hoover, and MacArthur lost patience with this strategy and decided to disperse the remaining 10,000 bonus marchers.

On July 28, after imposing restrictions on demonstrations, the district police, with army backing, began to disperse the B.E.F. Rejecting again police calls for gradual removals over several days to minimize violence, the district commissioners and President Hoover insisted on immediate action. When, during a scuffle, a panicked policeman shot and killed two veterans, Hoover ordered the army to clear the protesters from the center of the city.

MacArthur assembled nearly a thousand troops from infantry, cavalry, and mechanized units. When one of his aides, Major Dwight D. Eisenhower, questioned whether the Chief of Staff should be involved personally, the general responded that "there was incipient revolution in the air." When the nonviolent protesters retreated across the Anacostia river toward one of their encampments, MacArthur ignored suggestions by his aides to terminate the operation. His troops destroyed the squatters' camp and, in the process, inflicted more casualties on the marchers. Adding insult to injury, he told a late-night press conference that he had suppressed a mob "animated by the essence of revolution." The protesters were plotting to seize power and in another week the "whole government would have been severely threatened."[21]

Shortly after the so-called battle of Anacostia Flats, Democratic Presidential candidate Franklin D. Roosevelt disclosed his view of the general.

During a meeting with members of his "Brain Trust," FDR received a call from Senator Huey P. Long. The Southern populist berated Roosevelt for hobnobbing with Wall Street "blankety blanks" instead of plain folks. Later, as his aides chortled at Long's harangue, Roosevelt cautioned them to "remember all the time that he really is one of the two most dangerous men in the country." Economist Rexford Tugwell asked if the other culprit was Father Coughlin. No, Roosevelt replied, "the other is Douglas MacArthur." Many "Nazi-minded" Americans longed for a "symbolic figure—the man on horseback." "There was none so well endowed with charm, tradition, and majestic appearance as MacArthur." But, Roosevelt explained, he would try to "tame these fellows and make them useful to us."[22]

2
MacArthur and the New Deal, 1933-35

As part of his official duties, Army Chief of Staff Douglas MacArthur led the inaugural parade of President Franklin D. Roosevelt on March 4, 1933. Although the New Deal's social programs and the liberal administrators who flocked to the charismatic FDR had yet to make their impression in Washington, nothing about the new administration boded well for the general who had nineteen months remaining in his normal four-year tenure as head of the army. The departure of Herbert Hoover and other conservatives left MacArthur an awkward holdover. With the nation facing desperate economic straits, the military budget seemed a perfect target for further reduction.[1]

In the pre-New Deal America of 1933, the army represented one of the only federal agencies equipped to deal with major projects and large numbers of people. Naturally, Roosevelt turned to its leadership for help in launching his first mass relief program, the Civilian Conservation Corps (CCC). The administration wanted the army to recruit, transport, train, and deploy hundreds of thousands of unemployed youths. Although not enthusiastic about the program, MacArthur and his staff cooperated and played a major role in the early success of the CCC.[2]

Roosevelt displayed little gratitude when he examined the army's budget. In the spring of 1933, the administration proposed expenditures for the next year of about $277 million, requiring large personnel cuts and reductions in salary and pensions for many career officers. MacArthur visited the White House late in March to warn the president "bluntly" that he had put "the country's safety at stake." When FDR deprecated his statement, the general, in his own words, "spoke recklessly." If America "lost the next war," he predicted that a dying American boy "lying in the mud with an enemy bayonet through his belly and an enemy foot on his dying throat" as his final curse would "spit out" the name Roosevelt.

FDR demanded an apology and MacArthur, sensing his career had just ended, offered his resignation. But as the general rose to leave, the president said in a voice of "cool detachment which so reflected his extraordinary self-control, 'Don't be foolish Douglas; you and the budget [office] must get together on this.' " Despite the pride he felt in having "saved the army," upon leaving he "vomited on the steps of the White House."[3]

MacArthur claimed that from "that time on" Roosevelt "was on our side" and nothing diminished their "warmth of . . . personal friendship." Afterward, the general asked the president why he "often asked my comments" on New Deal legislation, but "never again on military affairs." The Chief of Staff recalled Roosevelt answering that he did not "bring these questions up for your advice but for your reactions. To me you are the symbol of the conscience of the American people."[4]

The response, if true, surely pleased MacArthur but did little to solve his budget problems. The only added support came when the administration agreed to combine the War Department's request for funds with the more pressing recovery program. Federal relief agencies earmarked a certain amount of money for the procurement of military equipment and for military construction projects. This alleviated some of the army budget cuts while creating jobs and stimulating production. However, few New Dealers expressed much sympathy with the army or its leader.

Interior Secretary Harold L. Ickes, a member of the administration's inner circle and a confirmed MacArthur hater since the Bonus incident, chortled at his own success in cutting army funding. During 1933 the general requested "hundreds of millions" of dollars from federal construction programs under the Interior Department. Ickes had limited allocation to only six million. After hearing the general demand more, Ickes characterized him as the "type of man who thinks that when he gets to heaven God will step down from the great white throne and bow him into his vacated seat, and it gave me a great kick to have him in and break the news to him."[5]

MacArthur found little relief outside his professional life in New Deal Washington. Professors, social workers, and progressive lawyers had replaced conservative businessmen as the social elite of the capital. His mother, now in her eighties and increasingly frail, demanded ever more attention. Even his love affair turned sour.

In 1931, the general had set up Isabel first in an apartment and later in a Washington hotel. Over time, both gradually grew dissatisfied. MacArthur must have found it difficult to balance the requirements of his job and the demands of his mother on top of a clandestine relationship. After 1932, Isabel sought other social and professional outlets, including travel, dating other men, and enrollment in art and law school. Early in 1934, MacArthur broke off the affair and sent Isabel (in a note signed "From the Humane Society") a return ticket to Manila. She refused and, eventually, made life miserable for her former lover.[6]

Ever since the Bonus Army incident, journalists Drew Pearson and Robert Allen had attacked MacArthur's alleged "dictatorial leanings" in their syndicated "Washington Merry-Go-Round" column. In the spring of 1934 the Chief of Staff sued them for $1,750,000, charging their writings slandered him as "dictatorial, insubordinate, disloyal, mutinous, and disrespectful of his superiors in the War Department." Privately, MacArthur also condemned the journalists for writing "disrespectfully" about his mother.

Pearson and Allen's muckraking had offended Roosevelt as well, despite their generally pro-New Deal slant. Pearson suspected that FDR secretly encouraged the general's suit in revenge for their criticism of some of his staff. Possibly Roosevelt relished the prospect of MacArthur and the journalists attacking each other. In any case, the suit had that result.[7]

Pearson searched for "dirt" that might force the general to drop the case. He turned first toward a wealthy stockbroker, Charlie Munn, who furnished details of a stag party hosted by MacArthur following an Army–Navy football game. The entertainment included a "dozen George White Scandals beauties dressed in Army uniforms and a dozen dressed in Navy uniforms, with MacArthur as the pivot." Although "a strict Army regulation about desecrating the uniform" might have been violated since "these young ladies had certain vital parts of their uniforms cut away," even Pearson realized the incident would hardly scare the general out of court. Then he heard about Isabel Cooper.[8]

MacArthur's first wife, Louise, had remarried but lost none of her animus toward Douglas. At Washington social gatherings she dropped hints about a Eurasian mistress now in financial distress after being jilted. Soon mutual acquaintances introduced the journalist and the damsel. Promising protection, financial security, and help in finding a job, Morris Ernst, Pearson's attorney, coaxed a sheaf of MacArthur's love letters from Isabel along with a detailed chronology of their affair. Ernst also agreed to represent Isabel in a suit against the general for at least $15,000 for unspecified damages. In December 1934, shortly before the Pearson case went to court, Ernst advised MacArthur's lawyer about the letters and his intention of calling Cooper as a defense witness.[9]

Pearson and Ernst, fearing for Isabel's safety, hid her in hotels and communicated with her using such dramatic code names as "the Flower of Cathay" and "Miss Ling." They provided selections from her intended testimony and excerpts from MacArthur's earlier letters to the general's representative, an officer in the Judge Advocate General's office. Isabel intended to make a deposition claiming that while amorously engaged the chief of staff denounced the secretary of war, called Roosevelt that "cripple in the White House," boasted that he had "eclipsed his father," in greatness and predicted "he would be the first man in history to be reappointed chief of staff." She also alleged that MacArthur took credit for "putting backbone into Hoover" when the time came to rout the bonus marchers.[10]

The threat of this testimony and the release of love letters addressed to "my own darling baby girl" from "Daddy" must have terrified Mac-

Arthur. Even if he could somehow explain away a liaison with a non-Caucasian woman, his alleged insults of his superiors could be damning. As Pearson expected, MacArthur beat a hasty retreat.

The journalists accepted an offer to settle the suit before trial with the condition that MacArthur pay all legal costs and provide money for Isabel's future. Through his two War Department aides, Captain T. J. Davis and Major Dwight D. Eisenhower, the general paid about $16,000 in legal fees and another $15,000 directly to Isabel. MacArthur got possession of his original letters, Pearson and Ernst promised that photocopies would not be revealed during MacArthur's lifetime, and all parties pledged to avoid further litigation. (Pearson, however, made multiple copies of the letters, telling Harold Ickes he feared "foul play" by MacArthur.)

Although a source of gossip in Washington for years, the scandal did not ruin MacArthur, as Pearson and Ickes had hoped. The incident took its harshest toll on Isabel. She soon left Washington, held a succession of jobs, and later worked as an actress in Hollywood. She died in 1960, a suicide.[11]

The strain this put MacArthur under, as well as his internal conflicts over sex and love, may have affected his subsequent behavior. At about this time, T. J. Davis (who had introduced Isabel to Douglas) rented a getaway apartment in Washington for his boss. As Davis retold the story years later, MacArthur used the apartment to entertain prostitutes. The general would sit in a chair and have the women "admire what a great man he was." On other occasions the two men "picked up one special girl" at a brothel in Baltimore. MacArthur would lavish favors and flattery on her, await her response, then "denounce her as a 'little whore' and walk away." Perhaps this was his symbolic revenge against Isabel.[12]

Now racked by self-doubt, the Chief of Staff complained to Davis how his mother's unrelenting pressure forced him to become a "glorious Apollo, Roland, and George Washington, all in one." He feared that at a critical moment he would panic and fail. Rather than await the event, he threatened suicide. In this drama, played out many times between 1934 and 1935, Davis would assure the general of his virtue and convince him to put down his pistol. One time, while traveling on a train near the scene of his father's great Civil War heroism, the general told Davis: "I've done everything I can in the army and life, my term as Chief of Staff is ending. As we pass over the Tennessee River bridge, I intend to jump from the train. This is where my life ends, Davis." Tired of replaying the scene, Davis wished him happy landings. MacArthur stormed out of the compartment and later apologized for being so emotional. After Pinkie's death in 1935, Davis never heard him make another suicide threat.[13]

MacArthur's position in Washington seemed even more anomalous. Outside of his work and, possibly, his forays into the sexual underground, the general found friendship among a group of conservative businessmen who later assisted his presidential aspirations. Robert Wood, then chairman of Sears, Roebuck and Co., had been at West Point with MacArthur and took his old friend under his wing. As leader of an anti-New Deal business lobbying group (the Committee for the Nation to Rebuild Prices and Purchasing Power), Wood introduced the general to such rich and powerful conservatives as James H. Rand—later MacArthur's employer. Although many of Wood's friends expressed support for New Deal critics like Father Coughlin, MacArthur avoided public attacks on the administration. However, his deputy, General George Van Horn Moseley, more than made up for his reticence. A notorious anti-semite, nativist, and admirer of Hitler, Moseley filled the office with denunciations of Roosevelt and the Jews he believed had seized control of the nation. MacArthur remained a close friend of his dyspeptic aide and, privately at least, shared some of his hates. But mostly, the general's staff recalled, he "lost himself in his work."[14]

Although not fixed by law, the typical four-year term as Chief of Staff would end in the fall of 1934. FDR came under pressure from liberals to announce a successor. Despite this, Roosevelt leaned toward extending the general's term. D. Clayton James notes that at this time the president and several members of Congress favored an increase in military expenditures and felt that MacArthur could help steer the army budget through a hostile legislature. In mid-November 1934 he announced a temporary extension of MacArthur's term and, a month later, an indefinite extension. (This came in the midst of the pretrial negotiations over the libel suit.)

Roosevelt's action permitted him to utilize MacArthur's talents in gaining congressional approval for a military budget increase, while placating liberals who opposed another term for the Chief of Staff. Given the trauma of the libel suit, MacArthur must have been happy to maintain his status a while longer before facing mandatory demotion in rank and probable reassignment to a corps area command.[15]

To MacArthur's great fortune, and Roosevelt's equally large relief, the Philippines provided a solution to the general's status. Once again the MacArthur family found its personal destiny tied to American policy in the Pacific.

Philippine Independence and MacArthur

The Great Depression prompted American leaders to reevaluate their commitment to the nation's large Pacific colony, the Philippines. For years both American and Filipino politicians had debated independence and all expected it to come eventually. Since the suppression of the insurrection early in the 1900s, the United States had gradually encouraged Filipino participation in both legislative and civil service. By the later 1920s, the Philippine legislature had gained a fairly large amount of power under the supervision of an American governor-general. Virtually all Filipinos favored the principle of independence, although they worried about its terms and implications.

Composed of more than 7,000 islands, some forty ethnic and eighty dialect groups, the Philippine archipelago was more a geographic expression than a unified society in 1935. Its nearly 14 million people were about 80 percent Catholic, 10 percent adherents to the Aglipayan or Philippine Independent Church (a Catholic offshoot), 4 percent Muslims and 4 percent pagan. Ethnically, it consisted of a mestizo elite (15 percent) that possessed mixed Malay, Spanish, and Chinese bloodlines. The remaining 80 percent were mostly pure Malays who worked either as tenant farmers or agricultural laborers. A smattering of Chinese and Japanese immigrants played important roles in urban commerce and commercial agriculture.

Since 1903, the United States had introduced modern education, legal, and political institutions but did little to change the entrenched patterns of landlord–tenant relations inherited from the Spanish. The mestizo elite continued to dominate the economic, political, and social life of a largely peasant population. The political order followed social structure. Loyalties and responsibilities stretched out from the nuclear family to the extended family, from local districts to more distant regions, from local political leaders to national officials. Filipinos, in the words of David J. Steinberg, lived according to "concentric circles of allegiance." Personal loyalties dominated political considerations. Values derived from the "sense of debt felt by each individual to the network of people who surround and help him." No Filipino dared ignore this social obligation or *utang na loob* (an internal debt of gratitude) for fear of being accused of *walang hiya* (shamelessness).[16]

The Nacionalista party served as the major vehicle for Filipino activism. Although dominated by Manuel Quezon, rivals like Sergio Osmeña and Manuel Roxas rent the party with factionalism. These oligarches monopolized the political arena from the 1920s through the late 1940s. Despite the facade of a western political organization, ideology took a back seat to the indigenous concepts of kinship and patron–client relationships. Political loyalties extended to individuals rather than to organizations or governmental institutions. National political leaders like Quezon built their power on networks of personal loyalties extending down into the countryside. Elite mestizo families in Manila and other cities depended on the support of local kingpins and, in turn, mediated between the rural landowners and the Spanish and, later, American colonial authorities.

The Manila elite, known as *illustrados* before 1898, were the key political actors under both the Spanish and American governors. They articulated the developing sense of Philippine nationalism and hoped that its conservative and evolutionary doctrine would speed power sharing with Spanish colonial authorities. When the American annexation drove Aguinaldo into revolt, the conservative elite shared the new rulers' fears of radical nationalism. Sensing this, Americans like Taft reached out to the elite and, through promises of power sharing, separated them from the rebels. The *illustrados* benefited from secular and land reform instituted by the

United States. Through purchasing church lands that the Spanish religious orders were compelled to sell, and by staffing local offices created by the American administration, they enhanced both their power and status. Their "collaboration" with conquerors went hand in hand with a continued espousal of evolutionary nationalism and American promises of freedom.

Mostly the United States cared about keeping the colony tranquil so that investments and profits would continue to grow. This seemed best achieved by working through local elites who controlled the masses effectively. Concentrating power in the hands of men like Quezon proved a simple and expedient way to oversee the colony. Neither democracy nor social justice thrived under such an arrangement.

Nevertheless, many Filipinos developed a significant sense of loyalty and obligation to the United States. The oligarchy, of course, benefited from their role as intermediaries and hoped eventually to inherit complete power. Ordinary Filipinos tended to follow traditional leaders, recognizing the relatively benign aspect of American rule and also anticipating benefits from independence.

Ironically, by the early 1930s, the United States rather than its wards took the initiative in proposing an end to colonial rule. A complex economic web enmeshed the two societies and the Depression made Americans, more than Filipinos, eager to cut the ties. In the words of Senator Arthur H. Vandenberg, Congress desired "not to free the Philippines from the United States, but to free the United States from the Philippines."[17]

Colonial trade policies bound the island's economy tightly to the United States. As a result of a free trade (Washington imposed no tariffs on exports or imports), 86 percent of Philippine exports went to the United States. Agricultural commodities like sugar, coconut products, and cordage fibers comprised most of the total. In effect, the Philippines became a specialized colonial economy where highly centralized land ownership and commodity exports serviced the American market.

Obviously, Filipinos and Americans who controlled this trade benefited from the dependency relationship. So did the several thousand long-term American residents who enjoyed the colonial ambience and special privileges granted them in Manila. However, other powerful interest groups hoped to sever the tie. American agricultural producers and marketers complained that Philippine cordage, coconut oil, and, especially, sugar, competed unfairly with domestic products. Of these, only sugar represented a significant commodity. In fact, Philippine sugar competed not so much with American produced sugar as against Cuban imports, in which American bankers had invested heavily. These investors hoped that independence would end the free trade relationship.

Racism also mobilized some Americans behind independence. Before the 1920s, immigration from the Philippines hardly existed. By 1932, nearly 60,000 single Filipino males had migrated to the United States. This concerned both overt racists as well as the American Federation of Labor, which condemned Filipinos as low-wage competition. Organized labor hoped that with an end to colonialism, the oriental exclusions provisions of existing immigration law would be applied to the Philippines.

Other Americans worried about the cost and dangers of protecting the Philippines against Japan. In 1932, the Japanese army had seized Manchuria. Tokyo, many Americans feared, had designs on the Philippines as well. The rapidly expanding scope of Japanese commercial agriculture in the southern Philippines was a source of growing concern. By casting off its colonial responsibility, Washington could unburden itself of a vulnerable Pacific target.[18]

While Americans thought of independence as a simple and definitive act, Filipinos favored continuation of trade preferences following political separation. This ran counter to the motives of those advocating decolonization. To make matters more complicated, in 1931 the Philippine legislature sent an advisory mission to Washington headed by Sergio Osmeña and Manual Roxas (the "OsRox Mission"), erstwhile rivals of Manuel Quezon. They, like Quezon, hoped to win personal credit for independence. As a result, each faction denounced anything the other endorsed.

A variety of independence bills wended their way through the American Congress during 1932. Different versions called for a transition period of from eight to twenty years; for immediate, gradual, or deferred imposition of agricultural tariffs; and for a plebiscite on independence either before or after the transition period. The compromise Hare-Hawes-Cutting bill (HHC) gathered the most support. It called for holding a constitutional convention and plebiscite at the start of a ten-year commonwealth period, the gradual imposition of tariffs and quotas on exports, the retention of American military bases and possible international negotiations to neutralize the Philippines. It also set strict immigration quotas limiting migration to fifty Filipinos per year during the commonwealth period and zero thereafter.[19]

Manuel Quezon opposed the legislation, in part to deny his political rivals credit. American agricultural interests condemned it for not imposing immediate tariffs on Philippine product. Given Japan's recent aggression against China, Secretary of State Stimson and President Hoover worried that Tokyo might misinterpret the grant of independence as signaling acceptance of a dominant Japanese role in the Pacific. Hoover vetoed the bill in January 1933, and Congress promptly overrode the veto.

In Manila, Quezon mobilized his followers in the legislature to reject the Hare-Hawes-Cutting bill. At one level, he could not abide the fact that his party rivals, Osmeña and Roxas, had succeeded in getting Congress to grant independence. He also worried about the economic consequences of the loss of free trade and the Japanese response to the provision that gave America bases in the Philippines but refused to guarantee the nation's security.

Quezon complained that the bill combined the worst of several plans by stretching out independence during a ten-year transitional commonwealth period while crippling the island's economy by imposing tarrifs and quotas. He also condemned the lack of American security guarantees. Of all these considerations, the Filipino leader probably cared most about embarrassing his rivals who championed the HHC bill. Quezon rallied those who preferred immediate independence, those who opposed any independence, and sugar exporters who feared economic disaster if free trade ended. In October 1933 the Philippine legislature rejected the bill, thus nullifying its terms.

The following month, Quezon led a delegation back to Washington where he expected to win concessions for faster independence while preserving free trade for fifteen years. He also wanted security guarantees without providing bases for American forces. Neither Franklin Roosevelt nor Senator Millard Tydings, the leading exponent of independence in the new Senate, expressed sympathy with these demands. In fact, they informed Quezon, any changes in the political or commercial timetables that Congress might consider would be for the worse. The administration conceded only one point. After consulting with military officials, FDR agreed to drop the demand for postindependence army bases. Instead, Washington consented to negotiate with the new Philippine government for naval bases only.[20]

With administration backing, these terms were included in the Tydings-McDuffie independence bill, approved by congress and the president early in 1934. The bill resembled closely its predecessor in calling for a transitional commonwealth period lasting ten years. Beginning in 1941, a progressive 5 percent tariff would be imposed on exports to the United States, jumping to 100 percent with the coming of independence in 1946. Although many Filipinos (and Americans) feared this would devastate the new nation's economy, Congress insisted on severing all links to Manila as quickly as possible.[21]

In May 1934, Quezon steered the Tydings-McDuffie Act through the Philippine legislature. Shortly afterward, his faction won control of the delegates elected to the July constitutional convention. The charter adopted in February 1935 provided for a presidential form of government that gave the chief executive tremendous power over the legislature and local government. With the approval of the National Assembly, the president could exercise "indefinite dictatorial power." A national plebiscite approved the constitution in May 1935 and the following September Manuel Quezon was elected as the first president of the nascent commonwealth. Despite claims by his opponents (Emilio Aguinaldo and Bishop Gregorio Aglipay) of massive fraud, Quezon's inauguration took place in November.[22]

Fundamental questions regarding the economic and military security of the islands remained unsettled. Roosevelt promised to consider a more gradual imposition of tariffs, thereby cushioning the blow to the critical export sector (this was done in 1937). However, in the years before the Second World War, the Philippine Commonwealth did little to rectify either the severe maldistribution of wealth or prepare for the realities of independence. When Roosevelt offered no additional security guarantees, Quezon turned to his old friend, Douglas MacArthur.

The general had formed a close friendship with Manuel and Aurora Quezon during his earlier tours of duty in the Philippines. Unlike many American officers, the general felt comfortable with elite Filipinos, enjoyed their company, and treated them as social equals. The Quezons reciprocated these feelings. When Governor-General Frank Murphy criticized their ties to a military figure, Aurora Quezon explained the relationship in a uniquely Filipino manner. "Douglas," she told Murphy "is our brother." The political implications of this "family bond" grew even more significant a few years later when MacArthur made the Quezons godparents to his son, Arthur. As "compadres," an intricate set of obligations existed between the families.[23]

When Quezon visited Washington at the end of 1934, he hoped to recruit an American military advisory mission. (During the 1920s, the United States sent such missions to several Central American countries, including the Dominican Republic, Haiti, and Nicaragua.) War, Navy, and State Department officials approved the idea, as did key members of Congress and the president. At Quezon's request, and with the understanding he was the leading candidate for the job, MacArthur drafted a bill amending current legislation to qualify the Philippine Commonwealth for such a mission.[24]

In his memoirs, MacArthur reported that he told Quezon that the Philippines could be defended if enough men, weapons, money, and time were available. Unfortunately, war came too quickly and "American help came too late and too little." Quezon remembered his response rather differently, as "I don't *think* that the Philippines can defend themselves, I *know* they can." Compared to the prospect of either a corps area command or retirement at age fifty-five, the prospect of returning to Manila to take charge of forging a new nation's security must have seemed heaven-sent.[25]

According to Major Dwight D. Eisenhower, then MacArthur's aide, the general fervently desired to guide the destiny of the Philippines. He told Quezon that he had concerned himself with the archipelago's security ever since his first tour of duty there. The "fervid spirit of nationalism" in the islands could be harnessed to train Filipinos to defend "every portion of the archipelago" against invasion. Any "hostile attempt at landing" would be instantly repulsed by "overwhelming strength." The economy of the islands could, MacArthur claimed, easily sustain the defending army and the population against even a prolonged blockade—all at a reasonable cost.

In chronicling the genesis of the military mission, Eisenhower also stressed the personal "attractiveness of the new offer" to the retiring Chief of Staff. Ike knew how MacArthur dreaded the prospect of "stepping down to a subordinate position in the hierarchy of the American Army." Besides its attraction as a professional challenge, Eisenhower noted that for MacArthur "the position [in Manila] might well prove to be the springboard for attaining an even more desirable position in official or industrial life."[26]

President Roosevelt also supported the Quezon–MacArthur arrangement. In December 1934, he told the general and Secretary of War Dern to make plans for a military mission headed by MacArthur and endorsed the necessary legislation, which Congress approved in May 1935. Roosevelt probably realized that sending MacArthur to Manila solved at least two problems. The administration partly relieved itself of direct responsibility for the arduous, probably futile, task of arranging for the future defense of the Philippines. The high symbolism of sending a retiring Chief of Staff as military adviser might cause the Japanese to take the emerging nation more seriously. On a different level, Roosevelt the politician had honorably exiled an ideological opponent before he could have an impact on the 1936 election.

The legislation authorizing the military mission, and the personal agreement approved by Quezon, permitted MacArthur to retain his army rank and pay while receiving extremely generous benefits directly from the commonwealth government. These included a salary of $18,000, a $15,000 expense account, and a luxurious suite in the Manila Hotel. The general may also have made a secret arrangement for even more compensation. Unsigned documents found in MacArthur's records requested the commonwealth to pay him substantial sums as a "commission" over a seven-year period for his work in drafting a defense plan. As Sidney Fine, Frank Murphy's biographer notes, these plans were drafted entirely by American government employees in their official capacities—making payments to MacArthur a violation of American law. As subsequent events showed, MacArthur and Quezon on at least one other occasion arranged a secret payment for the general's service.[27]

In addition to these financial wrinkles, MacArthur continued to express interest in a civilian post in Manila. He had Quezon insert a clause in his contract that permitted him to temporarily retire as military adviser if he were offered the position of high commissioner (the successor post to governor-general at the creation of the commonwealth). With this agreed to, the secretary of war issued orders in September 1935 for MacArthur to go to Manila while still serving as Army Chief of Staff. He would retire as chief in mid-December, after assuming his new post. The general felt this would enhance his status in the Philippines.[28]

These arrangements fell apart almost immediately. For a variety of reasons, American and Philippine officials wanted MacArthur to remain in the United States until after the new commonwealth government took office. Also, much preparatory staff work remained to be completed before the mission left Washington. In the interim, the general still lobbied with Quezon and FDR to secure simultaneous appointment as the new high commissioner. Frank Murphy, current governor-general, was expected to retire shortly. After discussing the question with MacArthur, Quezon apparently advised FDR to combine the posts of military adviser and high commissioner.

In September, the general broached the subject directly with Roosevelt.

The president tentatively endorsed the idea but MacArthur discovered that a law barred active duty officers from accepting civilian appointments. MacArthur asked that FDR sponsor a special congressional act to amend the provision. Meanwhile, he initiated a campaign against Frank Murphy who expressed renewed interest in remaining in Manila. The general predicted that Murphy would behave as a "Super President of Commonwealth," interfering with Filipino self-determination. In reply, the governor-general informed Roosevelt that MacArthur did not know what he was talking about.

Since FDR considered Murphy a valuable ally destined for future service in his administration, he sided with the governor-general. In late September, the White House announced that Murphy would stay on in Manila as high commissioner and the top American authority in the Philippines. Roosevelt's watchdog, in effect, would oversee the military adviser.[29]

Roosevelt may have had an additional motive for stringing the general along and delaying his departure for Manila. Since the president apparently desired to eliminate a MacArthur protégé from the list of generals eligible to serve as Chief of Staff, he had to keep the current chief in office until at least October 1, 1935. After that date the officer in question, General George S. Simonds, would be too near retirement to warrant appointment. FDR preferred General Malin Craig, whom he named to the post as soon as MacArthur left Washington. As noted below, this incident widened the breach between Roosevelt and MacArthur.[30]

Oblivious to these bureaucratic plots, MacArthur's staff labored over the actual defense plans for the Philippines. A War College committee, under the guidance of Majors Dwight Eisenhower and James Ord, drafted a plan based on the Chief of Staff's ideas. The group began working in November 1934 and continued right through the time the mission arrived in Manila a year later.

MacArthur's belief in a defensible Philippines ran counter to the past and present thinking of the entire American military establishment. Although politicians at the turn of the century boasted of the islands as a strategic outpost and asset, as early as 1907 Theodore Roosevelt privately described it as an "Achilles heel" threatened by Japan. Washington never appropriated sufficient funds to fortify the colony, hoping instead to assure its security through diplomacy. By the 1920s, as part of the treaties reached at the Washington conference, the United States secured promises of noninterference by Japan in return for agreeing not to fortify its major Pacific possessions.

In case of a Pacific war, the formal defense scheme, known as War Plan Orange, called only for holding Manila Bay while awaiting rescue by the U.S. Fleet. A 10,000-man garrison (half of whom were Philippine Scouts) and the 6,000-strong Constabulary were expected to resist up to 300,000 Japanese invaders for at least six months. The navy, in fact, estimated it might take two or three *years* for the fleet to fight its way to Manila, leaving the defenders little hope. Not surprisingly, in 1935 the head of the army's War Plans Division characterized the Philippines as a "military liability of incalculable magnitude."[31]

Even those drafting the "MacArthur Plan" questioned its logic. Typical among them was Major Eisenhower, who dreaded the prospect of going to the Philippines. MacArthur prevailed upon Ike to serve in Manila "for a year or so." Infact, the major remained in the Philippines until late 1939.

Eisenhower and Ord proposed a universal conscription scheme costing some 50 million pesos ($25 million) per year. MacArthur and Quezon resolved it must be cut in half. The next draft assumed that the U.S. Army would provide obsolete weapons at little expense, that few officers would be recruited, and that conscripts would receive almost no wages during their period of service. Even this proved to expensive for Quezon who imposed a 16 million peso ($8 million) ceiling on military expenditures.[32]

Ord revised the plan further downward, proposing a regular army of only 930 officers and 7,000 enlisted men. Each year, about 7,000 conscripts would be trained as a reserve force. Unlike their boss, MacArthur's staff dismissed the idea of insulating the Philippines from attack or defeating an aggressor without outside help. If Japan invaded, local forces would begin a war of attrition, holding out long enough for "powerful allies" (e.g., the United States) to come to the rescue. The plan also called for a small navy and air force to harass invaders, but no funds for such items existed. "Even though our Chief," Eisenhower recalled, "spoke and wrote in purple splendor," enthusiasm could not compensate for inadequate "amounts of money and men." They "would be scraping for pennies and pesos" while organizing a "skeleton force that some day might have flesh on its bones."[33]

During the long months of preparation, MacArthur refused to be burdened by any of these doubts. Failure to secure the high commissioner's post disturbed him more than the warnings of Ike and Ord. On the other hand, he was pleased that FDR had agreed that he should arrive in Manila still the American Army's Chief of Staff. As he left Washington on October 1, 1935, he emphasized the importance of this for impressing the new commonwealth government.

Even before the general reached the West Coast, Roosevelt yanked that rug from under his feet. During a train stop in Wyoming, the War Department notified him by telegram that the president had just named Malin Craig as Chief of Staff, effective immediately. MacArthur would now arrive in Manila a former head of the army, reduced from his temporary four stars to the two stars of his permanent rank of major general.

Upon receiving this blow, Eisenhower recalled, the general launched into an "explosive denunciation of politics, bad manners, bad judgment, broken promises, arrogance, unconstitutionality, insensitivity and the way the world had gone to hell." If, as James Farley claimed, the president had tricked MacArthur into delaying his departure until October to assure the appointment of Craig, the ploy had ugly ramifications. No one in the White House considered the consequences either for the Philippines or Mac-Arthur's pride. This indifference spoke volumes about the status of the impending mission and the future relations between Roosevelt and MacArthur.[34]

3
The Field Marshal of the Philippines

MacArthur's passage to the Philippines began inauspiciously and seemed to get worse as he approached his destination: during the ocean voyage Pinkie, now in her eighties, fell terminally ill and died a few weeks after reaching Manila. Even in death she exercised a hold on her son who had Pinkie's body preserved in the local morgue for over a year until he could escort it back for burial in America.

Despite (or because of) these strains, MacArthur met a women during the voyage with whom he immediately fell in love. Jean Faircloth, a 35-year-old, unmarried Southerner, had booked passage on the *President Harding* to visit friends in Shanghai. During the otherwise somber trip MacArthur found himself charmed by and drawn to her. She accepted his invitation to visit Manila and soon took up residence there. They married in April 1937 during a trip to the United States—the general's last until 1951. By all accounts, they adored each other and their happiness grew when Jean bore a son, Arthur IV, in 1938. Unlike the strong willed and tempestuous Louise or the furtive romance with Isabel, Jean devoted herself to making a home for Douglas and never competed with the army for his affection. Even twenty-five years after his death, Jean still referred to her husband as "mah general."

MacArthur's romance compensated for numerous professional disappointments in Manila. Although his selection as military adviser seemed a popular one, many Filipino and American officials had questioned the choice even before he assumed the post. Tensions had emerged as early as the summer of 1935 when the general tried to get FDR to appoint him to the new post of high commissioner. Murphy grew furious when he heard rumors that MacArthur had criticized his job performance. According to Murphy, MacArthur had struck a deal with Manuel Quezon to undermine the terms of the Tydings-McDuffie Act, allegedly encouraging Quezon to treat commonwealth status as merely a euphemism for de facto independence. And later, when Roosevelt decided to keep Murphy in Manila, the general tried to convince Quezon that the high commissioner would play only a ceremonial role.

The dispute surfaced publicly when MacArthur and Murphy clashed over the honors to be extended to the Philippine president at his November 1935 inaugural ceremony. Quezon demanded a twenty-one-gun salute (traditionally reserved for a sovereign chief of state). The high commissioner maintained he warranted no more than nineteen guns, appropriate for a state governor. MacArthur backed Quezon, leading Murphy to conclude that the military adviser would gladly compromise national honor in order to curry favor with Filipino politicians.[1]

Although it appeared to resemble the squabbles satirized by Jonathan Swift in *Gulliver's Travels,* the argument elicited an incredibly long and bitter debate among all the principals. In September, a bemused Roosevelt agreed that Quezon might have his twenty-one guns. Murphy and Secretary of War George Dern then went on the offensive, convincing FDR, just before the inauguration, to reverse himself. When Quezon threatened to boycott the ceremony, one of Murphy's allies assured the administration that "nothing . . . could have kept the *Presidente* from that occasion except his own demise." Roosevelt also informed Quezon that the high commissioner outranked all other American officials, including the military adviser.[2]

In addition to this imbroglio, Murphy and many senior War Department officials judged MacArthur's Philippine defense program as grandiose, inappropriate, and even dangerous. Suspicious of the general's initial plans, they were astounded by his formal "Report on National Defense in the Philippines," which he submitted to the commonwealth government in April 1936. The report opened a window into MacArthur's thoughts on both the Philippines and America's future in Asia. It was, he asserted, a "general exposition" on events of "international significance."

The new military adviser's report on an Asian country sounded remarkably like that of his father's from the 1880s. In many respects, MacArthur declared, the new commonwealth comprised "the most important section of the great and vaguely defined region known as the Far East." The "attention of thoughtful statesmen and businessmen throughout the world" would focus on that area for the "next century and more." Events there would have a "universal influence upon the prosperity and welfare of all civilized peoples."

Like earlier chroniclers who had traced the "science" of the rise and fall of civilizations, MacArthur saw commerce, power, and civilization unfolding according to inexorable patterns. He warned that the industry of the western nations had reached a state where "production far exceeds consumption." Already, "every existing market has been feverishly competed for and exploited," yet none could "absorb the overages that world industry constantly produces." Without vast new markets, an economic calamity in the West would "result in great upheaval" fatal to "governments and industry alike."

Commercial opportunities in the Orient, he predicted, could cushion the blow to the West and "postpone indefinitely" the coming crisis. Out of self-interest, America and Western Europe would have to "assist rather than merely to exploit" Asian states. Washington should acknowledge the connection between its own prosperity and security and that of the Philippines.

The report spoke of organizing a vast reserve of "citizen soldiers" led by a small regular army. Trained from primary school on, stationed throughout the archipelago, augmented by a small flotilla of torpedo boats and an air force, the Philippine army would deter or destroy any invader, "no matter how ruthless and powerful." Progress toward this goal, MacArthur claimed, had already "exceeded original anticipation". By following his plan, the Philippines would achieve a "peace which upholds Christian virtues and defies the threat of rapacious greed—a peace that will mean continued happiness and freedom for God-worshipping and democratic people." He would, as he had promised Quezon in 1934, "forge you a weapon that will spell the safety of your nation from brutal aggression until the end of time." [3]

This kind of talk made people sit up and take notice of MacArthur. Unfortunately, the facts were less compelling than the rhetoric. As he issued this declaration, not a single conscript had been called into service, not one training camp existed, no weapons were available for the proposed army, and virtually no military professional (aside from himself) thought the plan practicable. While the general and his defenders later blamed Washington for providing insufficient support and damned Tokyo for attacking "too soon" (in 1941 instead of 1946!), these excuses did not alter the fact that his inviolable Philippine force collapsed like a house of cards in December 1941.

MacArthur compared his program to the Swiss model of national defense. (Ironically, he would also use the Swiss analogy to describe a totally *disarmed* Japan in the late-1940s!) Reserve units scattered throughout the Philippines would provide the muscle to deter or defeat any aggressor. Critics argued that dispersing forces, weapons, and training camps would simply make it easier for a highly mobile enemy to subdue the archipelago piecemeal.

Although MacArthur barely mentioned the problem, the most pressing danger in the Philippines came from *internal* rebellion by disaffected political leaders in outlying areas. They saw the scattered arsenals and units as potential resources for their contest with the central government. During the early 1930s, for example, the Sakdalista movement gained great strength by organizing peasants against both colonialism and the landed oligarchy. The radical wing of this movement staged a bloody, if brief, uprising in May 1935, which the constabulary crushed.

In December 1935 nearly the entire General Staff endorsed a report by the Army War Plans Division, which debunked MacArthur's assumptions. It described the Philippine army as "wholly ineffective" in protecting the country against Japan. Ridiculing the Swiss analogy, the General Staff noted the complete geographic, social, and economic differences between the Philippines and Switzerland. Japan could easily overrun small, isolated army elements through its vastly superior firepower and mobility. Unlike Switzerland, the Philippines lacked the economic base to sustain any viable military establishment or to survive a prolonged blockade. The army envisioned by MacArthur had "little or no value as an end in itself." Its sole purpose might be to supplement "military (including naval) measures the United States might be induced to take for the defense of the Philippines."

The war plans staff feared that the proposed defense program risked bankrupting the new nation while, possibly, provoking Japanese aggression. These critics urged concentrating American defense efforts "purely and simply for the purpose of maintaining internal order." They favored gradually expanding the Philippine Constabulary (the existing national police force) instead of the building a large, expensive army.[4]

Self-confident and buoyed by President Quezon's strong initial support, MacArthur dismissed his critics. During 1936 he pushed his staff and the commonwealth government to implement the concepts he had articulated. The military adviser saw his role as giving overall direction and vision to the emerging nation. His staff and the Filipinos were expected to find the means to realize the dream.

From the beginning, the general's assistants such as Dwight Eisenhower and James Ord questioned the wisdom of their assignment. For example, Ike complained that he and Ord "learned to expect from the Filipinos with whom we deal a minimum of performance from a maximum of promise." Commonwealth officials cheered the defense plan, but displayed indifference toward its implementation. The War Department delayed sending surplus equipment, apparently preferring to postpone action until after the November 1936 presidential election. "From the President on down," Eisenhower complained "each official seems to act individually and on the spur of the moment with respect to any detail in which he is interested and without regard for possible effects upon other activities or upon the army as a whole." Only when patronage appointments came up did Quezon's staff evince much concern with national defense.[5]

In February 1936, MacArthur told Major Eisenhower that he might be named field marshal in the Philippine army. Ike rejected the suggestion that he, too, should solicit a promotion and tried, without success, to dissuade MacArthur. The advisory missioned depended on the good will and assistance of the American army in the Philippines. Officers in the Philippine Department already considered MacArthur's effort to develop a national army ridiculous. Flaunting an exalted Filipino rank, Eisenhower predicted, would further alienate the American military, while also dissipating the authority they possessed as American officers in dealing with the Filipinos.

Why, Ike wondered, did a man who had achieved a general's rank in the United States, want a "banana country" giving him a field marshalship? Eisenhower urged his boss to avoid the pomp and instead visit Washington to plead for more equipment. MacArthur ignored this counsel. He told his staff he would not go home to lobby until the late fall, after the 1936 election. The reason for this became clear shortly, when MacArthur announced his certainty that the Republicans would oust Roosevelt from the White House.[6]

The results of the first draft call in April 1936 surprised the American military advisers. An unexpectedly high total of 150,000 Filipinos responded. Upon closer scrutiny, however, the numbers appeared inflated, with many unqualified registrants included. MacArthur dismissed this evidence and, "quite suddenly," altered plans for the coming year. Eisenhower and Ord believed they could not train more than 6,000 conscripts during 1937. The general insisted that 40,000 be enrolled, half in January and half in July—even though his office lacked equipment, facilities, instructors, and money for the vast increase. The national assembly, he indicated, would appropriate the needed funds once the expansion began.

Eisenhower realized that even if the money came in later, it would not be possible to select and build the 125 training camps needed for the new recruits. When he and Ord strongly urged MacArthur to reconsider his order to call up 20,000 men in January, Ike remembered that "he gave us one of his regular shouting tirades." The general simply ignored the fact that "we cannot possibly get the money for some time yet," nor even begin to construct the massive infrastructure required for such an undertaking.

More worrisome, a distance began to develop between MacArthur and Quezon. The commonwealth president would make decisions and "we know nothing of them," Ike complained. "If the General would take the trouble to see Quezon weekly," these surprises might be avoided. But, MacArthur "apparently thinks it would not be in keeping with his rank and position for him to do so."[7]

The military adviser ignored these problems but decided, in mid-1936 to accept the title of field marshal, even though Ord and Eisenhower refused simultaneous promotions to the rank of general in the Philippine Army. The honor, in Ike's opinion, left MacArthur "tickled pink" and certain that "he's made a lot of 'face' locally." (Only later did Eisenhower learn that the general, not Quezon, conceived of the unique title.) On August 24, at an elaborate ("rather fantastic", in Ike's words) ceremony at Malacañan Palace, Aurora Quezon presented a new field marshal with his gold baton. MacArthur designed a flashy sharkskin uniform for the occasion, covered with braid, stars, and flashy lapels.[8]

The field marshal's optimism extended to (or derived from) his certainty that Republican nominee Alf Landon would sweep Roosevelt out of office in November. On the strength of the notorious poll conducted by the *Literary Digest* (predicting a Landon victory), he announced to Eisenhower and Captain T. J. Davis that "Landon is to be elected, probably by a landslide." When his aides expressed their belief that "Landon cannot even carry Kansas, . . . he got perfectly furious" and repeated his "almost hysterical condemnation of our stupidity." Only "fearful and small minded people," he admonished, would ignore the "evidence at hand."

MacArthur revealed that he and Quezon had based their future plans and pending trip to Washington "on the theory that Landon will be elected." The general admitted he had "bet several thousand pesos" on the election. After FDR's overwhelming victory, MacArthur backpedaled rapidly." He "accused the *Literary Digest* of crookedness" and feared that word of his enthusiasm for Landon would get back to Roosevelt. In postelection correspondence, the general insisted he had expected the president's reelection. The "victory was a great one," he wrote a friend in Washington, and "must have satisfied even the most vaulting ambition." MacArthur never bothered, however, to express regrets to his aides for what Ike characterized as "his awful bawling out."[9]

Roosevelt's victory, expected or not, boded ill for MacArthur's program. The War Department evinced even less sympathy than before, continued to charge high prices for surplus weapons and did nothing to expedite deliveries. Actually, the root of the problem was political. Officials in Washington, including the president, expressed growing disbelief in the Philippine defense plan.

Chief of Staff Malin Craig and his subordinates complained that while MacArthur peppered them with requests for more equipment, he never bothered to consult them about his policies. Shortly before the presidential election, Craig, Acting Secretary of War Woodring, and several State Department officials urged Roosevelt to call MacArthur home for consultations. When Woodring cautioned Roosevelt that MacArthur might respond by quitting ("unfortunate from the purely political view" right before the election), FDR decided to defer any action until after November.[10]

These critics feared that MacArthur's proposed arms buildup actually imperiled the Philippines and precluded a possible neutrality agreement with Japan. Diplomats considered it possible to make a deal whereby neither Tokyo nor Washington would further militarize their Pacific islands.

When Assistant Secretary of State Francis Sayre broached the idea with FDR in November 1936, the president seemed supportive. Roosevelt told Sayre (and on another occasion his cabinet) that an agreement with Japan probably represented the only hope of protecting the Philippines. Even if the islands remained under direct American protection, he predicted, they would be overwhelmed quickly by a Japanese attack. If war came, FDR explained, he would evacuate what forces he could and, after two years, "would gradually move westward in an island hopping operation" to recapture Manila. Although nothing came of the neutrality proposal, Roosevelt's forecast of Pacific war strategy proved uncannily accurate.[11]

Before retiring as high commissioner at the end of 1936, Frank Murphy also warned his superiors of the "growing menace" of the military mission in Manila. Sending MacArthur was a mistake from the beginning, he argued, and it should now be "abandoned as soon as practicable."

Others took up the chorus. Harold Ickes attacked MacArthur as a virtual mercenary, "greedy for power," wandering about Asia in search of a personal army to command. At a staff conference, General Malin Craig characterized the military adviser as a "lone horse" who should be recalled. War Plans Chief, General Embick, concurred, but predicted that a recall would leave the government with a "difficult situation on its hands" for MacArthur would "propagandize" against the administration when he returned. Nearly all concerned officials urged Roosevelt to bring the general home.[12]

Months earlier, the general had arranged to visit the United States where, he expected, a Republican president would heap rewards upon him and his mission. Now he would journey home as a supplicant. He and Quezon put a brave face on their mission by announcing their intention of visiting America to attend the swearing in of Paul McNutt as the new high commissioner to the Philippines.

Quezon and MacArthur crossed the Pacific by way of Japan, arrived in California, proceeded to New York, and waited, in vain, for an invitation to the White House. Quezon had proclaimed publicly his intention of demanding full commonwealth independence in 1938, seven years ahead of schedule. Roosevelt would probably have relished unburdening himself of the Philippines, if it could be done without domestic or international repercussions. But in going public without consulting the administration, Quezon had thrown down a challenge that gravely offended Roosevelt.

Finally, MacArthur requested a private audience with the president. The general pled for two hours the case for receiving Quezon. FDR relented, but brushed aside summarily Quezon's talk of early independence. Congress expressed similar lack of sympathy. Eventually, the administration agreed to phase in certain duties on Philippine exports more gradually, but balked at additional concessions. MacArthur also failed to coax more military supplies from the War Department.[13]

Shunned in Washington, the travelers soon left on a visit to Mexico. From there, Quezon began his journey home, via an extended tour of Europe. MacArthur opted to take care of personal business in America. First he arranged the burial of his mother's remains. With Pinkie finally interred, he felt free to marry. On April 30, 1937, he and Jean Faircloth wed in a civil ceremony in New York City and spent their honeymoon returning to the Philippines.

Back in Manila, MacArthur tried to salvage his dream. His staff spent the bulk of their energies teaching recruits the rudiments of sanitation and literacy, not military tactics. Only half way into 1937 they had spent all current funds and began to borrow against projected 1938 revenues. The Regular American Army in the islands, called the Philippine Department, became increasingly uncooperative with the mission. Even commonwealth officials discussed cutting the military budget and terminating the defense program in 1941, years ahead of MacArthur's schedule. Eisenhower and

Ord dismissed the operation as hopeless and suggested that the program train Filipinos only to suppress a domestic revolt or to slow an enemy advance until the United States came to the rescue.[14]

When, during 1937, the usually docile Manila press echoed these doubts, MacArthur lashed out at members of his staff. In July, he announced his readiness to "dispense, at a moment's notice, with the service of any or all members of the mission." He accused Davis, Ord, and Eisenhower of "conceit" and "looking out only for" themselves. Ike felt his chief resented the staff "simply because we labor under the conviction that someone ought to know what is going on in this army and help them over the rough spots."[15]

In August, Army Chief of Staff Craig notified MacArthur that his tour in Manila would end on December 31, two years after it began. Although the official record does not indicate what part the president played in this decision, Craig told Robert Eichelberger (then an officer on the General Staff) that FDR wanted MacArthur both out of Manila and out of the army. Allegedly, Roosevelt agreed to allow the general to retire while still in the Philippines with the understanding that he would "return to his home in Milwaukee and engage in cultural pursuits."

At first MacArthur informed Craig he was prepared to assume any "duty the War Department may have decided I should now undertake in the service of my country." But, on September 16, the general reversed himself. Pleading ill health, the completion of his work, and a desire to make way for the promotion of more junior officers, he applied for retirement as soon as possible. He told Craig that agreeing to be sent to a lesser command would be as if "President Roosevelt were required to go back to his former functions as Assistant Secretary of the Navy." Without comment, the army chief and Roosevelt accepted the retirement, effective December 31, 1937.[16]

MacArthur had little desire to leave Manila and still less to take up cultural pursuits in Milwaukee. Jean was pregnant, expecting a child early in 1938. The couple enjoyed the social life of the capital and their luxurious suite in the Manila Hotel. Douglas became a partner in the corporation owning the hotel and probably dabbled in gold mining ventures. In America he had no home nor any immediate employment prospects.

Manuel Quezon solved his dilemma by asking his friend to stay on as military adviser employed directly by the commonwealth. The president introduced a bill to that effect in the National Assembly in November. The assembly promptly voted a resolution of thanks to MacArthur for his "invaluable services in organizing the Philippine Army and national defense," but deleted a clause inserted by Quezon authorizing his retention as military adviser with the rank of field marshal. Some members objected to his proposed salary ($36,000), which reflected the total he had received annually from both the American and Philippine governments. Undeterred, Quezon issued an executive proclamation retaining MacArthur. The junior members of the mission continued their service as regular American army officers.[17]

Ultimately, MacArthur's defense program depended on strong support from Washington, close cooperation with his staff, Quezon's faith and backing, and deferral of the Japanese threat for several more years. The diminished support of the Roosevelt administration was apparent by the end of 1937. Each of the remaining pillars eroded in turn, leaving the program—and MacArthur's vision—a hollow shell.

MacArthur's alienation from Eisenhower, Davis, and Ord—the key members of his staff—accelerated in January 1938. The field marshal, Eisenhower recalled, hoped to encourage support for his program by staging a gala parade in Manila. Ike and Ord protested that the proposal would bust the budget, taking "money that was desperately needed for more important purposes." But MacArthur insisted on going forward.

Quezon, it turned out, knew nothing of the plan and exploded at the news when he quizzed Ike about proposed troop movements. Eisenhower and Ord were "astonished" that their chief had not cleared the idea with commonwealth officials. When Quezon criticized MacArthur, he tried to blame his staff. MacArthur raged at them that "he had never meant for us to proceed with preparations for the parade . . . he only wanted us to investigate it quietly." He also denied that he had "given us an order— which was certainly news to us." Even though Ike stayed on for another year and a half, he felt that "never again were we [he and MacArthur] on the same warm and cordial terms." Ord's strained relationship with the general ended altogether at the end of January when he died in an air crash near Baguio. For Ike, this tragedy eliminated the last solace of service in Manila.[18]

Even though Quezon agreed, in 1938, to become *compadre* (godfather) to the general's new son, the president began to drift away from his field marshal. During the summer of 1938 Quezon took a secret trip to Japan, which MacArthur only learned of belatedly through press reports. Possibly, the Filipino leader explored a neutrality agreement with Tokyo. Like many elite Filipinos, Quezon feared Japanese power even while admiring that nation's wealth, discipline, and the respect accorded it by the West.

In mid-1939 Quezon induced the National Assembly to separate the Constabulary (with its internal security functions) from the new army, as MacArthur's critics had urged. this deprived the army of important talent and equipment. The usually soft-spoken vice-president, Sergio Osmeña, also spoke against the military adviser. Probably with his boss's encouragement, he complained that MacArthur's program could not succeed and that "if pursued long enough it might well create conditions of false security" against Japan. Quezon ordered several staff changes in the army, replacing MacArthur loyalists with officers who voiced doubts about the force's ability to withstand a Japanese assault. During the next year, Quezon expanded his personal authority over the armed forces, limiting

MacArthur's freedom to take action independently of the president. Increasingly, the commonwealth president avoided speaking directly with the field marshal and, instead, made arrangements through subordinate staff.[19]

In 1939 and 1940 Quezon encouraged the Assembly to cut defense funds, making it impossible to create a small navy or air force. In these years draft registration and actual conscription fell precipitously. Those who appeared for service received only the most rudimentary military instruction, often without equipment. By 1940, the regular army consisted of only 468 officers and about 3,700 enlisted men. The impressive reserve, of some 130,000 men, existed mostly on paper.[20]

By late 1939 President Quezon referred openly to his waning faith in MacArthur. He told an audience celebrating the commonwealth's founding that "even if every last Filipino were armed with modern weapons" the islands could not be defended. In October, when Francis Sayre succeeded Paul McNutt as High Commissioner, Quezon broached with him the idea of firing MacArthur. Allegedly, he told Sayre that he now considered the whole MacArthur defense program "idiotic." Although Quezon decided not to dismiss his military adviser, he insisted that all further communications between them pass through the presidential secretary. In 1941 Quezon allies told journalist Clare Boothe Luce that the commonwealth leadership "was sick of doling out funds to MacArthur."[21]

Following the death of James Ord, Eisenhower resolved to get out of Manila. During 1938, he spent four months of leave in the United States and decided to terminate his work in the Philippines as soon as possible. With the outbreak of war in Europe in September 1939, Eisenhower insisted on leaving. MacArthur valued Ike's work and urged him to stay—until he discovered that Quezon had also intervened.

Quezon offered Ike a "blank contract for [his] services" and told the lieutenant colonel to name his price. Although flattered, Ike told Quezon that "no amount of money" would change his mind. T. J. Davis left about the same time, the last of the original staff of 1935.[22]

Clearly, Quezon's efforts to retain Ike in Manila drove a wedge between the general and commonwealth president. However, it undermined fatally the relationship with Eisenhower. Upon returning to the United States in 1939, Ike told Robert Eichelberger (who would serve under MacArthur in the coming war) that MacArthur had a nearly paranoid fear of subordinates "stealing his publicity." This "led eventually to the severance" of their ties. To make matters worse, after leaving Manila, Ike apprenticed himself to General George C. Marshall, the new Chief of Staff and MacArthur's rival.

Through the Second World War and beyond, MacArthur made frequent references to "that traitor Eisenhower." In 1942, he told a British confidant that years before, when he had "his difficulties with Filipino politicians," Eisenhower sided "with the latter and built up a strong personal position with them." MacArthur also claimed that Quezon and Ike conspired secretly to boost the lieutenant colonel's salary, whereupon Mac-

Arthur decided he was not "wholly loyal" and terminated his job. (In fact, it was MacArthur who lobbied with Quezon to boost his own pay; Eisenhower insisted on leaving Manila, despite MacArthur's pleas that he stay.)[23]

MacArthur's isolation increased steadily after 1938. When an old friend wrote encouraging him to return home as a Republican candidate for president in 1940, he gave an evasive but unenthusiastic reply. Interestingly, he coupled his remarks about lacking any political ambition with a winsome phrase that he asked his friend to pass on to the ailing General Pershing. He recalled the verse of the "famous old Philippine veteran's song that 'old soldiers never die, they just fade away.' " MacArthur savored the line for the next thirteen years.[24]

By the summer of 1939 MacArthur practically threw himself on Roosevelt's mercy. When Paul McNutt retired as high commissioner, the military adviser begged for appointment to the office twice denied him. On July 14, 1939, MacArthur cabled presidential aide Steve Early (whom he had befriended during World War I) of his interest in replacing McNutt. Roosevelt, he recalled, had discussed the post with him in 1935 and "he would be very glad to have the President's favorable consideration" at this time. As high commissioner, he "could render the President and the country valuable service in the Philippines." Ignoring this plea, FDR appointed Francis Sayre to the post.[25]

MacArthur and Sayre had little use for each other over the next two years. Admiral Thomas C. Hart, Commander of the Asiatic Fleet based in Manila, commented upon MacArthur's alienation from nearly all American and Filipino officials. In late 1940 he observed that "Douglas seems to have sunk to an unenviable official position and relationship" in which the "dignified thing to do would be to resign." A few months later, the admiral noted that practically no one spoke to the military adviser any longer. If it were not for his exorbitant salary, Hart believed, MacArthur would "duck out, because he doesn't appear to be too happy in the job."[26]

In December, 1940, Theodore White, then a young journalist beginning a career with *Time*, visited Manila. Officers in the Regular Army ridiculed openly the "the Napoleon of Luzon." According to one army major, the retired MacArthur "cut no more ice in this U.S. Army than a corporal." When White interviewed MacArthur, the general, "paced and roared, and pointed, and pounded and stabbed with his cigar, and spoke with an intelligence and magniloquence and a force that overwhelmed. He was holding himself, he said, in readiness to command the American expeditionary force in Asia when the war broke out." Dismissing the Japanese army as "not even second class," he predicted that his Philippine Army would soon be able to hold its own against Japan.

White's article for *Time* praised MacArthur as the best western military commander in all Asia. When the young journalist sent the general a draft of his piece, he received an invitation to come to MacArthur's suite. While the older man paced the floor in his old West Point bathrobe, the two of them "rejoiced" (in White's words) "that we alone understood the Japanese peril to America." Like his father, MacArthur reflected, he always felt his "fate and Asia's were intertwined." As the sun set over Manila Bay, he turned to his young companion and said: "It was destiny that brought us here, White, destiny! By God, it is destiny that brings me here now."[27]

In spite of his bravado, MacArthur had tentatively decided to leave Manila and return home as a civilian, unless he secured a civil commission in the Philippines. In March 1941 it appeared Sayre might return to the State Department. The general wrote again to Steve Early, making "application for appointment" as "he understands there may be a vacancy in the office of high commissioner. MacArthur informed Early that since his work as military adviser was "now completed," he intended to return home shortly. However, if offered Sayre's position, he would stay.

Besides enjoying the "the complete confidence of the Filipinos," MacArthur boasted, "From Vladivostok to Singapore I am thoroughly familiar with the most intimate details, political, military and commercial. I have a personal acquaintance with everyone of importance in the Orient and I believe no American holds the friendship and respect of this part of the world more than myself." He would help Roosevelt solve America's "Pacific problems" by responding "to any call here or elsewhere." No longer plagued by the age and infirmity he previously listed as reasons for retiring, MacArthur assured Early of his "robust health" and eagerness to serve FDR—"not only our greatest statesman but what to me is even more thrilling, our greatest military strategist."[28]

The next month presidential aide General Edwin "Pa" Watson replied that Roosevelt could offer nothing definite but hoped MacArthur would stay in Manila. In discussions of new assignments, Watson assured him, MacArthur's name was "most seriously considered." Steve Early informed the general that although he did not know "whether the President can utilize your services," Roosevelt's face showed "evident pleasure" when he read MacArthur's effusive praise. Now more hopeful of returning to active duty, the military adviser notified Early on May 11 that he stood "ready to respond to any call" from the White House.[29]

As time passed and he heard nothing, a despondent MacArthur wrote to Steve Early and Chief of Staff Marshall that he intended to close down the adviser's office and return to America. "Destiny", it seemed, had pretty much played itself out. But then Japan's thrust into Southeast Asia, also justified as destiny, rescued MacArthur from the shadows.[30]

4
The Road to War, 1937–41

MacArthur's Philippine defense program faced an overwhelming obstacle that not even his optimism could overcome. Japan resumed its military expansion in 1937, undeterred by American or European resistance. Tokyo intended to incorporate all East and Southeast Asia into its "Greater East Asia Co-Prosperity Sphere," regardless of developments in Manila. Ironically, MacArthur's last minute success, late in 1941, in convincing Washington to augment his strength, made the Philippines a primary military target.

The road to Pearl Harbor began with Japan's invasion of China and soon escalated when Tokyo sought to extend its hegemony over colonial Southeast Asia. At first indifferent to the contest in Asia, Americans experienced a panic after the string of German and Japanese advances in 1940–41. Roosevelt and the War Department turned belatedly toward MacArthur when all other pillars of resistance in the Pacific collapsed.

Since the 1920s Tokyo had feared a united China, be it led by Mao Tse-tung's (Mao Zedung) Communist movement or Chiang Kai-shek's (Jiang Jieshi) Nationalist, or Kuomintang, party. By 1937, the combined pressures of the world depression and an anti-Japanese united front within China convinced Japan's leaders they must dominate their continental neighbor.

On July 7, 1937, Japanese forces stationed in Manchuria began their assault on north China and the fighting soon escalated into a full scale, if undeclared, war. Within a year the invaders controlled the rich, eastern third of China and much of the coast. The Chinese government (in effect, the Nationalist party) as well as Communist forces, retreated inland, trading space for time. Chiang Kai-shek counted mostly on holding out until Japan blundered into war with America. The Communists picked up the cause of resistance and reform to build a peasant guerrilla army and a rival government.

Although most Americans sympathized with China's plight, Washington preferred to avoid a confrontation with Tokyo. Gradually, however, the specter of Japanese brutality and the prospect of that nation dominating Asia and the Pacific aroused the United States. Late in 1938, the likely collapse of organized Chinese military resistance convinced Roosevelt and his aides to begin economic and, eventually, military assistance to the Nationalist regime. American strategists hoped to tie down and weaken Tokyo's legions in a war of attrition in China, thereby buying time or deflecting danger away from American or European possessions.[1]

Between 1939 and 1941 Roosevelt followed up the first small loan to China with escalating assistance and selective embargoes on sales to Japan. The trade weapon had to be used delicately against Tokyo as any precipitous cutoff might cause Japan to seize mineral-rich Southeast Asia. Eventually, FDR hoped, the Chinese, Americans, and Europeans would grow strong enough to deter or throw back any Japanese military challenge. Neither he, nor anyone else in Washington, considered the Philippines a significant element in this scheme.

Roosevelt first counted on Britain, France, and Holland (the major Asian colonial powers) and the Soviet Union (which had given Chiang military aid and skirmished with the Japanese in 1937–38) to help restrain Japan. Stalin, however, feared the Western powers as much as the fascist states and signed nonaggression pacts first with Germany (1939) and then Japan (1941). Germany's conquest of western Europe by June 1940 left besieged England the only European ally able to support American policy in Asia and the Pacific.

The Japanese responded to Germany's blitzkrieg by occupying northern French Indochina, demanding increased access to petroleum in the Dutch East Indies and entering the Axis Alliance. Frustrated by three years of stalemate in China, Tokyo hoped London and Washington would be too preoccupied to resist its pressure on Southeast Asia. The festering war on the Asian continent still vexed the Japanese. But the main impediments to their strategy came from the Pacific Fleet at Pearl Harbor and American forces in the Philippines.

Up to this time, War Plan ORANGE (the standing plan for fighting Japan in the Pacific) presumed a minimal defense effort in the Philippines. American forces around Manila were expected to withstand a Japanese siege for up to six months while awaiting relief. In 1938 the chief of the War Plans Division concluded that "our present concept" no longer favored a prolonged defense since, in case of war with Japan, it seemed "highly improbable" that any relief expedition would be sent. Even if forces were sent, no estimate could be made of "when they would arrive."

President Roosevelt enhanced overall American preparedness by appointing, in September 1939, George C. Marshall Army Chief of Staff and, in June 1940, Henry Stimson Secretary of War. Yet neither man saw much value in dispersing scarce resources to Manila. During 1940–41 army and navy planners debated whether and how to protect the archipelago. Near the end of 1940 the War Department urged withdrawing all American forces, but just two months later, recommended reinforcing the Philippines. It hardly mattered, though, since no one bothered to implement either policy.[2]

Far Eastern military strategy remained a muddle during early 1941. Discussions in Washington among British and American military planners resulted in the so-called ABC-1 plan, affirming that, in case of war with Germany and Japan, the nascent allies would give priority to the war in Europe. Although only a contingency arrangement, these discussions set the course followed during the Second World War.

In April, American naval officers conferred with their Dutch and British counterparts in Singapore to consider committing greater resources to Southeast Asia in case of a Pacific war. But their superiors in Washington rejected this. Specifically, they decided "not to reinforce the Philippines except in minor particulars." The update of War Plan ORANGE that month further limited the American mission in the commonwealth. Army forces were expected to defend the access to Manila Bay, but they might have to hold out for two years before the navy relieved them. American war planners totally disregarded the contribution of Filipino forces organized by MacArthur over the previous several years.[3]

Ignoring these estimates, Field Marshall MacArthur radiated self-assurance. In an interview with the young journalist, John Hersey, the general predicted that if Japan entered the war, "the American, the British, and the Dutch could handle her with about half the forces they now have deployed in the Far East." The stalemated war in China had so sapped Japan that about "half" its army had been "reduced in effectiveness from first class to third class standing." In addition to Tokyo's weakness and fear of the western powers, MacArthur asserted, Germany would restrain Japan from starting a war. In hindsight, it is hard to imagine a less accurate prediction. Still, Hersey stressed, MacArthur's exuberance caused listeners to "go out feeling a little brisker . . . a little more cheery and more confident about things."[4]

European developments quickly outpaced the general's faith. Germany invaded the Soviet Union on June 22, 1941, inflicting such heavy defeats on the Red Army than many observers wondered whether Russia would survive. Even if it did, Moscow would no longer play much of a deterrent role in Asia.

Roosevelt told Harold Ickes that the "Japs" were "having a real drag-down and knock-out [sic] fight among themselves" over whether "they are going to jump-attack Russia, attack the South Seas (thus throwing in their lot definitely with Germany), or whether they will sit on the fence and be more friendly to us." The president sought a way to restrain Japan without driving it toward an invasion of Russia or the oil-rich East Indies. For the present he insisted on keeping control of the Atlantic by preserving "peace in the Pacific." He simply did not have "enough Navy to go around—and every little episode in the Pacific means fewer ships in the Atlantic."[5]

The day after Roosevelt spoke, Emperor Hirohito presided over an imperial conference that ratified plans for an "Advance into the Southern Regions." Since resources, especially petroleum, were more accessible in Southeast Asia, the Japanese deferred action against Siberia. They hoped a combination of intimidation and negotiation would secure Southeast Asia. If not, Japan would strike militarily.

This decision followed prolonged negotiations between Secretary of State Cordell Hull and Japan's ambassador in Washington, Admiral Nomura Kishisaburo. Although both men actually favored a diplomatic solution, neither fully acknowledged the chasm dividing them. Moreover, many Americans doubted whether the hardliners in Tokyo would accept a settlement achieved by Nomura. The admiral complicated the situation still further by misleading his superiors into expecting greater American flexibility as the talks dragged on from March through June. On June 21, Hull insisted that any overall settlement between Washington and Tokyo required the withdrawal of Japanese forces from China and Indochina. Interpreting this as new American intransigence, and excited by Germany's apparently successful attack on the Soviet Union, Japanese officials made the preliminary decision on July 2 to strike south.[6]

The Roosevelt administration had some awareness of the plans afoot since naval cryptoanalysts (starting in 1940) had broken Japanese diplomatic (but not military) codes. The messages between Nomura and Tokyo, as well as other sources, convinced intelligence analysts that the prospects for a peaceful compromise had diminished. From the summer of 1941 on, Roosevelt and his advisers understood that an end to negotiations would soon lead to an act of war. However, they had little evidence about the timing, nature, and location of an attack. The best informed experts anticipated a Japanese move into Southeast Asia, possibly including the Philippines. Few considered the fleet at Pearl Harbor a likely target.[7]

In July, Tokyo demanded and obtained from Vichy authorities air and naval bases in southern Indochina. Washington scurried to retaliate against this and took actions it hoped might deter further expansion. On July 17 Marshall sought to bolster the small American force in the Philippines by recommending to FDR that he call the Philippine Army into national service and also appoint the most senior officer in the region, Douglas MacArthur, as overall commander of "Army Forces in the Far East."

Late in May, while MacArthur languished in Manila, Marshall had mentioned to Secretary of War Stimson that "in case of trouble out there" (in Southeast Asia), they ought to "recall General MacArthur into service again and place him in command." The Chief of Staff, like Stimson, knew how few options existed. They had no reserves of troops, ships, or weapons to bolster American power in the Pacific. With Europe and the Atlantic considered the crucial front, strategists in Washington hoped to stall the Japanese through bluffs and symbols, at least until the European war stabilized and American strength increased.

Under these circumstances, the administration turned to MacArthur. Even after retirement, he remained one of the nation's highest profile military commanders. He knew the region better than most of his senior colleagues (actually, there was little competition) and had spent the past few years boasting about the impregnable Philippines. Even if many in the War Department knew better, to the public and to the Japanese his appointment would symbolize Washington's determination to resist Tokyo's pressure.

In order to stop MacArthur from leaving Manila in June, Marshall sent a personal message that the administration would probably name him army commander in the Far East in case of a crisis. Roosevelt had given the "impression," Marshall reported, that he "will approve." The military adviser now had a straw to keep him afloat.[8]

Anticipating Japan's move into southern Indochina, in mid-July Roosevelt ordered his deputies to draft plans for freezing Japanese assets in America. A major consequence would be the restriction of petroleum and gasoline exports. Harold Ickes believed that FDR continued to fear the consequences of "draw[ing] the noose tight, and still thought it better to slip the noose around Japan's neck and give it a jerk now and then." Ickes, like many of his colleagues, thought the president would impose on Japan only a partial trade freeze, allowing some petroleum sales and thus avoiding an imminent showdown with Tokyo.

This effort to fine-tune policy collapsed as the State, Treasury, and Interior departments jockeyed for control. Due to bureaucratic rivalries and vague wording, officials in both Washington and Tokyo interpreted Roosevelt's "freeze order" as mandating a total oil embargo following Japan's actual move into southern Indochina. By the time the president and his top staff realized their blunder, they hesitated to resume oil shipments since the Japanese might interpret a reversal of the embargo as a softening of American resolve.[9]

On July 24, after Japan's southern advance, Secretary of War Stimson urged Roosevelt to take "all practical steps" to "increase the defensive strength of the Philippine Islands. Besides supporting the idea of reappointing MacArthur, he recommended that a large shipment of the most powerful bomber in the American arsenal, the B-17, be sent to the Philippines.[10]

Roosevelt issued the new trade restrictions on July 26. At the same time he closed the Panama Canal to Japanese shipping and called the Philippine army into service. Immediately, the War Department established a new Far Eastern Command under major General Douglas MacArthur. Designated as commander of "United States Army Forces in the Far East (USAFE)," he now led both the federalized Filipino army as well as the regular forces of the American army's Philippine Department. Effective July 27, MacArthur received a promotion to the temporary rank of lieutenant general.

A few months before, MacArthur had seemed a virtual outcast in Manila. Now American and Filipino leaders rallied to him. President Quezon declared his confidence that the general would "attain in this difficult assignment the same success that has crowned your every endeavor in the past." (This from the same man who recently called MacArthur's defense plan "idiotic.") Washington's recognition of MacArthur convinced many Filipinos that the Untied States now considered their country important enough to lavish men, money, and materials upon. If MacArthur could bring them these things, they wanted MacArthur.[11]

The general threw himself totally into the task of preparing the Philippines for the Japanese onslaught. He appointed Colonel Richard Sutherland and Lieutenant Colonel Richard Marshall as chief and deputy chief of staff, respectively. He named Lieutenant Sidney Huff, Colonel Charles Willoughby, Colonel William Marquat, and Lieutenant Colonel LeGrande Diller to other headquarter posts. This group formed the nucleus of the what became known as the "Bataan Gang," the circle surrounding MacArthur for the rest or the war and beyond. MacArthur later placed Brigadier General Jonathan Wainwright in command of ground forces, while Major General Lewis H. Brereton, who arrived in November, commanded the growing air force. Significantly, neither of these two key officers was close to the general and both incurred the wrath of the inner circle. MacArthur assumed personal control of the Philippine Department and the Philippine Army in addition to his job as USAFE commander.

In theory, the general led an impressive military force. All told, the army numbered about 100,000 troops. His air force soon included one of the largest concentrations of advanced planes outside the United States. The Asiatic Fleet, based in Manila, supported his forces. His actual strength was less imposing. Philippine Army reserves, a force existing largely on paper, comprised nearly three-fourths of his army.

MacArthur confidently notified Washington on August 30 that defense preparations had progressed "by leaps and bounds," and the mood among Filipinos had changed from "a feeling of defeatism to the highest state of morale I have ever seen." Deliveries of additional supplies would insure the "development of a completeley adequate defense force," before a Japanese attack he anticipated coming in April 1942. (His previous estimate of Japan's timetable predicted an attack in 1946!)

The general seemed self-assured almost to the point of cockiness. He told Admiral Hart they had "plenty of time" remaining before war erupted. He declared that he was "not going to follow" or be bound in any way by "whatever war plans have been evolved, agreed and approved in Washington." Strutting about and pounding his hands together, he told Hart that he would soon lead in an expanded army of 200,000 men in a "glorious land war." If the Japanese came ashore, he would crush them. MacArthur brushed off Hart's suggestion that the two commanders coordinate their efforts, declaring that the "Navy had its plans, the Army had its plans and . . . we each had our own fields." His "primary interest was in the land battle he was going to wage." Although furious at MacArthur's condescension and egotism, even Hart admitted that the general radiated a contagious optimism.[12]

Unfortunately, the Philippine Army remained a glimmer in MacArthur's eye. Mobilized during August and September, its diverse units were hastily assembled in training camps where most recruits spoke different languages, hailed from rival ethnic and religious communities, and lacked any sense of common purpose. The American instructors possessed so little equipment that most troops had never fired a weapon before they saw combat. To alleviate these shortages, MacArthur ordered vast quantities of equipment from the United States. Relatively little arrived before the war began.

Although MacArthur could not be blamed for causing many of these deficiencies, he either ignored or misunderstood the facts. As late as October he wrote to an old friend about the "electrical" impact that American policy had on the Philippines. Since July 26, an "immediate and universal feeling of confidence and assurance" had dispelled the "pessimism and defeatism" of the past. Morale had risen to "a point almost of exuberance." Filipinos of all types, including all the "labor leaders in the islands," pledged "me their personal allegiance." His appointment brought "jubilation," MacArthur claimed, "throughout the Netherlands East Indies, Malaya, and China." It left Japan "dumbfounded and depressed." Soon, he would be able to repulse "any effort that may be made against us." In a remarkably expansive mood, MacArthur even lavished effusive praise on Roosevelt and Marshall, to whom he pledged "loyal and devoted service."[13]

In spite of standing policy to defend only Manila Bay, the new commander planned to defend the entire archipelago by repulsing the invader on the beaches. Such a strategy would have tested even a vast, fully equipped, and superbly trained army. With its many large and innumerable small islands, the Philippines had a coastline longer than that of the United States. Given the kind of army he commanded and the enemy he faced, it hardly mattered if MacArthur had 100,000 troops or twice that number. Nevertheless, as D. Clayton James observed in his study of the problem, the general's "overconfidence and unjustified optimism as to the abilities of himself, his staff and the untried Filipino soldiers . . . became a contagion which ultimately affected even the War Department and the Joint Army and Navy Board."[14]

In October, MacArthur learned that the latest army-navy defense plan (Rainbow-5) called on him to defend only Manila and Subic bays, not the entire archipelago. Denouncing this strategy as practically appeasement, he insisted that Washington give him more men and equipment and permit him to defend the entire archipelago. The next month the Joint Board relented, granting him the leeway he sought.

MacArthur proved an incredibly successful lobbyist for his own position, swaying the formerly skeptical Stimson, Marshall, and Roosevelt. He all but guaranteed his ability to deter or defeat the Japanese if only given a bit more. Moreover, his grandiose claims seemed buoyed by the emergence of a major new weapon—modern air power in the form of the B-17 heavy bomber.

Beginning in July, increased numbers of these so-called Flying Fortresses became available. Air war strategists placed a priority on sending these bombers and modern pursuit planes to the Philippines. The army allocated over 300 B-17s and two fighter groups of 130 planes during the autumn, although, due to delays in production, slow shipping schedules, and the need to fly the bombers via a circuitous route, only about 74 heavy and medium bombers (including a dozen B-17s) arrived by early December. Air commander General Brereton also possessed 175 pursuit planes and 55 miscellaneous aircraft. However, this total represented the strongest American air force outside the continental United States.

In theory, the B-17 could deliver a heavy payload with great accuracy along a 2,000 mile radius, or attack warships far out to sea. Operating from the Philippines, the plane might sink a Japanese invasion fleet or hit enemy strong points elsewhere in Asia. Some air force planners even suggested that the B-17 could bomb the Japanese home islands if a shuttle run were established between the Philippines and Soviet bases in Siberia. Some members of the White House staff actually devised a scheme to launch B-17 raids against Tokyo from China—in peacetime—using "private" American pilots.

Unfortunately, no one had tested the plane under realistic conditions. During the first year of the war a host of technical defects rendered it almost useless against ships. Nevertheless, Henry Stimson speculated that the B-17 would change the balance of power in the Pacific. Roosevelt boasted of the plane's "great effect . . . upon the Japs." Officers in the War Plans Division asserted that "American Air and ground units now available or scheduled for dispatch to the Philippine Islands in the near future have changed the entire picture of the Asiatic area." These almost literal flights of fancy rivaled those of MacArthur's imagination.[15]

Even when the small number of "Flying Fortresses" reached the Philippines, no one knew how to deploy them. Since only Clark Field on Luzon could accommodate the B-17, all were concentrated in one location. Even after Del Monte Field in Mindinao became partially operative, most of the B-17 force remained at Clark, despite the absence of effective defenses.

Unaccountably, neither MacArthur (who seemed unimpressed by the new weapon) nor the air power enthusiasts paid much attention to these operational problems. The general focused nearly all his attention on his land forces, paying little heed to cooperation with either air or naval commanders. Instead of developing a joint defense plan with the Asiatic Fleet commander, Admiral Hart, he took to ridiculing the naval officer and delivering patronizing advice. At one point he blurted out, "Get yourself a real fleet, Tommy, then you will belong."

The admiral complained to the navy leadership that MacArthur's behavior had grown increasingly "erratic" during the autumn. He offered an even more dismal assessment to his wife. "The truth of the matter is," he wrote, "that Douglas is, I think, no longer altogether sane. [H]e may not have been for a long time."[16]

It remains a testament to MacArthur's powers of persuasion that, despite Hart's fury, even the admiral half believed the general could use his ground and air strength to repulse the enemy. In case of war, the small Asiatic Fleet had orders to leave Manila and join Anglo-Dutch units to the south. Hart urged the Navy Department (without success) to change these orders so that he could engage Japanese naval units near the Philippines where, he expected, MacArthur's B-17s would join the attack.[17]

The same pressures that prompted a change in American strategy drove Japanese policy in new directions. The stalemated China war (characterized by growing American military aid to Chiang's forces), the uncertain military outcome in Europe, and the July 26 freeze order all hung like a sword over Tokyo. Given its rapidly shrinking oil reserves, Japan faced the specter of slow strangulation unless America resumed sales or Japan seized the sources of production in Southeast Asia.

The Japanese army and navy, through both tradition and intimidation, put tremendous pressure on the civilian imperial government. The armed services had a virtual veto over diplomatic policy and enjoyed privileged access to the emperor through the Imperial General Headquarters. Increasingly, civilian officials played a secondary policy role.

The extremely nationalistic army justified imperialism on the basis of economic and racial dogma. In the aftermath of the Depression, army leaders attacked liberal values, big business, Marxism, and Chinese nationalism for hurting Japan. Often downplaying the importance of material factors, they promoted a military doctrine based on spiritual power, courage, and duty. Since classical doctrine postulated fighting a major war on the Asian mainland, the army was ill prepared for amphibious, tropical warfare as would be experienced in Southeast Asia and the Pacific.

The navy more closely resembled its American counterpart. Since 1907, Tokyo's plans for a Pacific war had resembled Washington's War Plan ORANGE. In the event of war with America, the navy would seize Guam and the Philippines and lie in wait for the American fleet sailing to the rescue. It planned to destroy the United States fleet in a decisive battleship engagement.

As technology evolved and after Japan acquired many mandated islands in the Pacific, these plans changed. Operating from forward island bases, submarines and other small craft would harass the American battle fleet en route to Manila. The navy developed huge battleships and excellent naval aircraft, but paid too little attention to building aircraft carriers, improving anti-submarine techniques, or perfecting methods of amphibious warfare. Again, the American navy shared these deficiencies.[18]

The escalation of tension in the Pacific in mid-1941 prompted both Japan and the United States to adopt a more offensive profile. By September, Washington was augmenting land, sea, and naval forces on both Hawaii and the Philippines. The Japanese army and navy made preparation to seize Malaya, the Philippines, and the East Indies at the outset of war. However, success in this effort to break out of an economic stranglehold depended on crippling the striking power of the fleet at Pearl Harbor.

On August 28, Prime Minister Konoye proposed a meeting between himself and Roosevelt to avert war. The president responded that Japan must first abandon its alliance with Germany and withdraw troops from China. Military leaders in Tokyo dismissed this as an arrogant ploy to drag negotiations on while the strategic oil reserve ran dry. Early in September the Japanese government resolved that if "by the early part of October there is still no prospect of being able to obtain our demands, we shall immediately decide to open hostilities against the United States, Great Britain, and the Netherlands." The emperor, still hoping to avoid war, chided the military for insisting on an October deadline. He wondered how they expected to defeat the Western powers when they had failed to achieve a military solution in China after four years of war.

Nevertheless, Japanese negotiating terms virtually insured a deadlock. Tokyo insisted that Washington halt the growing military assistance program to the Chinese Nationalists and press Chiang to make his peace with Japan. The Americans and British were to cease strengthening their own military forces in the Far East while, at the same time, assuring Japan adequate supplies of strategic raw materials. In return, Japan pledged that its bases in Indochina would only be used in the war with China, not to attack nearby Western colonies. A withdrawal from Indochina or guarantee of Philippine neutrality depended on reaching a comprehensive economic and military settlement with the United States and Great Britain. Even following all this, Tokyo intended to retain control of Manchuria.

When he learned of these terms late in September, Roosevelt felt justified in rejecting the earlier summit proposal. Parlaying with Tokyo, he felt, would only undermine American, British, and Chinese resolve. Intelligence intercepts made American leaders even more skeptical, since they revealed that the Japanese army opposed withdrawing large numbers of troops from China even if civilian leaders agreed. When Washington rejected Tokyo's latest terms, Konoye resigned (on October 16) and General Tojo Hideki became prime minister.[19]

Despite their anxiety, Roosevelt and his advisers still preferred to postpone a military showdown. On November 5, for example, General Marshall and Admiral Harold Stark urged the president to avoid any dramatic moves until at least mid-December. After that, growing American "air and submarine strength in the Philippines" would become a "positive threat to Japanese operations south of Formosa." If a confrontation were delayed until February or March, the B-17 force in the Philippines might actually become a deciding factor in deterring Japanese operations." A premature war in the Pacific would divert critical resources away from the European front. Roosevelt's military chiefs felt it important that only a direct attack upon American, British, or Dutch territory, Thailand, or selected Pacific islands, justified going to war with Tokyo.

On November 7, the president informed Prime Minister Winston Churchill that he hoped that the steady American buildup in the Philippines, British reinforcement of Singapore, and mounting military assistance to

China would deter war with Japan for several months longer. Of course, what Roosevelt described as defensive or deterrent buildups appeared quite provocative from the Japanese perspective.[20]

Army and navy leaders in Tokyo intended to act before the Philippine or Chinese military buildups increased. If they postponed action until the spring, the oil embargo would threaten to choke off the civilian economy as well as military operations, and the growing American bomber forces might negate Japanese striking power. Plans to seize Southeast Asia in the fall of 1941 lapsed when the emperor ordered that the October deadline for negotiations be extended. But planning for a preemptive attack continued.

Rather than awaiting the response of the American navy to operations in Southeast Asia, Combined Fleet commander Admiral Yamamoto Isoroku proposed initiating war with a carrier-based aircraft strike against the fleet at Pearl Harbor. In mid-October, when General Tojo became prime minister, planning for the attack accelerated. An imperial conference of November 5 decided that war would begin by early December unless Tokyo and Washington reached a diplomatic accord.

During November *pro forma* negotiations continued as Ambassador Nomura presented several plans to American officials. Japan, in essence, demanded a free hand in China and guaranteed access to Southeast Asian raw materials. Washington must stop assistance to China and resume oil sales. Besides being unacceptable to the United States on their face, American officials had other reasons to question Japanese intentions. Intelligence intercepts revealed that the military opposed Nomura's efforts and had no intention of withdrawing from China even after a settlement.[21]

At one point, hoping to avert war, Roosevelt contemplated a short-term deal in which both sides would make no military moves for several months and America would resume some trade with Japan. The president abandoned the idea of this *modus vivendi* when British and Chinese leaders warned it would be interpreted by their people as a form of surrender. Also, intelligence reports revealed accelerated Japanese troop movements toward Southeast Asia, placing the region in great jeopardy.[22]

On November 26 Cordell Hull handed Nomura the formal reply to the earlier Japanese proposals. The note asserted that peace in the Pacific required that Japan evacuate all its forces from China, Manchuria, and Indochina; pledge no further aggression; and accept the colonial status quo throughout Southeast Asia. American leaders understood that Japan would probably respond by attacking in Southeast Asia.

During the last days of November, American intelligence analysts learned of large Japanese troop convoys leaving Shanghai in the direction of Thailand and Malaya. On November 27, both MacArthur and the commanders in Hawaii received a "final alert" from the War Department and a "war warning" from the navy. The news did not overly concern the general. In a meeting that day with Admiral Hart and Commissioner Sayre, he paced the floor confidently, puffed his cigar, and told his colleagues that the "existing alignment and movement of Japanese troops convinced him that there would be no attack before the spring."[23]

At about the same time, *Life* magazine received the draft of an article on MacArthur written by the publisher's wife, Clare Boothe Luce. She had interviewed the general in Manila during October and written an extremely flattering portrait. However, in a note to the magazine's editor, she apologized for the exaggerated "buildup." There were, she admitted, "quite a few unpleasant things to be said about the man," but she avoided them "for a good reason"—America's military prestige "in the whole Far East" was "now indissolubly linked with the military (and personal) prestige of the General." Anything that "might tend to destroy his prestige in the eyes of his troops, the Filipinos, or for that matter the British at Singapore, the Dutch and the Japanese, will inevitably weaken our military effort." United States power in the Pacific, she suggested, depended on praising the emperor's—or MacArthur's—new clothes.[24]

On December 6, Roosevelt sent a personal appeal to Emperor Hirohito, urging some action to avert war. Even as FDR dictated the message, American cryptoanalysts intercepted thirteen parts of a fourteen-part cable from Tokyo to Ambassador Nomura announcing the termination of negotiations. It indicated an imminent decision for war. Still, most evidence suggested a Japanese thrust against Southeast Asia, possibly bypassing the Philippines. (Although FDR had given informal assurances to the British and Dutch to come to their assistance in case of Japanese attack, he had made no binding commitments.) In order to avoid ambiguity and minimize dissent, the president and his aides insisted that Japan would have to fire the first shot. At dawn, December 7, carrier-launched planes from a Japanese task force obliged.

5
Retreat in the Pacific

At dawn on December 7, 1941, Secretary of War Stimson speculated about the still undelivered fourteenth section of the message Ambassador Nomura was to bring to the State Department later that day. "Everything in MAGIC" (the secret code breaking project), he noted, "indicated they had been keeping the time back until now in order to accomplish something hanging in the air." Certainly the Japanese were "planning some deviltry," but all wondered "where the blow will strike."

The attack upon the fleet at Pearl Harbor brought war with dramatic suddenness. Plans to strike the Philippines simultaneously with planes from Formosa were delayed by bad weather. The events of the day terrified most Americans, accustomed to thinking of the Pacific Ocean as a great, protective moat. Fortunately, the Japanese failed to destroy the navy's aircraft carriers or critical logistic and repair facilities in Hawaii. These proved vital in subsequent battles. When, at the end of December, the Soviet army held the Germans at the very gates of Moscow, Japan lost its gamble that the United States would become so involved in Europe as to be paralyzed in the Pacific. Aroused as never before, the American nation rose to fight back on two fronts.[1]

On the other side of the international date line, first word of the Pearl Harbor raid came to Manila on December 8 when radio operators of the Asiatic Fleet picked up news reports at 2:30 in the morning. Admiral Thomas Hart and his staff began combat preparations at once, although naval personnel neglected to inform the commanding general of the army forces. An hour later, army staff learned of the debacle from a commercial radio broadcast and General Richard Sutherland, MacArthur's chief of staff, broke the bad news to his commander.[2]

The events of the next few confusing hours sealed the fate of the Philippines. Startled and possibly disoriented by the crisis (MacArthur had insisted war would not come before April), the general closeted himself in his office, communicating only through Sutherland. When air force chief

Brereton requested permission to launch a B-17 strike against Formosa, the general stalled. By the time MacArthur gave the go ahead (about nine hours after the Pearl Harbor attack), waves of Japanese aircraft had descended on Clark Field. In minutes, over half the B-17 bombers and modern pursuit planes (the strongest American air concentration outside the United States) were destroyed on the ground. As Admiral Hart noted ruefully, the "highly efficient" Japanese had had a "grand Round One."[3]

The reasons for MacArthur's delay in approving the strike against Formosa have never been explained to anyone's satisfaction. All the principals offered mutually exclusive accounts. However, sometime later Manuel Quezon told Dwight Eisenhower that when the "Japanese attacked Pearl Harbor MacArthur was convinced for some strange reason that the Philippine Islands would remain neutral and would not be attacked by the Japanese." Fearful that an air strike would compromise this neutrality, he hesitated to permit Brereton to attack. As a result, "all our planes were still on the ground." The general, perhaps, believed his own claim that no nation would dare challenge the Philippines.[4]

Unopposed in the air, the Japanese began landing their main force at Lingayen Gulf, northeast Manila, on December 22. MacArthur and Hart counted on the submarines of the Asiatic Fleet to impede the invasion. They, too, proved a disappointment, failing to sink even stationary ships. (The cause lay in a faulty torpedo mechanism that had never been adequately tested.) With little difficulty, the Japanese secured their beachhead. As for the Philippine Army, it folded in about forty-eight hours, despite the fact that it outnumbered the invading force.

General Brereton soon flew his handful of operational bombers to Australia while Admiral Hart ordered the Asiatic Fleet to retreat south. By December 24, MacArthur declared Manila an open city and ordered a military evacuation. Forced to resurrect the long-despised War Plan Orange, the general instructed American and Filipino forces to withdraw to the Bataan Peninsula and Corregidor Island. There, possibly, they might withstand a siege and hold Manila Bay.

Previously, MacArthur had ordered supplies put in forward depots. Now, the defenders scurried to recover as much as they could before the Japanese overran them. Regular American army units (including the Philippine Scouts) prevented the retreat from becoming a rout. By early January, about 80,000 soldiers, 26,000 civilians, and substantial amounts of material reached Bataan.

In spite of the confusion, the American defenders in the Philippines resisted the Japanese far longer than had the British and Dutch. Although the five-month defense of Bataan and Corregidor had only limited impact on the overall Japanese battle plan, it transformed Douglas MacArthur into the first American hero of the war. Hardly a day passed between January and May of 1942 without a dramatic account in the newspapers or radio about the heroic resistance in the Philippines. In his three months on Corregidor, MacArthur's headquarters released over 140 press releases, many of which he wrote himself. A large majority referred to the general exclusively. Singlehandedly, it seemed, he parried enemy thrusts and frustrated Tokyo's entire war plan. Neither fellow officers nor front line troops received much credit for anything. MacArthur, or his public relations officers, were so determined to stress the positive that their releases frequently reported victories in imaginary battles.

As the Japanese tightened the noose around Bataan and Corregidor, the general became convinced that enemies in the Roosevelt administration and the army had chosen to sacrifice him. In fact, most military and civilian officials strongly supported MacArthur and believed, initially, that he could hold off the Japanese. Even the alleged leaders of the cabal (FDR, Harold Ickes, George Marshall, and Dwight Eisenhower) suspended their disbelief for a while and succumbed to the MacArthur magic. They "abandoned" the general only when the dimensions of his defeat and the desperate need to bolster the European theater became apparent.

During the first weeks of war, virtually no one in Washington blamed MacArthur for the success of Japan's air strikes. (He suffered none of the shame and punishment inflicted on the commanders at Hawaii, for example.) Administration officials rallied to the cause of defending the Philippines and heaped praise on the general. For example, Harold Ickes stated that MacArthur (with his "magnificent" Filipino troops) had given "an excellent account of himself in the Philippines," driving the Japanese "back into the sea." The normally critical interior secretary "thanked our stars . . . that General MacArthur has been in charge" and now hoped the entire archipelago might be saved "despite the long established expert military opinion here that we could not hold onto them in the event of the war with Japan."[5]

Secretary of War Henry Stimson also voiced faith in MacArthur's leadership. Fearful that the weakening of the Pacific fleet and the "defeatist" attitude of the navy might lead to writing off the Philippines, he reported that he, General Marshall, and President Roosevelt were determined to "make every effort at whatever risk to keep MacArthur's line open." Stimson even threatened to resign unless still greater efforts were made to assist MacArthur.[6]

In the weeks following the Pearl Harbor attack, Dwight Eisenhower, by then the second in command of the War Plans Division, struggled to get supplies and reinforcements to the Philippines. Although uncertain whether the islands could be saved, like Stimson, he argued that the "Far East was critical and no other sideshows should be undertaken." Ike hoped to utilize Australia as a way station for assisting MacArthur and for organizing a counteroffensive. By mid-January, however, Eisenhower realized that the losses in Hawaii and the Japanese blockade of the Philippines made it virtually impossible to deliver supplies to MacArthur. (Even then, the War Department sent Patrick Hurley to Australia with money to hire private blockade runners—few of which got through.)

When MacArthur realized that little assistance was forthcoming, he lashed out. As early as January 2, 1942, Stimson complained of receiving messages of a "most harassing and agonizing character" from the Far Eastern commander. When the general (incorrectly) berated Eisenhower for delaying reinforcements, a beleaguered Ike wondered why MacArthur had not "made a better showing at the beaches" or "saved his planes" on December 8. To the public MacArthur might still be the "hero," but Ike considered him "as big a baby as ever." Nevertheless, the War Department somehow had "to keep him fighting."[7]

American weakness in the Pacific and the determination of the British and American leadership (confirmed at the ARCADIA conference late in December) to defeat Germany before Japan made it impossible to do much for the Philippines. The fall of Singapore and the Dutch East Indies during February eliminated any strategic reason for reinforcing MacArthur. Eisenhower voiced the opinion of most military experts when he declared: "We've got to go to Europe and fight, and we've got to quit wasting resources all over the world—and still worse—wasting time."[8]

By mid-January, when the Japanese attacked Bataan, MacArthur sensed that no relief would arrive. His troops suffered almost as much from disease and hunger as from battle. Many Filipino units began to melt away. Even when fighting slowed during February and March, disease, malnutrition, and desperation continued to extract a toll on the defenders. The situation on the Bataan peninsula deteriorated when MacArthur ordered some of its food stocks transferred to Corregidor, for use after the anticipated loss of the peninsula.

The general's decision to visit Bataan only once in three months also undermined the morale of the front line troops. Irate soldiers composed derisive songs about "Dugout Doug," safe in his underground shelter on Corregidor Island. In truth, danger and discomfort dogged MacArthur there almost as much as on Bataan. But, inevitably, when soldiers compared his behavior with the exaggerated claims of official press releases, they judged him a fraud. The epithet "Dugout Doug" stayed with him for the rest of the war.[9]

Desperate for support, MacArthur even looked toward the Soviet Union. He issued effusive praise of the Red Army's defense of Moscow (declaring that the "hopes of civilization rested on . . . the Red Army) and urged Stalin to begin a campaign in the Pacific. The general also sent what Eisenhower called a "flood of communications" to Washington in which he and Manuel Quezon demanded that naval convoys come to their rescue. Ike feared that MacArthur "refused to face facts," was "losing his nerve," and propounded strategic ideas more suitable for "plebes at West Point" than for real war.[10]

In addition to the military debacle in the Philippines, the desertion of so many old friends among the Manila elite stung MacArthur personally. Even President Quezon had wavered when MacArthur declared Manila an open city. The commonwealth president proposed remaining at his post to mediate between the Japanese and his people. Although MacArthur convinced him to go to Corregidor, Quezon reportedly told subordinates that both he and the American commander agreed they might cooperate with the Japanese so long as they did not formally pledge allegiance to Tokyo. (MacArthur later denied this.)

As soon as they took Manila, Japanese military authorities called upon Filipino officials to remain in their jobs. The conquerors promised early independence as a reward for collaboration. Leading members of the oligarchy responded by organizing a Japanese-approved administrative council, which became the nucleus of a puppet regime.[11]

Shocked by the collaboration of so many of their friends, Quezon and MacArthur both sought a way to reaffirm their allegiance to each other. On January 3, 1942, President Quezon issued a secret Executive Order awarding $640,000 in commonwealth funds (on deposit in the United States) to General Douglas MacArthur ($500,000) and members of his staff, including Richard Sutherland ($75,000), Richard J. Marshall ($45,000), and Sidney L. Huff ($25,000). The American commander waited almost six weeks before accepting the award.

In theory, this represented special payment for services rendered by the office of military adviser. However, it included the period July–December 1941, after MacArthur had reentered active service. Acceptance, therefore, violated various army regulations.[12]

Historian Carol Petillo, who first revealed the transaction, speculates that the payment reflected a cultural and psychological relationship. Quezon hoped the gift would confirm his faith in MacArthur and obligate the general to press Washington harder for assistance to the Philippines. In his mind the money may have been symbol of reciprocal obligation and bonding so common in Philippine culture, not a bribe.

MacArthur, who had pressed since 1935 for a higher salary, may simply have considered the $500,000 an overdue payment for past services. Possibly, he failed to realize that compensation for July–December 1941 violated army rules. Whatever the case, he must have found this generous expression of Filipino faith in him emotionally reassuring.[13]

By early February, Quezon and MacArthur grew so despondent with what they called America's "inactivity" and "timidity" they contemplated a separate peace with Japan. After discussing his plans with MacArthur, the Philippines leader informed FDR bluntly that the situation had grown "desperate" and that all realized that "no help will reach us from the United States in time" to rescue the trapped forces or to protect the rest of the commonwealth. Washington, he insisted, had broken faith by declining to send the "slightest assistance" to relieve the "death, misery and devastation" of the Philippines. No longer concerned with past loyalties or his peoples' "ultimate destiny" under American rule, Quezon intended to "preserve the Philippines and Filipinos from further destruction" by reaching a settlement with the Japanese.

He proposed that Washington grant the commonwealth immediate independence and permit him to negotiate the simultaneous withdrawal of

Japanese and American forces. The new nation would be neutral in the Pacific War. Since America intended to grant independence in a few years and Tokyo had also "publicly announced its willingness to grant the Philippines her independence," Quezon urged Roosevelt to agree quickly.

Although MacArthur did not formally endorse this proposal, he raised no objections. His accompanying message confirmed Quezon's charge about the "violent resentment" felt by Filipinos toward the United States and agreed that Tokyo's promise of independence made a "powerful impression" upon most Filipinos. Given these facts, he suggested Quezon's plan might be the "best possible solution of what is about to be a disastrous debacle."[14]

Not surprisingly, Quezon's demand shocked and offended American officials. Roosevelt, Marshall, and Stimson judged the Filipino a near traitor and expressed outrage at MacArthur's implicit support. In disbelief, they wondered how the top military commander in the Far East could even contemplate assenting to a deal with Japan abandoning American territory in the wake of the Pearl Harbor attack and invasion of the Philippines. It seemed especially ironic that this same general had described the islands as virtually invulnerable and able to resist indefinitely the Japanese threat. After meeting with Marshall and the president, Stimson ridiculed the "wholly unreal message" which took "no account of what the war was for." How on earth, the secretary wondered, could MacArthur even consider a "virtual settlement with the Japanese."

FDR and Marshall assigned Eisenhower to draft a response to Quezon. It left no doubt that even if Filipino forces surrendered, American soldiers would defend the flag "to the death." The president authorized MacArthur to arrange for the evacuation from Corregidor of Quezon (who, Washington feared, might deal with the Japanese if left behind), Commissioner Sayre and his family, and the general's own wife and son.[15]

Startled by the suggestion of disloyalty and the vehement rejection of the proposal, the General informed the president he would arrange for the evacuation of Quezon and the Sayres, conditions permitting. (They left by submarine a few weeks later.) However, his own family would "share the fate of the garrison." Now distancing himself from the neutralization scheme, he described it only as a vague contingency that Quezon thought Roosevelt might wish to consider. Both Filipino and American troops, he assured Washington, would hold "steadfast to the end."[16]

Following this latest rebuff, MacArthur decided to accept Quezon's January 3 offer of a large cash award. On February 15, he wired instructions for the War Department to transfer $500,000 from the holdings of the Philippine commonwealth to his personal account in a New York bank. Smaller sums went into the accounts of the other recipients. Pending confirmation of the transfer, MacArthur asked Manuel Roxas, Quezon's aide, to loan him 1.2 million pesos worth over a half million dollars. (He returned this after the transfer was confirmed.) Of all the honors and awards he received, this was the only one the general kept secret. Besides implying that he knew about the legal complications, MacArthur's convoluted behavior suggested he did not intend to "share the fate of the garrison." [17]

In fact, the president and Marshall had already told him they wanted him to leave Corregidor and assume command of troops assembling in Australia. But he insisted that the president issue a direct order that he leave, lest anyone challenge his motive. FDR and Marshall, Eisenhower felt, had not decided on this new command on the basis of military logic but had caved in to "editorials and public opinion."

Ike objected to the decision of February 22 pulling MacArthur out. The siege of the Philippines seemed "made to order" for the general. He could not handle more "complicated situations." Corregidor had "all the essential of drama" MacArthur craved and made him the "acknowledged King on the spot." In another command, public opinion would force him into a position where his "love of the limelight may ruin him."

After stalling for two days, MacArthur accepted Roosevelt's order that he leave the Philippines. Still, he insisted upon postponing departure until the right psychological moment. Not much more was heard from Corregidor until March 11, when MacArthur announced suddenly his decision to leave. [18]

Since January a "MacArthur craze" had swept America. This reflected the drama of a siege, MacArthur's dramatic press releases, and the fact that soldiers in the Philippines were practically the only Americans actually fighting the enemy. The anti-Roosevelt press played up this struggle. William Randolph Hearst, Roy Howard, Colonel Robert McCormick, and Henry Luce all considered MacArthur and his plight a weapon against the New Deal. Their publications described the general as the "Lion of Luzon" and "Hero of the Pacific," singlehandedly frustrating the Japanese war machine. Imagine what he could do with a real army! Even the liberal *Nation* magazine praised MacArthur effusively.

Cities and towns throughout America renamed bridges, streets, and buildings in honor of the general. Patriotic societies voted awards to both him and his dead parents. The National Father's Day Committee named him "Father of the Year" and numerous state legislatures rushed to declare special holidays in his name. The Polish government-in-exile granted him a medal for bravery.

Political leaders of both parties urged that he receive special awards. Republican Wendell Willkie delivered a speech in February calling upon FDR to bring MacArthur to Washington and make him "commander of the Army" and coordinator of "all the armed forces of the nation." Senators Millard Tydings (Democrat of Maryland) and Arthur Vandenberg (Republican of Michigan) denounced the administration for "abandoning" the hero. The latter condemned FDR for pursuing his "social revolution" in the midst of a "private war." While Roosevelt rescued the British and Russians, he made MacArthur "fight alone." If the general "gets out alive," Vandenberg pledged, "I think he will be my candidate for President in 1944." [19]

MacArthur, his family, and personal staff (including the beneficiaries of Quezon's largess) left Corregidor for the safety of a world that judged them heroes. They traveled for two grueling days on small patrol boats toward the southern island of Mindinao. After a short rest, on March 17, B-17 bombers flew them to an airfield near Darwin, Australia. The party finally reached Melbourne on March 21.

En route to Melbourne, MacArthur addressed a group of reporters anxious for a statement. The general declared that Roosevelt had ordered him to Australia to organize the "American offensive against Japan, a primary object of which is the relief of the Philippines. I came through and I shall return." Despite belated efforts by the Office of War Information to change the "I" to "We," the phrase became a catchword of the Allied cause. Among those remaining in the Philippines, however, it became a bitter joke. Troops ridiculed it by saying, at the appropriate time, "I am going to the latrine, but I shall return."

Quite apart from MacArthur's pronouncement, the Joint Chiefs had decided to bolster Australia for use in a future offensive against Japan. By March, about 80,000 troops had been sent there, with an additional 200,000 scheduled to go later in 1942. Although MacArthur expressed initial disappointment over the paltry resources at his command, these forces, with accompanying aircraft, actually represented the largest concentration of American power outside the Western hemisphere during the early stages of the war.[20]

The Australian government and people (the bulk of whose troops were fighting in the Middle East and who feared Japanese invasion) were relieved by the arrival of MacArthur and American forces. Prime Minister John Curtin emerged as one of the general's most vocal champions in Allied councils. Later, many social and racial frictions developed between American servicemen and their hosts, tarnishing the good feelings. Still, the United States soon replaced Great Britain as Australia's protector.

About the time MacArthur arrived, the American and British military chiefs (known as the Combined Chiefs of Staff, or CCS) agreed upon a global division of responsibilities, with the United States undertaking the major role in the Pacific. Largely because Admiral Ernest J. King, chief of naval operations, opposed placing the Pacific Fleet under army control, the Joint Chiefs divided the Pacific into two separate theaters of operation. They assigned Admiral Chester Nimitz to command the Pacific Ocean Areas (POA), meaning the naval and amphibious campaign in the central and southern Pacific. MacArthur assumed command of the Southwest Pacific Area (SWPA), consisting of Australia, New Guinea, the Bismark Archipelago, Solomon Islands, and Netherlands East Indies (minus Sumatra). By mid-April the foundation was laid for MacArthur's operations during the next three years.[21]

The administration continued to reveal an ambivalence toward the new theater commander. Before MacArthur left Corregidor, George C. Marshall weighed the idea of awarding him a Congressional Medal of Honor as a morale measure. When Henry Stimson and Dwight Eisenhower disputed the basis for the award, Marshall argued he must find some way to "offset any propaganda by the enemy directed against his leaving his command." A Medal of Honor, the army chief told Stimson, would have tremendous "popular approval" and "constructive morale value"—presumably to dispirited troops and civilians reeling under the unbroken succession of Japanese victories. Marshall may also have wanted to short circuit efforts by several members of Congress who had submitted bills to award a medal to MacArthur.

In mid-March, as the general prepared to leave Corregidor, George Marshall took the initiative of writing a citation based on information supplied by General Richard Sutherland, MacArthur's chief of staff. It described the commander's bravery in personally supervising defensive and offensive operations on the Bataan Peninsula and referred to his disregard for his own safety while exercising command. At best, this strained credulity since the general had visited Bataan but once. Certainly, the troops there who called him "Dugout Doug" did not share Sutherland's opinion. Nevertheless, on March 25, President Roosevelt approved without comment Marshall's recommendation.[22]

In other contexts, Roosevelt made no secret of his disappointment with the general. In March, he told Admiral Thomas Hart that MacArthur had assured Washington he could easily repulse a Japanese attack on the Philippines anytime after December 1, 1941. "If I had known the true situation," Roosevelt mused, "I could have babied the Japanese along quite a while longer." On April 10, the president discussed the matter with former High Commissioner Frank Murphy, now a Supreme Court Justice. Roosevelt charged that the general had repeatedly misled the War Department about the situation in the Philippines after the Japanese attack. The Philippine Army had actually "dropped out by the hundreds" when fighting began, while the retreat to Bataan and Corregidor was "more a rout than a military achievement." Why, Murphy asked, did FDR approve the Medal of Honor? Roosevelt defended his action as "pure yielding to Congressional and public opinion." Two years later when asked to sign a special citation for the medal, Roosevelt claimed he could not recall what heroism the award was actually based on.[23]

The decision to give MacArthur his own theater also reflected, in part, public and political pressure. Henry Stimson remarked how news of the general's safe arrival in Australia had a "great effect on the United States and all over the world." The public interpreted his escape as a victory over Japan. It was "the first good news we have had for so long" that people "inclined to exaggerate" its importance. The Allies needed MacArthur as a symbol, Stimson told himself, even if he presented "difficulties" for the administration's efforts to maintain "strategic control of events."

A few days later, the secretary of war declared himself in "shock" over the "emerging problem" of MacArthur. Stimson discovered that when leaving Corregidor the general ordered that all forces remaining in the

Philippines should remain under his direct command rather than that of his successor, General Jonathan Wainwright. MacArthur interfered with the War Department's efforts to communicate with the officer it supposed was the new commander. Stimson, among others, considered this a terribly dangerous division of authority.[24]

Eisenhower complained that the "newspapers" and the public had created a "hero out of its own imagination." Everyone expected miracles from MacArthur—and demanded that the War Department give him the resources for a miracle. The SWPA commander would probably attempt too much and "could be ruined by it." If "we tie up our shipping for [the] Southwest Pacific" in support for the hero, Ike lamented, "we'll lose this war."[25]

Harold Ickes at first lauded the decision to give MacArthur a theater— far away from home. It would have been a "terrible mistake" to bring him to America but he was the "right man in the right place in Australia." His pluck made "everyone's spirit rise." Like Eisenhower, Ickes hoped, the public would "not make the mistake of expecting too much, too soon." In Australia, fortunately, he was "thousands of miles away from American newspapers."[26]

MacArthur's talk of a major Pacific offensive soon enraged administration officials. Ickes and Assistant Secretary of War John J. McCloy now complained that the general was not "entitled" to the Congressional Medal since he had "not been on the field of battle in the fight for the Bataan Peninsula, but had stayed under cover at Corregidor." Roosevelt actually ordered the War Department to "make[up] an incident" to justify the award, McCloy claimed. Ickes fumed about the press covering MacArthur so favorably that papers were printing pictures of "the Hero's Son" (Arthur) getting his hair cut. Other generals actually had to win battles but this only had to "live and breathe and let his press agent pound the typewriter." MacArthur, the interior secretary informed his colleagues, had begun a campaign for the presidency as "our man on horseback."[27]

General Marshall and Eisenhower as well as Secretary Ickes raised their eyebrows over MacArthur's accepting the $500,000 gift. In June 1942, Quezon arrived in Washington and attempted to give Eisenhower a similar, if smaller, "honorarium." Ike checked it out with Marshall (who warned him to stay clear lest it "destroy" him) and asked the Filipino leader to instead present him a "written citation of thanks" for service as a military adviser. This, he assured Quezon, "would be of great and more lasting value to me and my family than any amount of money his government could possibly present to me."[28]

Roosevelt made no recorded comments about the payment but, by May, complained to aides about the "trouble MacArthur was making" with his public demands for a bigger Pacific campaign. In June, FDR remarked to his press secretary, William D. Hassett, about the SWPA commander's "constant playing to the grandstand." He "seems to have forgotten," Roosevelt noted, that his record in Manila resembled that of Admiral Kimmel and General Short who "face court-martial on charges of laxity at Pearl Harbor." MacArthur's leadership at Corregidor was "criminal," not heroic.[29]

Congressman Lyndon Johnson, then a young protégé of FDR, fueled these doubts with his report on a trip to the Southwest Pacific. Temporarily on duty in the navy, Johnson looked forward to his June trip as an opportunity to "get into the danger zone" and to "look into charges against General MacArthur" on Roosevelt's behalf.

Besides almost getting killed while flying on a combat mission, Johnson had a long meeting with the general who lectured him "for two hours and ten minutes." MacArthur then awarded LBJ a medal for bravery and sent him home.

In an effort to evade rigorous SWPA censors, *Time* correspondent Robert Sherrod handed Johnson a long report challenging the "official line" delivered by MacArthur. Sherrod described the general as brooding, aloof, and remote, a "poser and an extreme egoist." MacArthur's staff spoke "bitterly against Roosevelt," charging that he starved SWPA in order to "dampen the MacArthur political boom." Outside of the sycophantic headquarters staff, most officers "hated MacArthur almost beyond description."

Yet, even cynics, Sherrod admitted, felt uplifted when listening to the general make a speech. MacArthur might not be very successful on a real battlefield, but "he talks the best war I have ever heard. He has a greater sense of drama—ham acting, his officers call it—than Churchill or Roosevelt." Perhaps this contributed to what Sherrod called the "amazing . . . stories we hear about MacArthur's overwhelming heroic elevation in the U.S." The people were "hungry for a hero" and, as a fellow correspondent who escaped from Bataan put it, "we put him there, knowing better."

Although Johnson later told Sherrod the report was "so hot I put it in the toilet," he actually kept multiple copies. During July the congressman met with FDR and Harold Ickes, both of whom, no doubt, pumped Johnson for information.[30]

The End in the Philippines

Once MacArthur took command in Australia, the public's breathless concern with the fate of the Philippines waned. Still, the defenders held out for nearly two more months. When he left, the general created four separate commands (all under his control) hoping the separation might prevent a total collapse when the Japanese overran Bataan or Corregidor. He expected that units on outlying islands would survive and merge with Filipino guerrillas. Still, for all his public anguish, MacArthur expressed more concern with planning a new offensive than with alleviating the suffering of his comrades.[31]

Ultimately, no one could offer the besieged Americans much help. The

Japanese resumed their offensive against Bataan on April 3. From Australia, MacArthur ordered the broken men to counterattack. Instead, the ravaged units quickly collapsed. On April 9, over 70,000 sick, wounded, and starving troops surrendered. MacArthur reacted with rage, arguing that the Bataan force should have broken through Japanese lines and begun a guerrilla campaign.

The no surrender order issued by Roosevelt in February still applied when Bataan fell, although Stimson and Marshall had considered revoking it. Roosevelt quickly withdrew his order that the garrison on Corregidor fight to the end. After a heavy bombardment, Japanese landing forces established a position on Corregidor on May 6. Later that day, Wainwright capitulated.[32]

Wainwright's surrender covered all Americans in the Philippines, dashing MacArthur's hope that troops in the south might continue guerrilla resistance. The general condemned Wainwright's decision, unaware that the Japanese commander had threatened to massacre everyone on Corregidor unless he complied. The Mindinao forces surrendered on May 10 and most other units followed quickly. (Although MacArthur embraced Wainwright at the war's end, he actually opposed efforts to award him a Medal of Honor while he languished as a Japanese prisoner. In 1945, Stimson and Marshall insisted that Wainwright receive the medal.)[33]

Defeat in the Philippines had an especially depressing impact on Eisenhower, who had both served in Manila and led the effort to assist the beleaguered troops there. MacArthur's effort to shift the blame of defeat to others enraged him. When Corregidor fell, Ike poured out his feelings in a diary entry.

> Poor Wainwright! He did the fighting in the Philippine Islands, another got such glory as the public could find in the operation. . . . General MacArthur's tirades to which TJ [Davis] and I so often listened in Manila would now sound as silly to the public as they then did to us. But he's a hero! Yah.

Nothing revealed more clearly the gap between the public's MacArthur and the general known to his peers.[34]

6
The Pacific War
and American Politics

From virtually the moment of his arrival in Australia until late 1944, Douglas MacArthur focused his energies on returning to the Philippines. In practical military terms, this meant pushing the Japanese out of New Guinea and utilizing that island as a stepping-stone toward his goal. This operation had to compete for resources with the navy's plan for an offensive across the central Pacific, with efforts to build up China's military capability, and ultimately, FDR's "Europe First" strategy.

MacArthur feared that institutional and personal enemies in Washington opposed his drive for ignorant or selfish reasons. To counter this opposition he worked hard to sell his strategy and accomplishments to the American public and relied on the influence of anti-New Deal political leaders and newspaper moguls. He alternatively cajoled and threatened, begged and demanded so much that the president and Joint Chiefs committed greater resources to the Southwest Pacific than they thought wise. His wartime accomplishments were as much a testament to his public relations abilities as to his military skill.

To a large degree, the Allied determination to open a second front in Europe and to defeat Germany before Japan dictated the pace of the Pacific War. Army and navy strategists also questioned whether the best route to Tokyo lay through New Guinea and the Philippines. A central Pacific offensive promised to secure islands from which to stage heavy air attacks on Japan. A successful Chinese offensive would neutralize millions of Japanese troops. Hitler's early defeat, of course, would free immense American, British, and Soviet resources for the Pacific. Even so, during 1942 the Joint Chiefs agreed on the need to halt Japan's advance and develop the foundation for a counterattack. As of December 1943, they had actually committed more troops and supplies to the two Pacific theaters than to Europe.

Still, the military chiefs could do little to counter the general's frequent assertion that "some people in Washington would rather see MacArthur

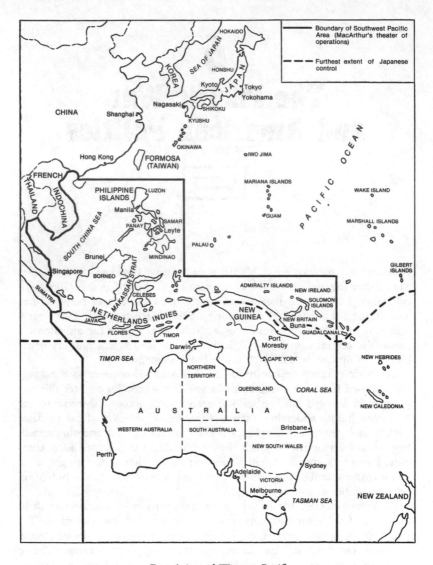

East Asia and Western Pacific

lose a battle than America win a war." In November 1943 he wrote political backer Robert Wood that he was forced to fight Japan on "a shoestring." A few months later he told another friend that sufficient "resources have never been made available to me for a real stroke." MacArthur complained that the judgment of civilian and military officials were "handicapped by politics" and "even the most basic decisions are corrupt." From "the beginning" his enemies had denied him vital resources. He had done his best, "but no commander in American history had so failed of support."[1]

During the second half of 1942, crises in the Mediterranean and North Africa had forced the Joint Chiefs to postpone promised deliveries to MacArthur, delaying his operations. General Marshall and Admiral King sympathized with the SWPA commander and tried to hasten the resumption of supplies. Roosevelt had set the priorities, however, and these remained in the west. If a major second front in Europe was not possible in 1942, the president declared, "then we must take the second best [North Africa], and that is not the Pacific." Marshall tried to put it more delicately. "You should be aware," he wrote to MacArthur,

> that the pressures to meet the growing dangers of the situation in the Aleutians, to build up again the depleted air force in Hawaii, to meet the debacle in the Middle East, not to mention Russia, China, the losses in ocean tonnage, and the urgent necessity of creating new air squadrons sufficiently trained and equipped to go overseas, make our problem exceedingly difficult and complex.

But this struck the SWPA commander as patronizing and untrue.[2]

MacArthur interpreted Marshall's leadership and Eisenhower's operations in North Africa as slights against himself. General Robert Eichelberger, a corps commander and later head of the Eighth Army, recalled that throughout the war MacArthur raged "about his dislike for FDR and his statements about General Marshall and General Eisenhower were rich, rare and racy." MacArthur "had a fetish" that Marshall was "working against him at all times" and still accused Ike of having sabotaged the work of the military adviser in Manila by "stealing publicity." Eichelberger discovered quickly that to steal "any publicity from MacArthur was like driving a dagger into his heart."[3]

The SWPA commander insisted that liberating the Philippines would both redeem American honor and shorten the war by severing Japan's economic lifeline to Southeast Asia. Critics countered that the first order of business must be stopping the Japanese naval advance through the South Pacific (the enemy's string of victories continued through the early summer of 1942) and then bolstering the central Pacific and China theaters. Although not discounting the value of the Southwest Pacific, most planners in Washington doubted the wisdom of giving everything to MacArthur.

In fact, MacArthur could not even begin his New Guinea offensive until the navy had scored decisive victories in the Coral Sea and at Midway. In May, intelligence intercepts revealed that the Japanese intended to seize

Port Moresby on the southern coast of Papua, New Guinea. MacArthur intended to use the port as a key base areas. That month, at the Battle of the Coral Sea, American naval forces beat back a Japanese task force and saved the thinly held Australian positions at Moresby.

Fearful that Australia would be Japan's next target, MacArthur urged that "the Atlantic and Indian oceans should be temporarily stripped" of naval and air forces to augment his command. If this were not done, he predicted, Australia would be jeopardized and America would "face a series of such disasters and a crisis of such proportions as she never faced in the long years of her existence."

Instead, the Joint Chiefs dispatched these forces to Hawaii. Military planners in Washington knew, from reading intercepted enemy messages, that the Japanese navy planned to seize the Midway islands (near Hawaii) in hopes of drawing the U.S. fleet into a decisive engagement. This critical information prompted the navy to set a trap during the first week in June. At the dramatic battle of Midway, American ships and planes sank four Japanese aircraft carriers and the best of the imperial navy's pilots. In effect, the navy broke the enemy's offensive striking power and forced Japan to begin what its strategists dreaded most—a war of attrition with a far more richly endowed adversary.[4]

Encouraged by these victories, MacArthur pushed for authority to attack the large Japanese base at Rabaul, on the island of New Britain. The Joint Chiefs again refused, believing it too formidable a target for the limited forces currently available. Instead, planners in Washington decreed that the navy and marines should attack Guadalcanal, in August, while MacArthur's forces moved on the southeast New Guinea coast. Both campaigns proved far more difficult and bloody than predicted, and lasted for months rather than weeks.

By July, MacArthur turned his attention from Rabaul to Buna, a tiny mission station in Papua, the southeastern part of New Guinea. He proposed to seize the town, construct airfields, and begin a leapfrogging advance along the New Guinea coast. (The northwestern tip of New Guinea lay only a few hundred miles from the southern Philippines.) The general ignored warnings from naval intelligence that the Japanese also planned to take Buna, as a base for an overland assault against Port Moresby on the southern coast, facing Australia.

On July 21, 16,000 Japanese troops moved into Buna and launched an expedition across the Owen Stanley Mountains toward Moresby. Only extraordinary efforts by Australian forces, who fought the enemy along the treacherous mountain path called the Kokoda Trail, stymied the attack. Subsequent Japanese plans to capture Port Moresby, as well as Milne Bay at the eastern tip of Papua, were postponed because of the intense battle for Guadalcanal to the east. Still, because of his delay, MacArthur would now have to fight his way into Buna.

As the summer lapsed into autumn the general grew frustrated by his comparative inaction while other commanders fought in North Africa and

Guadalcanal. The appointment of General George Kenney as SWPA air commander improved prospects for gaining air superiority over New Guinea. But the troops training in Australia would not be ready for a major campaign for several months.

When General Henry H. Arnold, commander of U.S. Army Air Forces, visited MacArthur late in September, he found his host "obsessed by a plan he can't carry out, frustrated, dramatic to the extreme" and even "shell shocked." The SWPA chief complained about inadequate support and warned the enemy might take all New Guinea, the Aleutians, and "move into Siberia." Dismissing the value of a second front in Europe or Ike's operations in North Africa, MacArthur argued that everything in the American arsenal ought to be sent to the Pacific. He would make increased aid to the Soviets conditional on their willingness to fight Japan. Failure to centralize command, presumably under himself, would result in the "Japs . . . control[ling] the Pacific for 100 years."[5]

In mid-November, Australian and American forces began the battle for Buna in earnest. MacArthur established a field headquarters at Port Moresby and delegated Major General Edwin F. Harding to direct the fighting. Unfortunately, some 8,000 Japanese defenders (not the 2,000 estimated by SWPA intelligence) stopped the mainly Australian attackers in their tracks. While disease and casualties thinned the SWPA ranks, MacArthur pressed for an all-out offensive. When it failed he relieved Harding.

Blaming the delay on Harding and his Australian units, not the ability of the Japanese defenders, MacArthur ordered his new commander, General Robert Eichelberger, to win a quick victory. He was to "take Buna or not come back alive." If victorious, MacArthur promised to "announce" Eichelberger's name publicly, making him a "very prominent figure in the United States."

At the front, Eichelberger found his men hanging on by a "shoestring" against superb Japanese troops. On several occasions he feared complete defeat from either the enemy or tropical diseases. At least a third of the attacking force was wounded in combat, and more would have been struck down by bullets, Eichelberger noted, "if the malaria mosquito hadn't removed so many from the field of battle." Among frontline units, nearly half the men were wounded or killed.

Skillfully revitalizing the offensive, the new field commander secured Buna by mid-January. MacArthur, however, waited a week after the victory to award Eichelberger his promised medal. The latter felt he did this to insure the public associated MacArthur's name, not Eichelberger's with the first victory in New Guinea. This interpretation seemed confirmed when MacArthur reacted to several press reports praising Eichelberger by threatening to "reduce you to the grade of Colonel tomorrow and send you home." Outshining his boss, the combat general wrote his wife, could "prove more dangerous than a Japanese bullet."[6]

The Buna victory made possible MacArthur's subsequent sweep along the New Guinea coast. Besides the campaign's operational importance, the way SWPA public relations personnel portrayed the battle set a pattern that continued throughout the war. Since leaving Corregidor, MacArthur had not actually commanded troops in the field. He would not do so, in any significant way through Japan's surrender. As a theater commander he planned overall strategy and coordinated the efforts of talented subordinates like Generals Eichelberger, Walter Kreuger, and George Kenney. Yet, he had a "special talent," in the words of one critic, to "make people forget who his field commanders were."

MacArthur's public relations staff manipulated facts in such a way as to give the "impression that he was the field commander himself, in tactical command of every jungle and on every beachhead." The public's image was shaped by what one correspondent called "the most rigid and dangerous censorship in American history." Colonel LeGrande Diller, MacArthur's press officer, insured that no news left the theater unless, in the words of General Kenney, it "painted the General with a halo and seated him on the highest pedestal in the universe." Only "favorable news, reflecting complete credit on an infallible Mac," got through the censors.

The world learned of events in SWPA through a release of the notorious "MacArthur Communiqué." Distinguished by their "lush and elaborate prose," which imitated the general's spoken rhetoric, they possessed a "special air of fantasy." Always, "they came back in the end to the General himself." Besides their egotistical focus, an "inevitable, chronic and ineradicable inaccuracy" distinguished the releases. They portrayed MacArthur as the sole architect of victory, continually, surprising, outsmarting and utterly destroying the enemy. His "brilliance" demolished opponents while sparing American lives.[7]

For example, at Buna MacArthur announced that he had minimized casualties by insisting on a cautious attack in which "no attempt was made to rush the [enemy] positions." He boasted that "no campaign in history" had achieved such decisive results "with so low an expenditure of life and resources." It is impossible to reconcile this description with the general's "win-or-die" order to Eichelberger, his demands for frontal assaults, and the resulting casualties.

As D. Clayton James concluded in his study of the New Guinea campaigns, MacArthur's assertion "appears fantastic." Of the approximately 33,000 Allied troops in Papua, some 8,546 were combat casualties and 3,095 died. The death rate was actually three times greater than that in the notorious Guadalcanal campaign. Although many subsequent SWPA offensives proved more economical with human life, the official army history of the Papua campaign concluded that "the victory there, proportionate to the forces engaged, had been one of the costliest of the Pacific War."[8]

During the following eighteen months SWPA communiqués routinely exaggerated the numbers of Japanese ships, aircraft, and soldiers destroyed. When officials in Washington (especially naval officers) questioned these figures, MacArthur accused them of jealousy. Later, during the occupation of Japan, a preliminary investigation of Japanese records revealed gross errors in MacArthur's accounts. The general, by then Supreme Occupation Commander, blocked further research.[9]

Disillusioned soldiers, like those who had served at Bataan, expressed outrage at SWPA headquarters for denying any credit to enlisted men or for proclaiming victories when the troops knew that difficult "mopping up" operations still remained. GIs circulated critical poems and songs about the hyperbole in "Doug's Communiqué." The last verse of one such poem serves as an example.

> And while possibly a rumor now,
> Someday it will be fact
> That the Lord will hear a deep voice say
> "Move over, God—it's Mac."
> So bet your shoes that all the news
> that last great Judgment Day
> Will go to press in nothing less than
> DOUG'S COMMUNIQUÉ!

During the almost two years that passed between the Buna victory and the invasion of the Philippines, MacArthur's headquarters publicly reported killing between 150,000 and 200,000 Japanese. In the same period, they put Allied losses at 122 killed, 2 missing, and 529 wounded. Truly, this was remarkable leadership.[10]

The only difficulty, one journalist noted, was that while MacArthur "had to win the war every morning," he also had to convince "the public that Roosevelt and the dastardly Chiefs of Staff were withholding from him the weapons that were rightfully his." Whatever the factual errors in SWPA publicity, it convinced most of the American public that not only was MacArthur's theater doing most of the fighting against Japan, but that he singlehandedly "was the one who was licking the Japs." Years later, most Americans still credit Douglas MacArthur with Pacific victories won not only by his forces, but by the navy and marines. As one war correspondent remarked, few generals have achieved a "comparable public relations coup."[11]

Recent assessments of MacArthur's military leadership have called into question the popular wisdom that he accomplished very much with very little. In the words of Sidney L. Falk, "during most of his campaigns he controlled resources far more extensive than those of the enemy he faced." The disparity in air, ground, and naval strength was often "overwhelming." Up through 1944, his casualties remained low in part because Washington "prevented him from embarking on excessively wild or dangerous operations." On those occasions when the Japanese faced him on equal terms, he either could not defeat them or suffered casualty rates comparable to that of other theater commanders.

Perhaps his greatest defect, Falk concludes, was the degree of "strategic parochialism" and "inflated sense of the value of his own operations" that surrounded the general. He could neither admit the validity of other strategies nor the competence of other commanders. MacArthur disputed the honor of anyone who disagreed with him. Ultimately, SWPA played only a supporting role in defeating Japan. Its greatest contribution came in assisting the central Pacific offensive. At some level, MacArthur realized this fact and it pained him in his soul.

During the last few months of 1942 and into early 1943, MacArthur felt trapped in Australia, awaiting the outcome of the first offensive in New Guinea. At this time he conferred regularly with Lieutenant Colonel Gerald Wilkinson, a British liaison officer who reported directly to Winston Churchill. In spite of the general's professed contempt for Great Britain, Wilkinson was the first of several British officials (including Alvary Gascoigne in Tokyo) to whom he frequently unburdened himself.

MacArthur blamed many of his troubles on Eisenhower who, he charged, was "enhancing his own position by feeding the White House with anti-MacArthur data." Early in 1943, he criticized bitterly Ike's North African victories as a sham." Japan, not Germany, was the "main enemy." Even if Hitler won a partial victory, "life under a civilized race . . . would be tolerable." In contrast, the Japanese were a "total menace to civilization" and "life under her would be impossible." (Charles Willoughby went further, telling Wilkinson that Allied strategy should be to stop fighting in Europe, "let Nazism mellow," and encourage the "Germans and Russians to kill each other off.")

MacArthur described the Joint Chiefs of Staff as bunglers under the thumb of Roosevelt. Any sensible strategist would divert all resources from Europe to SWPA, he asserted. In 1943, after leaving Australia, Wilkinson penned a candid portrait of MacArthur which summed up the general more precisely than any contemporary assessment.

> He is shrewd, selfish, proud, remote, highly strung and vastly vain. He has imagination, self-confidence, physical courage and charm, but no humor about himself, no regard for truth, and is unaware of these defects. He mistakes his emotions and ambitions for principles. With moral depth he would be a great man; as it is he is a near miss which may be worse than a mile. . . . His main ambition would be to end the war as Pan-American hero in the form of generalissimo of all Pacific theaters. . . . He hates Roosevelt and dislikes Winston's control of Roosevelt's strategy. He is not basically anti-British, just pro-MacArthur.[12]

The War at Midcourse

During 1943, American and British planners concentrated upon the roads to Berlin, not Tokyo. At such summit meetings as Casablanca (January), Washington (May), Quebec (August), and Cairo (December), the British pushed for expanding operations on the European periphery, including North Africa, the Mediterranean, and Italy. Above all, they sought to secure their colonial lifelines and avoid a repetition of trench warfare. In contrast,

American strategists urged opening a second front in Western Europe as soon as possible. Any resources not needed for that operation, the Americans argued, should go to the Pacific. The British chiefs of staff favored only defensive, holding action in the war against Japan.

In the American camp, the navy pushed hard for a major central Pacific offensive, largely bypassing the Philippines. More sympathetic army planners barely succeeded in preventing the navy from placing MacArthur's theater under the control of Admiral Nimitz. In a compromise decision reached at the Casablanca conference, MacArthur and Nimitz were authorized to continue their operations. However, the Joint Chiefs committed less resources to SWPA than MacArthur had hoped, possibly setting back his plans to attack the Philippines.

Despite these uncertainties, MacArthur's forces made real progress during the first eight months of 1943. Increasingly, American dominance of the sea and air, the arrival of more and better weapons, and Japan's inability to reinforce its isolated garrisons tipped the scales. MacArthur's field commanders developed great skill at coordinating amphibious landings and used their technological edge to the utmost advantage. Despite their tenacity, Japanese forces in New Guinea and the Solomons found themselves overwhelmed by American firepower. By the spring, MacArthur's troop strength had grown large enough to create the Sixth Army, under General Walter Krueger. As his American resources expanded, the SWPA commander turned away from cooperation with the Australian ground forces which had formed a substantial part of his manpower. By mid-1944 he cut these allies out of almost all SWPA operations.

Progress on and around New Guinea did not convince military planners in Washington that SWPA held the key to victory in the Pacific. In April–May 1943, Joint Chiefs of Staff (JCS) strategy committees drafting plans for the defeat of Japan envisioned a tight naval blockade, heavy bombing, and probably, an invasion of the home islands. MacArthur's operations were expected to keep the enemy off guard, but the navy's advance across the central Pacific remained the crucial element. Even the projected air offensive using a new generation of heavy bombers would be launched from either China or captured Pacific islands, not the Philippines. This "Strategic Plan for the Defeat of Japan" was on the table when FDR and Churchill convened the Washington conference (TRIDENT) in May.

At TRIDENT, the two delegations struck a compromise. In return for London's approval of a May 1944 cross-channel invasion, the Americans agreed to a small offensive in Italy. In order to strengthen Nimitz's central Pacific drive, operations in Burma would be postponed and SWPA would receive fewer resources than promised.[13]

As the August QUADRANT conference in Quebec approached, MacArthur feared that growing support for Nimitz would further marginalize his operations. The continued refusal of the Joint Chiefs to approve an attack upon the large Japanese base at Rabaul, on New Britain, upset the general especially. Anxious to engage the enemy, he insisted that Rabaul must be captured to protect his advance toward the Philippines. However, JCS planners considered Rabaul far too strong to attack directly and argued that air strikes could effectively isolate and neutralize it. Despite MacArthur's howls, the plan imposed by the Joint Chiefs proved successful and eliminated Rabaul as a threat to SWPA operations. (After the war, the facts showed that MacArthur had grossly underestimated Japanese strength on New Britain. By then he claimed that bypassing the base had been his own idea.) [14]

At Quebec, Roosevelt, Churchill, and their advisers affirmed the course set earlier in Washington. The invasion of France (OVERLORD) would take priority over Italian operations. The central Pacific offensive would proceed, along with a later campaign in Burma. Airfields were to be built in China and on Pacific islands for mounting long-range attacks on Japan with the new B-29 bomber.

The British almost convinced Roosevelt to put all of MacArthur's operations on hold before General Marshall and Admiral King warned that this might wreck the tenuous interservice balance between the SWPA and POA theaters. Still, the American service chiefs agreed that the central Pacific thrusts "promise[d] more rapid advance" and should receive the bulk of military resources. These decisions threatened to delay an invasion of the Philippines until 1945 or 1946.

By December 1943, when British and American leaders met at Cairo, a coalition of air force and naval planners pressed successfully to make conquest of the Mariana Islands a priority. MacArthur could advance along the New Guinea coast, striking at nearby targets of opportunity such as the Admiralty Islands. Although he had permission to plan for future operations in the Philippines, the Joint Chiefs placed their big bets on a thrust across the central Pacific. [15]

MacArthur appealed these decisions to Secretary of War Henry Stimson. The general hoped that as an old "Philippine Hand" (Stimson had served as governor-general in the 1920s) the secretary might support an early invasion of the archipelago. MacArthur claimed he could take the Philippines with a fraction of the resources that Nimitz required. He condemned the navy for pounding away at Pacific islands like Tarawa, resulting in "tragic and unnecessary massacres of American lives." MacArthur predicted that the operations scheduled against the Marshall and Mariana Islands would prove even bloodier. But if Roosevelt gave him "central direction of the war in the Pacific, " he assured Stimson, "I will be in the Philippines in ten months." [16]

(The comparison between SWPA and POA losses were dubious, at best. MacArthur stressed how, during 1943–44, his forces advanced 1500 miles along the New Guinea coast, losing only about 1,600 men and killing 26,000 Japanese. Nimitz's assault against the Marianas and Palaus claimed six or seven times as many American and about 46,000 Japanese. On Saipan alone, some 3,400 Americans died. During and after the war, Mac-

Arthur argued that these figures proved his ability to advance faster and farther at far less cost. He neglected to add that the Japanese defended New Guinea much more lightly than the strategically located central Pacific islands. Because the enemy realized these islands could be used for strategic bombing, they mounted a major defense. Most American strategists felt the importance of securing these island bases justified the high casualty toll.[17])

Concerned that talk of a MacArthur presidential candidacy might have alienated Roosevelt, the SWPA commander asked an assistant to the secretary of war to "Tell Mr. Stimson I have no political ambitions. I have but one ambition; to return to the Philippines, to save the Philippine people . . . from their present agony; to restore the prestige of the United States. That is all that is moving me." The navy's "pride, position and ignorance," he added, had misled the president and brought a "great tragedy to our country."[18]

In the spring of 1944, it appeared the bulk of the Philippines might be bypassed or liberated only late in the war. Meanwhile, Nimitz would push on to the Marianas, the Palaus, and possibly, Formosa. MacArthur knew that once the navy and army air forces possessed these bases of operation, the military relevance of the Philippines would diminish further. Apparently, enemies in Washington would succeed in frustrating his mission. Gradually, the SWPA commander came to believe that only political leverage—exercised by supporters or by himself—could force these adversaries to clear his path back to the Philippines.

MacArthur's hostility toward the Roosevelt administration and interest in challenging it surfaced in discussions he held with Lieutenant Colonel Wilkinson, Churchill's liaison with SWPA. During 1942–43, the general lectured Wilkinson about his dissatisfaction with Allied policy, knowing the fact that the lieutenant colonel transmitted his views directly to FDR's close comrade, Winston Churchill. MacArthur might have used these confessionals to send warnings, indirectly, home.

Although MacArthur claimed to have no presidential ambitions, his remarks to Wilkinson raised the British officer's eyebrows. On the eve of the 1942 congressional election, the general gloomily predicted a Democratic sweep. (In fact, the Republican party nearly recaptured legislative control.) Once that occurred, MacArthur thought FDR would "find some pretext to relieve" him from command. Nothing could be done about it, he added, since removal of any general for any reason was a "presidential prerogative."

The SWPA commander told Wilkinson that, although he never thought about the political implications of his acts, he would "nevertheless respond to a nationwide appeal" from the American people to "lead them as president in this time of trouble." Early in 1943, he averred that "one day" a "voice from the fighting services" would have to be heard in the White House. Incredibly, MacArthur asked Wilkinson to invite Churchill to visit

Australia so he and the prime minister could consult on Pacific strategy. Churchill, he added, was his "only ally" against the anti-MacArthur forces in Washington![19]

The Politics of War

The startling gains achieved by the Republicans in the 1942 congressional elections heartened all New Deal critics. The GOP gained forty-seven House and ten Senate seats, revealing the weakness of a Democratic ticket not headed by FDR. Since Roosevelt had not committed himself to run in 1944, the prospects of capturing the White House whetted Republican appetites.

Among Republicans, Wendell Willkie and New York governor Thomas Dewey led in public opinion polls held from 1942–44. Voters surveyed evinced great enthusiasm for Douglas MacArthur as a military leader, but only 20 percent rated him as a top choice for president. The most reliable poll, by Elmo Roper in March 1944, showed FDR the overwhelming favorite of voters asked to compare the incumbent's abilities to those of Dewey, Wilkie, and MacArthur.[20]

The general's supporters fell, broadly speaking, into two, somewhat overlapping categories: old isolationists and midwestern Republicans. The isolationists, as historian Justus Doenecke observed, came in two varieties: Liberal reformists who feared that war would short circuit domestic reform and conservatives who believed that war would extend socialistic tendencies of the New Deal and undermine capitalism. Some liberal isolationists, like Wisconsin's Senator Bob LaFollette, as well as conservatives, such as Chicago *Tribune* publisher Colonel Robert McCormick and America First Leader General Robert E. Wood, rallied to MacArthur. Following the outbreak of war, both types became convinced that "Europe First" made the United States the "unwitting agent of either [British] colonial despotism or [Soviet] revolutionary terror."[21]

Politicians like the conservative and staunchly isolationist Senator Arthur Vandenberg (R–Michigan), who recoiled at Roosevelt's dalliance with Churchill and Stalin, championed MacArthur as a nonpartisan hero who could bring new voters into the Republican tent. The senator did not care that, before the Pearl Harbor attack, the general had been a moderate inverventionist favoring aid to the Allies. Vandenberg hoped that the very uncertainty surrounding MacArthur's politics could become an advantage. The general's simple—or simplistic—patriotism permitted him to seem all things to all people, while his lack of an organization base would make him beholden to professionals able to secure his nomination in a brokered convention.

Even before MacArthur left Corregidor, rumors swept Washington that the White House had, for strategic or political reasons, abandoned the general. In February 1942, for example, Vandenberg complained about "Roosevelt's private war," in which he maintained that Roosevelt's "social revolution units" helped England and Russia and left MacArthur to "fight alone." If the general got out of the Philippines, Vandenberg promised, "he will be my candidate for president in 1944."[22]

Newspapers published by Joseph M. Patterson (New York *Daily News*), Colonel McCormick (Chicago *Tribune*), Cissie Patterson (Washington *Times-Herald*), and the Hearst and Gannett chains all demanded that the administration do more to help MacArthur on Corregidor. After the general reached Australia, they urged that Roosevelt put him in charge of *all* American military operations or, at least, the war against Japan. Later, these conservative publishers charged that the navy and White House vindictively split the Pacific in two and starved MacArthur's theater of vital resources. Colonel Lloyd A. Lehrbas and Brigadier General LeGrande A. Diller, the general's press officers, nurtured reporters who circulated these charges.[23]

In October 1942, Joseph C. Harsch of the *Christian Science Monitor* suggested that MacArthur had not been given command of a unified Pacific theater in retaliation against moves by conservative supporters who had launched a "MacArthur for President campaign." The general dismissed speculation about his candidacy, saying that all claims to the contrary "must be regarded as merely amiable gestures of good will dictated by friendship." He had "started as a soldier" and would, if he survived the war, "finish as one." His only ambition was to defeat Japan and "return to that retirement from which this great struggle called me."[24]

Several members of the president's cabinet doubted this. Harold L. Ickes charged that the general was "taking himself every whit as seriously as some of us have thought." The interior secretary set about stockpiling scandalous ammunition (gathered from Drew Pearson and the first Mrs. Douglas MacArthur) to use in case the general sought political office. Secretary of War Stimson also pondered the SWPA chief's motives. The day after the general denied all political aspirations, Stimson wrote in his diary:

> MacArthur, who is not an unselfish being and is a good deal of a prima donna, has himself lent a little aid to the story by sending people here who carry a message from him that he was not a presidential candidate, thereby playing into the hands of the people who would really like to make him a candidate.

A man of true honor, like George C. Marshall, would "never say a word on the subject."[25]

Although MacArthur never committed himself publicly to a campaign, several of his closest staff acted as intermediaries on his behalf. They listened to his attacks on FDR and served as a liaison with Republican officials who organized a low-key movement to draft the general. During 1943 and early 1944, the SWPA commander told General Eichelberger that the "only reason I want to be President . . . is to beat that S.O.B. Roosevelt." In June 1943, the "Chief" discussed "the Republican nomination for next year" and, Eichelberger concluded, "I can see that he expects to get it, and

I sort of think so, too." During a visit to the United States in the spring of 1943, SWPA chief of staff Richard Sutherland and air commander George Kenney discussed MacArthur's political prospects with Congresswoman Clare Boothe Luce and Senator Vandenberg. The latter denounced Secretary of War Stimson for publicizing a directive forbidding regular army officers from campaigning for or accepting political office.[26]

On April 13, MacArthur had an aide hand deliver a confidential message to Senator Vandenberg. It thanked the senator profusely for his "complete attitude of friendship" and hoped, someday to "reciprocate." Although circumstances prevented total frankness, he wanted Vandenberg "to know the absolute confidence I would feel in your experienced and wise mentorship." Vandenberg characterized this first (and only) direct message from the general as one which "might be supremely historic." He assumed MacArthur was responding to the encouragement given Sutherland and Kenney and to the senator's denunciation of the War Department's attempt to prohibit officers' candidacies. "Mac," Vandenberg decided, "certainly is not 'running away' from *anything*."[27]

During the summer of 1943 the SWPA commander dispatched General Charles Willoughby, his intelligence chief, to discuss his prospects with Senator Vandenberg and other prominent Republicans, such as Robert E. Wood. Vandenberg and Wood (who spoke with MacArthur in April in Australia and agreed to fund a behind-the-scenes campaign) proceeded to organize a small movement of conservative Republicans who would communicate with the general through his staff. They sought to develop popular interest in a MacArthur candidacy while keeping the general out of politics. Vandenberg considered it vital that the movement appear a "spontaneous draft" of a hero without the "appearance of any connivance on his part." The senator hoped Willkie and Dewey would split the party, deadlock the nominating convention, and allow MacArthur's supporters to put the general forward.[28]

The group around Vandenberg and Wood (which included publishers Frank Gannett, Roy Howard, and Robert McCormick, as well as some veteran Republican officials) worried that leaders of several "MacArthur for President Clubs" in the Midwest would wreck their plans by running the general in state primaries. This "scares me almost to death," Vandenberg told his allies, since without a large organization, the general might be swamped. The Senator and Wood did all they could to stifle independent boosters threatening the "under-cover movement." As Vandenberg put the issue to General Willoughby in August, their goal was:

> a matter of developing a situation in which, in the first instance, produces a national Convention in which a majority of the delegates are "uninstructed" so that they are free to make the proper decision when the time comes: then to develop through general consciousness that "our man" is the best answer under all the circumstances. . . . tell my friend to just "get on with the war" and to forget this whole political business back here in the States. None of us wants him to do anything else.

For the remainder of 1943, Vandenberg and Wood believed, they ought to work quietly and try to suppress the MacArthur for President Clubs. In November, Vandenberg wrote Wood that he found it "quite inexplicable" that servicemen home on leave from the Pacific were "not enthusiastic about our friend." Perhaps some "diabolical arrangement" insured that "only anti-MacArthur veterans are furloughed home." Wood speculated that the negative reports came mostly from sailors, imbued with their service's dislike of the general.[29]

Besides utilizing his staff to communicate with his supporters, MacArthur benefitted from, and often encouraged, a series of heroic biographies appearing after 1942. For example, the general encouraged his friend, journalist Frazier Hunt, to prepare a substantial "campaign biography." Hunt enjoyed special access around SWPA headquarters and MacArthur actually edited much of the journalist's prose. (The biography did not appear until after the general withdrew from the campaign.) About this time the general also adopted his signature corncob pipe. Previously, he enjoyed posing with a cane, but abandoned it when an aide remarked that the public might think him "feeble."[30]

Meanwhile, Roosevelt responded cautiously to these maneuvers. The president made clear to his aides that they must avoid doing anything that might make a "martyr" out of the general. In January 1944, for example, Secretary of War Stimson announced that the SWPA commander did not have to retire later that month when he reached the statutory retirement age of 64. Instead, the War Department hurried to award him a medal, citing no special reason. In February, Roosevelt declared that regular officers could accept political nominations they did not solicit.[31]

In private, administration officials like Ickes and Assistant Secretary of War John J. McCloy, as well as Drew Pearson, continued to denounce MacArthur as a "Man on Horseback" threatening democracy. The three compared potentially embarrassing information about the general's private life, connections to Filipino fascists, and acceptance of the 1942 payment from Quezon. They anticipated that the SWPA commander would try to force FDR to "discipline him" so he could "pose as a martyr." (For example, in September 1943 MacArthur took time to entertain a group of conservative junketing senators in Australia, but snubbed Eleanor Roosevelt when she visited wounded soldiers.) Ickes and McCloy discussed ways of leaking damaging material from army files that "would not put MacArthur in a good light."[32]

Early in 1944, by chance or design, the War Department got involved in an intense flap with MacArthur. In January, writer John McCarten published an iconoclastic account of the general's career in the (then) liberal *American Mercury*. McCarten commented only parenthetically on the Pacific war but, in a style worthy of H. L. Mencken, debunked nearly all the heroic legends surrounding the general. He ridiculed the strange coterie of ultra nationalists, anti-New Dealers, "Yorkville Bundists," and right-wingers from Vandenberg to Gerald L. K. Smith and Father Coughlin, who hoped the general could unseat "that man in the White House." McCarten insisted there was precious little in MacArthur's record to justify public adulation or elective office. The "sole object" of those promoting him for president, the journalist wrote, was to find someone able to "take the reins of government out of Franklin Roosevelt's hands" and to halt the trend "toward what they call 'internationalism.'"

McCarten observed tartly that MacArthur "never committed himself on any nonmilitary subject more controversial than the weather." Nevertheless, his supporters portrayed him as a selfless warrior, patriot, and victim of White House jealousy. They asked the nation to forget his bungled defense of the Philippines, the loss of his air force, and ultimate triumph of the Japanese invaders. "America needed a hero to take the edge off humiliation and MacArthur filled the bill with a margin to spare." In the process, questions about the general's role in the Bonus incident, military failure in 1942, and bad reputation among his troops went unasked.[33]

McCarten attributed the successful counteroffensive in the Pacific mostly to the navy, marines, and air power, rather than to SWPA forces. Yet, because of rigorous SWPA censorship and FDR's decision to handle MacArthur with kid gloves, few critical reports ever appeared in print. The article compared MacArthur to the Civil War general, George McClellan, noting that he had the grace to leave the field before challenging Lincoln.[34]

Within days of the article's publication MacArthur commented he "had never read such lies and misstatements, but that there was always a thread of truth which prevented his answering it." This attack was a "cross . . . necessary for him to bear."[35]

Given the limited circulation of the *American Mercury*, as well as continued censorship, the exposé seemed destined to end there. However, two months later, someone in the Army War College's library service listed the article as recommended reading for servicemen. On March 9, Vandenberg took the Senate floor to denounce the War Department's collusion in circulating "smear literature." MacArthur sent a blistering message to the War Department describing the article as containing false, scandalous, and even "libelous" material. Like Japanese propaganda, it sought to undermine troop morale and advanced the "enemy's sabotage efforts." Marshall and Stimson promptly withdrew it from official circulation and ordered a second critical article, scheduled to appear in *Harper's Magazine*, barred from publication "on the grounds of security."[36]

In March, MacArthur's chief-of-staff, Richard K. Sutherland, met with Herbert Hoover in New York to discuss the "political prospects of General MacArthur." The former president thought Dewey the likely nominee and suggested that MacArthur's handlers work to nominate him as "vice-president with the title of vice-commander in chief . . . in control of the entire war." Sutherland "thought MacArthur would be pleased to undertake such an appointment."[37]

In April, the MacArthur "bubble" burst. Thomas Dewey had already

A young MacArthur on the eve of the Armistice in Europe, November 1918

MacArthur as Major General, 1925

MacArthur and his first wife, 1925

The Military Adviser and Philippine Field Marshall relaxing in Manila, 1940

MacArthur in front of his personal airplane, 1944

MacArthur, President Roosevelt, and Admiral Nimitz during the strategy
conference at Honolulu, July 22, 1944

Ceremony restoring Philippine government during liberation of Leyte,
October 23, 1944

MacArthur stepping off his plane in Japan, August 30, 1945

MacArthur and his staff land in Japan, August 30, 1945

Signing the Japanese surrender agreement aboard the *Missouri*,
September 2, 1945

MacArthur and Emperor Hirohito meet for the first time, September 27, 1945

made great strides toward winning the nomination. In a mistaken belief that a public campaign would improve the general's chances, the independent MacArthur for President Clubs entered their man in the Wisconsin and Illinois primaries. As Vandenberg and Wood predicted, this proved a disaster. Dewey took most of the Wisconsin delegates on April 4, knocking out Willkie and the chance for a deadlocked convention.[38]

By April 10, Vandenberg considered it "all over but the shouting." He wrote a somber appraisal to MacArthur's headquarters, offering the small solace that "our Cabinet" might wait a few weeks before "joining the Dewey parade." A few days later, MacArthur's victory in the nonbinding Illinois primary briefly boosted Vandenberg's spirits. But political observers noted that the general had beaten a local nonentity.[39]

MacArthur, who had not publicly sought the nomination, could easily have avoided further embarrassment by saying no more about the activities of his "over enthusiastic" supporters. In fact, Vandenberg thought he might emerge "from this adventure with enhanced stature and with greater right than ever to the confidence of the country and a greater place than ever in current history."

Unfortunately, just as discussion of the general's candidacy began to fade, Congressman Albert L. Miller (R Nebraska) revealed several letters he and MacArthur had exchanged during 1943–44. The representative had written the SWPA chief to denounce "left wingers and New Dealism" and tell him that "you owe it to civilization and the children yet unborn [to] be our next president." MacArthur thanked Miller and agreed "unreservedly" with the "complete wisdom and statesmanship" of his remarks.[40]

The general made the disclosure worse by first protesting the "unauthorized publication" of his replies (as if that were at issue) and then by insisting that "any fair-minded person" would realize his letters were "neither politically inspired nor intended to convey blanket approval of the Congressman's views." MacArthur denied he intended "criticism of any political philosophy or any personages in high office." More remarkably, the general told a reporter that, although he was not a "presidential candidate," he could not rule out accepting a draft.[41]

Miller followed this up by sending more "public" letters pleading with his hero to drive the New Deal dragon from Washington. As this farce continued, MacArthur issued a "Sherman-like" statement. On April 30, he declared that since many people thought having a general considered for political nomination might prove "detrimental to our war effort," he wanted nothing done that "would link my name in any way with the nomination." "I do not covet it nor would I accept it."[42]

Privately, Senator Vandenberg expressed "shock" that MacArthur had "ever written the letters which Miller made public." If he "hadn't written them, Miller couldn't have used them"; more importantly, the general (and his boosters) could have "retired from the scene with dignity and honor" intact. MacArthur would suffer a final humiliation at the June Republican convention when an obdurate MacArthur delegate insisted on having his vote recorded. The tally was Dewey, 1056 votes to the general's 1.[43]

Franklin Roosevelt, Admiral Leahy noted, "did not show much interest in the announcement" of MacArthur's noncandidacy. One suspects he savored the farcical collapse of his rival's boomlet. Yet despite this, the general's political venture probably helped his campaign to return to the Philippines. Although a majority of military planners still believed this strategy would not hasten—and might even delay—Japan's defeat, MacArthur found an unexpected ally. For complex reasons, Franklin Roosevelt came to support MacArthur's plan, with major consequences for the general's immediate and postwar prospects.

MacArthur, FDR, and the Road to Manila

During the spring of 1944, the Joint Chiefs had authorized MacArthur to plan an assault on the Philippines. At the same time, they permitted Nimitz to launch summer and fall offensives against the Marianas and Palaus. They hoped that long-range bombers operating from these islands (and eventually from Formosa or the China coast) could carry the air war directly to Japan. Possession of these bases would make liberation of the Philippines a much lower military priority.

The SWPA commander criticized the navy's strategy in many messages to the JCS. He doubted the navy's ability to seize Formosa and belittled its importance. In contrast, he described Luzon as the ideal forward base for future operations against Japan. Control of the Philippines would allow the United States to sever Tokyo's economic lifeline to Southeast Asia and, as importantly, would avenge "our unsupported troops [who] were destroyed by the enemy" in 1942. "We have a great national obligation to discharge," he declared.

> Moreover, if the United States should deliberately bypass the Philippines, leaving our prisoners, nationals and loyal Filipinos in enemy hands without an effort to retrieve them at the earliest possible moment, we would incur the gravest psychological reactions. We would admit the truth that we had abandoned the Filipinos and would not shed American blood to redeem them; we would undoubtedly incur the open hostility of that people; we would probably suffer such loss of prestige among all the peoples of the Far East that it would adversely affect the United States for many years.

MacArthur also warned of "extremely adverse reactions" at home if he did not liberate the Philippines. Even though no longer a candidate himself, the general warned that others would attack Roosevelt for not redeeming MacArthur's pledge to return.[44]

Many strategists questioned these assertions. As George Marshall put it to Henry Stimson, the path through the Philippines struck most planners as the "slow way" to Tokyo. Marshall informed MacArthur on June 24 that strategists in Washington recommended an attack on Formosa or Japan directly. "We must be careful," he wrote, "not to allow our personal feelings and Philippine political considerations to override our great objective which is the early conclusion of the war with Japan." "Bypassing" part or all of the Philippines was not "synonymous with abandonment." The sooner the Allies crushed Japan, the sooner Manila would be free.[45]

As the summer of 1944 wore on, both the military and verbal battles intensified. The successful, though bloody, capture of the Marianas secured vital airfields and resulted in critical losses to the Japanese fleet. The cabinet of General Tojo fell in mid July, partly in response to the American victory. MacArthur and the conservative American press criticized these battles as needlessly costly and argued that an attack on the Philippines would achieve the same strategic advantage at a fraction of the cost in American lives. The general condemned plans for future operations against Formosa and the China coast and urged Washington to approve invasions of the south and central Philippines in the late fall. The navy replied in kind, ridiculing any major operation in the archipelago as a diversion. With little warning, Franklin Roosevelt threw himself into the strategic debate.

On July 21, (just after the Democratic convention nominated him for a fourth term), FDR decided to confer with MacArthur and Nimitz in Honolulu. The president may have wanted to show his concern for the "neglected" Pacific and be photographed with two popular military figures. Admiral Leahy (the president's chief of staff) made the trip, but since no other senior military advisers in Washington came along, no substantive debate over strategy seems to have been anticipated. MacArthur expressed private outrage at being summoned to Hawaii at the last minute with no idea of what he was to do.

Accounts of the meeting in Honolulu on July 26–28, 1944, resemble the classic Japanese story, Rashomon. Judge Sam Rosenman, Robert Sherwood, Admiral Leahy, and Douglas MacArthur all published or verbally related different accounts of what transpired. (No official transcript exists.) Although no formal policy decisions resulted from the conference, unofficially it set the agenda for future operations.

In his memoirs, MacArthur recounts with unusual grace the events of late July. The navy lobbied the president intensively but Admiral Nimitz demonstrated a "fine sense of fair play." The president listed to the general's cogent argument for bypassing Formosa and retaking the Philippines. In the end, this swayed FDR, much to the chagrin of the assembled admirals.

Roosevelt's close aid, Sam Rosenman, also described the conference as something of a spirited, but good natured, Army–Navy football game. Rosenman relished repeating how just as the president prepared to leave his ship (the *Baltimore*), "a terrific automobile siren was heard and there raced onto the dock and screeched to a stop a motorcycle escort and the longest open car I have ever seen." MacArthur sat majestically in the back then leaped out and "dashed up the gangplank." He stopped "halfway up to acknowledge another ovation, and soon was on the deck greeting the President." (MacArthur recalled that the crowd gave "me . . . an ovation about three times that which the President received.")

During the course of three days, MacArthur and Nimitz jockeyed for the president's attention. Occasionally they got to speak with him alone, more often while crowded into a car touring the island. On July 29, after the general had departed, Roosevelt spoke glowingly about their meeting. "We are going to get the Philippines back," he declared, and "without question General MacArthur will take part in it. Whether he goes direct or not, I can't say."

Two weeks later, FDR sent the SWPA commander a letter thanking him for the "splendid picture" presented at the "most successful conference." Roosevelt commended MacArthur for the "magnificnt job" he had done thus far and added "you would make more of a go as President than I would as General in retaking the Philippines." When it came time to raise the flag in Manila, the president told him, "without question I want you to do it."

MacArthur answered with equal warmth, telling his "old friend" how "nothing" in the course of the war had pleased him as much as "seeing you again." The general revealed an "even higher and more soul filling vision" about the Manila flag raising. He hoped his beloved Franklin would preside over the solemn ceremony!"[46]

Did the two mean really mean any of this? Harold Ickes thought not. The president, he maintained, "was only playing some politics because I know how little he thinks of MacArthur." Roosevelt knew the "political value in his [MacArthur's] name in some circles and was "shrewd enough to capitalize upon it." Privately, Ickes insisted, FDR considered the general "a fake."[47]

MacArthur at least partly concurred with Ickes's attribution of motive. The SWPA commander told several associates that FDR "felt it would be good politics to show himself intent on winning the Pacific war" and be photographed with his popular general. Roosevelt, he believed, came to the meeting with his mind set on bypassing the Philippines, but left realizing this was politically (if not militarily) unwise.

MacArthur spoke openly of his "amazement" at the president's physical deterioration. "His head bobbed up and down and his mouth sort of hung open." The former master orator suffered stage fright and allegedly depended on MacArthur to tell him what to say to "crowds of sailors and others" who assembled to see them. Roosevelt was a "doomed man" who would not live out his next term, MacArthur predicted. At some point, the general claimed, White House aide Edwin "Pa" Watson took MacArthur aside and offered him any job he desired in postwar Washington or Manila if only he would declare himself "with us" in the coming election. Leahy and FDR went still farther: they offered him the liberation of the Philippines.[48]

Several historians speculate that Roosevelt and MacArthur made a se-cret deal whereby the general would be permitted to retake the Philippines in return for a pledge to stop attacking the administration and provide "good news" on the eve of the 1944 election. In postwar conversations, MacArthur claimed he made just such a pact with his commander-in-chief. He told Robert Eichelberger (in 1947) and Herbert Hoover (in 1946) that the two men hatched the deal during a brief private meeting in Honolulu. MacArthur "told Roosevelt that if he wanted to show progress in the Pa-cific War before his election it could only be done by island hopping to the Philippines on the Southern route." He "guaranteed to show great progress and even to land in the Philippines before November . . . if his plans were carried out."

If the president balked, the SWPA chief predicted a political backlash. The public, MacArthur told FDR, had "forgiven" him for "what took place on Bataan," but would never reelect him if he approved a plan "which leaves 18 million Christian American citizens to wither in the Philippines under the conqueror's heel." The navy and JCS might not care but "polit-ically it would ruin you." Roosevelt, MacArthur reported, smiled and said "Douglas, you have nothing to worry about." In just ten minutes of bar-gaining he had "secured Roosevelt's approval." FDR had little interest in the "superior strategic argument," but evinced tremendous "interest in the political possibilities for himself," MacArthur recalled.[49]

Whether or not MacArthur really made this deal (formal approval for the Philippine operation did not come until early October), several factors began to break his way. Although the navy fought to confine SWPA op-erations to the southern Philippines—allowing Nimitz to attack For-mosa—military developments in the region undermined the navy's case.

In mid-September, SWPA intelligence discovered that very thin Japanese forces held Leyte, in the central Philippines. If so, it seemed logical to bypass Mindinao and attack Leyte earlier than planned. This would bring MacArthur much closer to Luzon. About the same time, the Japanese army in China launched a huge offensive (ICHIGO), which overran American-built airfields designed to support the attack against Formosa. With the Chinese Nationalist regime tottering and their airfields lost, it no longer made sense to devote American forces to clearing Formosa or the China coast. In theory, MacArthur's talk of making Luzon a major base now had more credibility. (In fact, it never played this role.) Meanwhile, the navy and air forces could launch direct attacks on Japan from B-29 bases in the Marianas.

During the Anglo-American summit conference in Quebec in late Sep-tember, the Joint Chiefs decided MacArthur should invade Leyte a few weeks hence. On October 2 the JCS approved a mid-December attack upon Luzon. The navy was to assist MacArthur's operations and then move against Okinawa, bypassing Formosa completely.[50]

Even before he received final approval for the Luzon operation, Mac-Arthur's communiqués took on a new look. Gone were the complaints about Washington "starving" SWPA. Now, upbeat reports emphasized how cooperation at all levels permitted the general's forces to make especially rapid progress. On October 30, for example, little more than a week after his troops landed on Leyte, MacArthur declared the battle virtually won. When skeptical journalists noted that the serious fighting had barely begun, a SWPA press officer remarked that "the elections are coming up in a few days and the Philippines must be kept on the front pages back home." Roosevelt and MacArthur had at last found a way to be "useful" to each other.[51]

7

The Philippines and Liberation Politics

As SWPA forces approached Southeast Asia, MacArthur for the first time had to deal with large civilian populations and the thorny issue of liberating colonial territory. His relations with the Australians had deteriorated so much that by 1944 he excluded them from any substantial policy role in SWPA and consigned their troops to mopping up and garrison duty around New Guinea. MacArthur also opposed British efforts to dispatch ground and naval forces anywhere near his theater, unless they were placed under his control. Like many American civilian and military officials, he sensed the British were more intent on recovering their lost colonies than with defeating Japan promptly or liberating the Philippines.

MacArthur's attitude toward colonialism and nationalism reflected his experiences in the Philippines. Since the 1930s, he accepted the principle of self-determination for colonial peoples, provided they were "ably" led. He acknowledged the intellectual and social equality of elite Filipinos who had proved themselves "fit" for independence.

This elitist critique of traditional colonialism voiced contempt for "decadent" European rule in Asia while also disputing the legitimacy of revolutionary nationalism. Although both men might have winced at the comparison, MacArthur and Roosevelt shared a similar attitude on the subject. They promoted the Philippines as a model of decolonization while decrying French rule in Indochina or the radical alternatives of the Vietnamese and Chinese Communists.[1]

MacArthur's ambivalence toward colonial issues (especially ones that affected his drive for the Philippines) became apparent in 1944 when he negotiated with Dutch officials in Australia about the status of the East Indies. For some time President Roosevelt and the State Department had pressed the Dutch government-in-exile to grant its colony a promise of postwar self-determination. Without bothering to consult Washington, MacArthur made an agreement with Dr. H. J. van Mook, lieutenant governor-general of the East Indies. In order to facilitate operations related to the Philippines, the Dutch granted him complete authority in the East Indies during any military operations. He, in turn, agreed to restore Dutch authority and administration as "rapidly as possible." In effect, the SWPA commander committed American power to the reimposition of colonial rule in the face of competing claims by Indonesian nationalists. MacArthur ignored the immensely complicated situation in the East Indies to facilitate his drive on Luzon.[2]

The general's global perspective revealed a curious blend of the grand and the petty. In conversation with Admiral Forrest Sherman on November 9, 1944, the MacArthur spoke of his growing anxiety over international developments. Admittedly in a sour mood—having just received word of Roosevelt's reelection—he complained that the "American people" had chosen "further regimentation and had forsaken the things for which our forefathers had fought." He went on to prophesy a postwar world split between the United States and the Soviet Union. The Russians, he declared, had already begun to "unify Europe economically" and would soon try to monopolize political control. Germany seemed "ripe for going to the left" and the French Communists might even win a predicted civil war. In Asia, MacArthur warned, the Soviets would "eventually tak[e] over North China and compet[e] for the hegemony of the Pacific."[3]

Following the invasion of the Philippines and Roosevelt's reelection, MacArthur resumed his attack upon administration policy. (Presumably, his bargain with FDR lapsed after election day.) During an "off-the-record" discussion he told reporters that "every mistake that supposedly intelligent men could make has been made in this war." He decried yet again Eisenhower's campaign in North Africa and Roosevelt's effort to help China as an "absolutely useless" waste of resources. Because of it the "Pacific war had been starved." These errors, MacArthur charged, arose from Washington's fundamental ignorance, if not "treason and sabotage." Other civilian and military officials were too blind or foolish to realize that

> the history of the world will be written in the Pacific for the next ten thousand years. . . . we made the same old mistake of intervening in European quarrels we can't hope to solve because they are insoluble. . . . Europe is a dying system. It is worn out and run down, and will become an economic industrial hegemony of Soviet Russia. . . . The lands touching the Pacific with their billions of inhabitants will determine the course of history . . . for the next ten thousand years.

While Washington worried over postwar Europe, Stalin wisely set his long-term sights toward dominating northeast Asia.[4]

MacArthur shared similar thoughts when Turner Catledge and Arthur Hays Sulzberger of the *New York Times* visited him on Leyte. (His revelations of impending operations worried the journalists since they would soon be traveling in a combat zone.) Catledge recalled how

> The general went through his whole life story. He told how Roosevelt had ordered him to leave Bataan [sic], how he got to Australia, how he came to

realize that his enemies in the War Department wanted to undercut him. He spoke of his conflicts with the Navy, how his name had been mentioned for the Republican nomination for President, and the problems this created between him and Roosevelt.

The general surprised the reporters by describing virtually all wartime events in "strictly personal terms." He spoke of "my infantry, my artillery, my men, my strategy, and so on. As he spoke, he was variously the military expert, the political figure, the man of destiny." Catledge and Sulzberger agreed they had "never met a more egotistical man nor one more aware of his egotism" and determined to match it with his deeds.[5]

Of course, officials in Washington shared many of the same anxieties about European strategy and future relations with the Soviet Union. Looking toward victory in Europe, Roosevelt and his advisers realized that the Grand Alliance might easily fracture, complicating the finish of the Pacific War. Anticipating hard bargaining with Stalin at the upcoming (February 1945) Yalta summit, Marshall and Stimson sought MacArthur's views on handling the Soviets and assuring their cooperation in defeating Japan.

In February, MacArthur told a visiting member of the General Staff that he considered it "inevitable" that the Soviets would take "all of Manchuria, Korea and possibly part of North China." The United States should press Russia to "pay her way by invading Manchuria at the earliest possible moment," thereby preventing Japan from using these forces against the Americans.

Shortly after the Yalta conference, MacArthur told Major General George A. Lincoln, one of Marshall's deputies, much the same thing. The War Department must make "every effort to get Russia into the Japanese war before we go into Japan." Although Stalin would demand compensation in the form of Manchurian rail and port privileges, MacArthur felt it would be "impracticable" to refuse given Soviet strength and America's desire that "they should share the cost in blood in defeating Japan."

(At Yalta, the Soviets promised to fight Japan approximately three months after Germany's defeat. The Western Allies agreed that the Nationalist Chinese should grant Russia Manchurian rail and port privileges and that the Kurile Islands should be taken from Japan and given to the Soviet Union. Stalin, in turn, pledged to support the Nationalist regime in China and not assist the Chinese Communists.)

The SWPA commander made the same argument at the end of February, speaking with visiting Navy Secretary Forrestal. Since China had proved a toothless ally, Washington "should secure the commitment of the Russians to active and vigorous persecution of a campaign against the Japanese in Manchukuo of such proportions (sixty divisions, he hoped) as to pin down a very large part of the Japanese army." Like many officers and enlisted men in the Pacific, he urged American leaders to make concessions to hasten Soviet assistance in the war against Japan.[6]

Then, in March, MacArthur told Robert Sherwood (officially an emissary of the Office of War Information but, more importantly, a close friend of Harry Hopkins) that the administration should condemn Soviet actions in Europe and make no deal (as had been done at Yalta) to assure Russian assistance in the Pacific War. He predicted the Japanese would be ready to surrender when he subdued Luzon. If they did not, he admitted, American forces would need Soviet help.[7]

These contradictions revealed how the ideologue and martyr inside MacArthur tripped over the pragmatic strategist. An innate loathing of communism and dislike of FDR led him to suspect the worst about any deal with Moscow. (Listening to the general and his aides, Sherwood reported, "one would think that the War Department, the State Department, the Joint Chiefs of Staff—and possibly, even the White House itself are under the domination of 'Communists and British Imperialists.' ") Yet, an awareness of Japan's residual strength compelled him to seek Soviet help in northeast Asia. Finally, his determination to redeem his pledge to liberate the Philippines made it vital to bring maximum force to bear on Tokyo as soon as possible.

Speaking to Sherwood, MacArthur seemed both anxious and self-assured, open and defensive. The general made his standard speech on his unique ability to comprehend the mysterious and wonderful Orient, the site of world power and markets for the next thousand or even ten thousand years. The visitor left "enormously impressed with the extent of his [MacArthur's] understanding of the Orient and the breadth of his views." At the same time, the "inaccuracy of the information held by General MacArthur and his immediate entourage" about the "formation of high policy in Washington" struck Sherwood as "shocking." He observed "unmistakable evidences of acute persecution complex at work" in the general's headquarters. This "strange misapprehension" (that "Communists and British Imperialists" controlled Washington), Sherwood reported to the president, produced an "obviously unhealthy state of mind." Yet, MacArthur's talk of using American power to end colonialism and promote "liberalism" in postwar Asia impressed Sherwood so much that he recommended to FDR the general's appointment as "Military Governor of Japan after surrender." As so often when astute visitors met the general, they dismissed his comments on American politics as naive ranting, but took seriously his grandiloquent prose about America's future in Asia. Even Sherwood, a sophisticated writer on the "left" of the New Deal coalition and someone who must have held MacArthur's political and economic ideas in contempt, surrendered his skepticism to "America's greatest expert on Asia."[8]

The Return

Throughout Japanese-occupied Asia, the war unleashed a tidal wave of social and political upheaval. In a few months, Japan destroyed the structures of Western colonialism built up over four centuries. Asians responded to the collapse of the old order in a variety of ways. Some collaborated with the Japanese, hopeful that the new masters would establish an order based on "Asia for the Asians." Other collaborators doubted Japan's sincerity but cooperated in order to develop political leverage that could be mobilized against the returning Europeans. Many simply hoped to aggrandize their own power. Probably, most collaborators operated from a combination of motives. "Puppet" officials in Burma, India, China, the East Indies, and the Philippines, for example, could be seen as patriots, villains, or opportunists, depending on one's perspective.

Resistance groups also spanned the political spectrum. At one extreme, the right-wing Chinese Nationalist regime fought as a recognized government, hoping to parlay its alliance with America into postwar strength. Most Asians opposed to Japan organized underground movements that combined political and military goals. The Chinese Communists, the Vietminh (in Indochina) and the Hukbalahap (in the Philippines), for example, fought the Japanese under the banners of nationalism and social revolution.

In the Philippines, the occupation intensified lingering problems that American promises of independence failed to address. The Japanese conquest fractured Filipino politics. Part of the old elite collaborated, while others resisted. Simultaneously, a peasant-based guerrilla movement challenged the Japanese, the old elite, and in some cases, American authority. MacArthur faced an explosive internal crisis even before SWPA forces landed at Leyte.[9]

Although MacArthur and Quezon fled Manila, most commonwealth officials remained behind. When he left the capital Quezon told his subordinates that he and MacArthur (who later denied it) wanted them to cooperate in routine matters with the Japanese so long as they did not pledge allegiance to Tokyo. For their part, the invaders promoted collaboration by extending the promise of rapid independence. Motivated by fear and hope, many leading Filipinos—in and out of government—organized an administrative council under Japanese direction.[10]

The Japanese cultivated several prominent Filipinos to serve as possible puppet leaders. At first they expressed interest in Manuel Roxas, a protégé of both Quezon and MacArthur, who remained behind after his mentors departed in 1942. Initially he refused to collaborate, but eventually joined the cabinet of the puppet government. Occupation authorities eventually selected José Laurel, another member of the prewar oligarchy, as their client.

More than a mere opportunist, Laurel took seriously Japan's pan-Asian rhetoric and viewed Tokyo's modernization under authoritarian rule as a model for his own country. After sponsoring a new constitution, Japan declared the Philippines a nominally independent republic in October 1943. Laurel resisted demands that his government declare war on the United States but bowed to this pressure in 1944.[11]

Tokyo encouraged similar movements in Burma, Indochina, and the East

Indies as well as in the Philippines, hoping to gain popular approval and encourage colonial subjects to participate in the "Greater East Asia Co-Prosperity Sphere." In spite of this, it proved difficult to integrate most of Southeast Asia into a regional economy.

Chronically underdeveloped, the Philippines produced a surplus of what Japanese industry needed least—sugar and copra. Dreams of large scale cotton cultivation suffered from lack of funds and improper planning. After 1944, the destruction of Japan's merchant fleet made it difficult to move commodities anywhere. Meanwhile, currency inflation disrupted the entire food distribution system within the Philippines. Despite these problems, most of the prewar oligarchy continued to collaborate with the Japanese.

The exiled commonwealth government consisted of little more than President Quezon, a few close aides, and the good will of MacArthur. The vice-president, Sergio Osmeña, had almost no political base at home and few important American contacts. In the late 1930s he had gotten on MacArthur's hate list by openly criticizing the general's defense program. In any case, as the most respected student of the period observed, both the puppet and Quezon regimes, "were served by the same oligarchic elite, which only divided into these two hostile units by the accident of history."[12]

To complicate matters further, a broad-based guerilla movement arose during the war. The product of Japanese cruelty, Filipino nationalism, economic privation, and at times, banditry, the guerillas operated mostly in small units, harassing the enemy and sabotaging military and economic targets. American or Philippine Army soldiers who had avoided capture in 1942 or escaped confinement led some units. Others formed spontaneously in response to Japanese oppression or because guerrilla resistance seemed a way to strike out against landordism as well as the new imperialists.

By early 1943, MacArthur recognized the potential advantages and dangers of the guerrilla movement. It could provide vital intelligence and support as well as a rationale for return—the rescue of loyal subjects. Guerrilla leaders approved by SWPA could erase the stain created by prominent collaborators and provide a successor to the dying Manuel Quezon.

At the same time, MacArthur realized, the movement could easily evolve into a radical insurgency. Without proper constraints, guerrillas might become bandits, religious or regional separatists, or even a revolutionary force. The growth of the Hukbalahap (People's Anti-Japanese Army), or "Huk," movement on Luzon epitomized this danger. Peasant organizers and farmers had created an extremely effective guerrilla army, which struck against the Japanese, the landholding elite, and the puppet regime. Such radicals could easily destabilize the Philippines after liberation. Worse, they might become a legitimate claimant for national power with no ties to the commonwealth, the Filipino elite, or MacArthur.[13]

In 1943, the SWPA chief established a special staff section solely to deal with Filipino guerrillas. He recruited Courtney Whitney, an attorney he knew slightly in prewar Manila, to take charge of the group. He and Whitney barred the Washington-based OSS from working with resistance fighters. Only groups sanctioned by SWPA would receive payment, supplies, and the right to compete for power when the Americans returned. In effect, MacArthur sought to license those fighting Japan and to extend his political reach in advance of his troops.

Several high administration officials objected violently to the general's political agenda for the Philippines. Interior Secretary Ickes (his department's Bureau of Insular Affairs supervised the commonwealth), Secretary of War Stimson, and several of their subordinates tried to preempt MacArthur's actions and have Washington play an active role in liberation. At least some of those concerned believed that his acceptance of the half-million dollar gift from Quezon in 1942 put him virtually in the pay of commonwealth officials. They also complained that he appointed to his headquarters several naturalized Filipinos, former Spaniards reputed to be wealthy supporters of General Franco. This, critics charged, showed the direction he intended to push Manila politics.

The question of political succession loomed large by 1943. Gravely ill with tuberculosis, President Quezon was not likely to resume power in Manila even after liberation. Ickes urged Roosevelt to allow Quezon's mandate to lapse when his term of office ended in 1943. But the president and Congress bowed to MacArthur's recommendation that the incumbent retain his office until liberation. The commonwealth president clung to his position until his death in August 1944. Vice President Sergio Osmeña, his successor, enjoyed modest support among Filipinos, American politicians, and SWPA headquarters. As a sympathetic observer commented, he could compete neither with a dead hero, like Quezon, nor with a live one, like MacArthur.[14]

Ickes and Stimson received some satisfaction early in 1944 when Roosevelt created a special committee on the Philippines chaired by Assistant Secretary of War John J. McCloy. This group urged that a new high commissioner and civilian experts be sent to Manila upon liberation to assist the nearly powerless commonwealth.

Furious at such meddling, MacArthur set about to frustrate these recommendations. He told one of the few civilian experts on Philippine affairs whom he trusted (Michigan political science professor Joseph Hayden) that any "elaborate" plans would "melt in the heat of battle." In August of 1944 he confidently informed Hayden that no one could "tell me a thing about the political situation" in the Philippines. He would exercise "full authority and responsibility" over Philippine affairs during liberation. The general pronounced himself "more competent than any other American, or than any other Filipino either" to run things in Manila. In the view of one specialist, he intended to "return as liberator and operate as power broker."[15]

MacArthur recalled vividly his father's dispute with the Taft Commission in Manila, nearly half a century before. The general promised to "fix things" this time so that a civilian high commissioner "couldn't do a thing.

He would be a prisoner of the Army." If Washington sent a "cloud of carpetbaggers," MacArthur warned, the Filipinos would revolt "and some of them [the Americans] would be killed." He would put the survivors "on a boat . . . home" with a "message to the President telling him why."[16]

As the October invasion date neared, Ickes brought Osmeña together with FDR to plead for assistance in reconstituting civilian authority. But the weak and vacillating Filipino made such a bad impression that, to Ickes's horror, Roosevelt suggested appointing MacArthur "High Commissioner at the end of the war." (MacArthur claimed this post was offered to him at the Honolulu conference.)

The interior secretary then let fly all his accusations about MacArthur's questionable "financial interests in the islands," support for Spanish fascists who would run Manila, and his alleged disdain for democracy. But, in light of the upcoming election and their deal, Roosevelt obviously had no interest in battling MacArthur over the Philippines. Instead, he empowered Ickes to designate an informal personal aide to assist Osmeña. Dejected, the interior secretary wrote in his diary that "the President is going to let MacArthur have all the lead that he wants."[17]

Aware of Roosevelt's attitude, the Joint Chiefs (through a special committee on the Philippines) conceded nearly all the authority the general desired. The JCS granted MacArthur power to determine the fate of all collaborators, appoint Filipinos to civil office in liberated territory, and decide when commonwealth authorities might resume administration. There would be no civilian oversight.[18]

Ironically, MacArthur wanted to project the image (if not the reality) of civilian authority. He insisted that Osmeña accompany him to the Philippines, even though the commonwealth president would be a figurehead, without authority, a staff, a budget, or even a jeep. The general would exercise real power, appointing civil officials and dispensing emergency relief. The government returned to Osmeña would actually be MacArthur's creation, loyal primarily to himself or his Filipino allies.[19]

The invasion of the Philippines began on October 20, 1944, on the beaches of Leyte. The general waded ashore to proclaim, emotionally, "I have returned." Because of an error in judging the tides, the pilot of the landing craft had to drop his passengers off in knee deep water. MacArthur announced his return after cursing the subordinate for dumping him in the surf. However, when he learned that the American public considered the scene heroic, the general made a habit of repeating his walk through the water during later landings.[20]

Liberating Leyte and the rest of the Philippines proved far more bloody and protracted than MacArthur predicted. Even on the main islands, fighting continued until Japan surrendered ten months later. The Japanese army and navy committed larger than expected forces to the campaign. At the battle of Leyte Gulf, the Japanese navy came close to endangering the entire invasion. It required over two months to secure Leyte, despite MacArthur's October 29 declaration of imminent victory. During four months of "mopping up" operations, the Eighth Army killed 27,000 Japanese. General Eichelberger described these "minor" operations as "physically the most terrible we were ever to know" in the war.[21]

Prolonged Japanese resistance on Leyte, including the introduction of kamikaze suicide attacks, prevented the island from becoming a base for further operations. This delayed the invasion of Luzon and set back the scheduled attacks on Iwo Jima and Okinawa. The invasion of Luzon finally began on January 9, 1945, at Lingayen Gulf, where the Japanese landed in 1941. Ostentatiously refusing to step onto a prepared dock, MacArthur again jumped into the surf the afternoon of the invasion for a suitably dramatic entrance.

On Luzon, for practically the first time since 1941, the general took charge of nearly all aspects of the battle. He seemed especially eager to retake Clark Field and Manila before his sixty-fifth birthday later that month. This proved an elusive quest.

As American forces approached, the Japanese commander, General Yamashita, decided to abandon Manila and withdraw his sizable army into the mountains of Luzon. From there he planned to fight a war of attrition. Although most of Yamashita's troops withdrew, some 20,000 soldiers and sailors stayed behind, either because they felt trapped or due to a collapse in communications and discipline. Barricaded within the old walled city, known as Intramuros, they fought to the death.

On February 5, General MacArthur announced that his forces had entered Manila and predicted the "imminent destruction" of the enemy. The battle, in fact, had only begun and it was the Filipino population of the city that also faced destruction. During the next several weeks Japanese troops went on a rampage against trapped civilians. One of MacArthur's top lieutenants suggested that his premature pronouncement about recapturing Manila reflected his determination to get some publicity for himself while "the big conference [Yalta] was going on in the Black Sea area" among the Allied leaders.

When the fighting ceased on March 3, some 1,200 Americans, 16,000 Japanese, and 100,000 Filipinos had perished. In Manila, General Eichelberger "found there was practically nothing that hadn't been entirely knocked down and in ruins." Most major buildings were leveled. Within Intramuros, he wrote his wife, "it is all just graveyard." Much of "Manila in effect has ceased to exist." The destruction took a great emotional toll on MacArthur. Before their death, Japanese troops pillaged his apartment in the Manila Hotel, ruining the one home he had really enjoyed. When the war ended, MacArthur insisted that General Yamashita be tried and executed for the actions of his troops, even though most observers acknowledged that the Japanese commander bore no responsibility for the tragedy.[22]

After capturing Manila, the general surprised both his field commanders and the Joint Chiefs by dividing his forces. As elements of the Sixth and Eighth armies cleared Luzon, MacArthur dispatched other units to subdue

Japanese troops on the innumerable islands in the central and southern Philippines. He justified the campaign to take the entire archipelago by citing a need to rescue prisoners of war and to centralize political control before regional separatism took root. However, the dispersal of forces made it more difficult to subdue the large Japanese army fighting in the mountains of Luzon. Yamashita continued resistance until ordered to surrender in mid-August. Even after losing 200,000 men, he still fielded nearly 50,000. Meanwhile, another 8,000 Americans died while "mopping up" these remnants. In any case, SWPA forces engaged in these scattered operations were unavailable for use against other enemy targets.

The major campaigns on Leyte and Luzon proved just as costly in American lives as had the battles waged by the navy and marines in the central Pacific, battles MacArthur criticized as "tragic and unnecessary massacres." To liberate Manila, the general relied on the same kind of "frontal assault" he ascribed to "mediocre commanders." Instead of six weeks, it required six months to secure most of Luzon. The victory there took the lives of some 200,000 Japanese. American forces suffered 47,000 battle casualties and nearly twice as many non-battle casualties. Proportionately, this ranked the Philippines as one of the bloodier operations in the war against Japan.[23]

MacArthur said little more about his prediction that Japan would surrender after he took the Philippines. Instead, in March, he decided to invade the East Indies. This time the Joint Chiefs refused to be bullied. They blocked what most military historians feel would have been a bloody and unnecessary assault on Java and nearby islands. MacArthur then turned his attention to Philippine politics and the competition to lead the invasion of Japan.

Philippine Politics

Three overriding questions affected the political future of the Philippines: When and how would the commonwealth resume functioning? What treatment should be meted out to thousands of collaborators? What role should the large guerrilla movement play in postwar politics?

MacArthur urged early restoration of commonwealth authority, even suggesting to Roosevelt that he grant the colony formal independence as soon as military operations ceased. Not only was this impractical (the commonwealth had no administrative capability), but Ickes and Stimson considered "independence" merely a smoke screen for manipulation of local politics. MacArthur, they suspected, planned to discredit and ruin Osmeña by transferring to him prematurely responsibility for civilian affairs even though the commonwealth government had no resources to minister to the needs of Filipinos. Then, one of the general's favorites would sweep Osmeña from office. As for MacArthur's interest in all this, his opponents in Washington suspected that since he might well remain in Manila after the war, he had deep concern with the internal power struggle.[24]

The general hinted at this in a discussion with Paul P. Steintorf, the newly arrived American consul general. Although he denied any desire to "dictate" a political agenda, MacArthur condemned Osmeña as "incompetent" and badly in need of "direction." He had sent Osmeña to Washington to plead for aid, but feared he might "falter when subjected to the influence" of Ickes and others.

The general justified his interest in maintaining internal order by claiming that the Philippines was "destined to become the principal staging area for the Pacific war, with possibly a million troops stationed here." (Of course, this was not true.) At the same time, he repeated his fanciful demand that Washington grant the commonwealth full independence by August 1945. In spite of the massive destruction wrought in clearing out the Japanese, MacArthur predicted an era of unprecedented prosperity for the islands. Partly because of "tremendous American military expenditures" (which he expected to administer) it would have a "totally new economy." The liberation he envisioned would mean the end for the "small wealthy group" that "dominated the country and exploited the common people." Certainly the Philippines needed this tonic; but nothing MacArthur did over the next few months seemed consistent with this goal.[25]

The treatment of collaborators was a case in point. During the war, Roosevelt declared it national policy to remove from positions of political and economic influence all "those who have collaborated with the enemy." Ickes demanded that the army "shoot or hang any Filipino who had anything to do with the puppet government, no matter what reason they may have had for collaborating."

Collaboration, of course, was a complicated phenomenon that could not be reduced to moral absolutes. For example, most of President Osmeña's family and friends supported the puppet Laurel regime. Accordingly, before he accompanied MacArthur home, Osmeña sought guidance from several American officials about adopting a flexible standard. But, as a prominent historian of collaboration noted, Osmeña found that "Stimson and Hull were friendly and vague; Ickes pontifical and vague; [Senator] Tydings mildly interested and vague." All "agreed that something should be done regarding collaboration," but looked to the president to set a policy. When Roosevelt spoke with Osmeña, he offered some homilies and sent him on his way.[26]

MacArthur's public statements sounded as absolute and unbending as those of Ickes. Insisting that nothing justified cooperating with the enemy, he condemned all collaborators as traitors to the United States and to him, personally. On the eve of the Luzon invasion MacArthur issued a proclamation ordering the apprehension by American forces of all Filipinos who had given "aid, comfort and sustenance to the enemy." All were to be removed from "positions of political and economic power."

In practice, however, MacArthur redefined collaboration in a manner that transformed this policy radically. He decreed that service in the puppet regime did not automatically constitute evidence of a crime. Collaboration, he explained in private, was less a "legal question" than one of motivation. Some Filipinos served the puppet regime in order to mitigate harsh Japanese rule; others were venal or traitorous. Any general purge, he feared, would ravage the fragile social order. MacArthur decided that American forces should detain suspects for an interim period. As soon as possible, all would be transferred to the commonwealth for judgment. In practice, the general wanted Filipino politicians to make painful choices while he exercised informal influence from the sideline.[27]

This meant that the hapless Osmeña would begin his rule by weighing the fate of a significant portion of the Filipino oligarchy, the same class from which he sprung. Any sign of softness would outrage the guerrilla movement and American liberals. A firm policy would alienate the landholding and political elite. MacArthur justified his action by describing it as a vote of confidence in the commonwealth.

Even while fighting still raged in the capital on February 27, the general held a ceremony at the Malacañan Palace transferring civil authority to Osmeña. The Philippine commonwealth, he declared, was "again at liberty to pursue its destiny." Manila stood once more as "the citadel of democracy in the East." MacArthur then drove off, leaving Osmeña to search for a jeep to take him home.

As American forces liberated towns and provinces, they appointed Filipinos to local administration. Technically, these appointees held office on an acting basis only, pending Osmeña's approval. In practice, he had little choice but to confirm the selections. MacArthur decided that those chosen should, in most cases, be drawn from "recognized" guerrilla groups, meaning those sanctioned by his headquarters or by Quezon. Other units were deemed illegitimate.

The general's flexible yardstick worked to the benefit of both sanctioned guerrillas and influential collaborators. MacArthur reached a critical juncture in April when he decided that even high level service in the Laurel regime could be excused. The case of Manuel Roxas exemplified the transition.

Although the Japanese evacuated José Laurel and some associates to Japan, many members of the regime remained behind. In mid-April, Roxas and three cabinet ministers surrendered to American forces. After jailing the others as collaborators, MacArthur brought Roxas to Manila, announced he had been "liberated," restored his army rank, and declared his power to pardon, unilaterally, anyone he chose. This decision transformed Philippine politics.

A prewar leader of the Philippine congress and ally of Quezon, Roxas had served as an officer on MacArthur's staff at Corregidor. The general argued that his friend had only feigned collaboration while serving in the puppet government. In reality, he had gathered intelligence as was "one of the prime factors in the guerrilla movement." Although he revealed no evidence for his claim, MacArthur appointed Roxas a brigadier general on Willoughby's intelligence staff.[28]

MacArthur acted out of many motives. As early as 1942 he envisioned Roxas as a successor to Quezon, a protégé so trusted by both that he had handled the details of the 1942 "gift" to the general. MacArthur knew that many of the oligarchs had not considered collaboration a betrayal of America but a pragmatic adjustment to circumstances. Transforming a traitor (real or feigned) into a patriot might erase MacArthur's own initial sense of betrayal by so many of his Manila friends. Also, by exonerating his protégé, he could rebuild the prewar bond between himself and the ruling class. In placing his mantle over Roxas, MacArthur was "pardoning, liberating, and encouraging Osmeña's great potential rival for the presidency." Most Filipinos interpreted the act as a "sign . . . clearly signifying support for Roxas over Osmeña."

In the aftermath of Roxas's "pardon," many local officials turned toward him as their patron. Hoping to reverse the erosion of his power, Osmeña made a pilgrimage to see Roosevelt in April 1945. FDR listened patiently to his complaints about MacArthur, promised to look into the matter, but died within a few days and before taking action. Osmeña begged other officials to support appointment of a high commissioner able to check MacArthur's power. He even approached Supreme Court Justice Frank Murphy, a former high commissioner, to take the job. At an earlier point Murphy had broached the idea to FDR of returning to Manila. But without presidential support, he doubted anyone could stop MacArthur. Osmeña returned home empty-handed.

MacArthur treated the guerrilla forces on the left as cavalierly as he had Osmeña. Even before liberation, he targeted the Hukbalahap for destruction. Organized among the peasants of central Luzon in 1942, the Huks shunned association with either the remnants of the old Philippine Army (the guerrillas loyal to MacArthur) or the oligarchs around Laurel or Quezon. Although General Willoughby described it as one of the "largest and most powerful guerrilla groups in central Luzon" during the war, he also condemned it as a rough organization of bandits and Communists. Willoughby told his chief that the Huks planned to create a "communistic government in the Philippines after the war on the early Russian model." He also believed the guerrillas had "connections with the communistic elements in China."[29]

The Hukbalahap and its leader, Luis Taruc, had strong Communist leanings in 1945, but were not Soviet stooges. They built a popular following among poor peasants who hoped to break the economic and political monopoly of the landed elite. At the same time, they fought bravely against the Japanese. However, MacArthur, Willoughby, and Whitney saw them entirely as a radical threat to the postwar order. Even before the invasion of Luzon, SWPA officials ordered the Huks to cease military activities and threatened reprisals if they disobeyed. Whether they obeyed or not, SWPA intended to disarm, isolate, and dissolve Huk units. As one student of the movement concludes, "MacArthur's principal concern was to put the lid on what he perceived to be a political tinderbox, and to do this as quickly as possible." He needed time to find a group untainted by collaboration (or one he could exonerate) and willing to cooperate with his plans for the Philippines. This required suppressing any fundamental challenge to the status quo.[30]

General Willoughby's Counter Intelligence Corps (CIC) took the lead in appointing Filipino officials on the local level. In most cases those chosen came from prewar elites who either joined approved guerrilla groups or had not openly collaborated with the Japanese. General Whitney wanted to disarm and dissolve all Huk units as a threat to law and order and the commonwealth. MacArthur preferred to move cautiously, concerned that if American forces tried to disarm the Huks, violence would erupt. However, he agreed to move politically against them.

As American units entered the villages of Luzon, they removed all Huk political officials and replaced them with leaders from "acceptable" guerrilla groups. Simultaneously, intelligence units arrested individual Huk leaders. Many were held for several weeks as interrogators probed for evidence of Communist links. Meanwhile, MacArthur encouraged other Philippine guerrilla groups to disarm the military wing of the Huks. These efforts were designed to prevent the peasant movement from competing for postwar power with ballots or bullets.[31]

During the spring, American forces also detained some 6,000 suspects accused of cooperating with the Japanese. Whoever assumed control of the Philippine government would, of course, decide their ultimate fate. Since assuming power in February, Osmeña had ruled by decree. Under pressure from MacArthur, he called into session the Congress elected in 1941—even though it had never sat and included many suspected collaborators. Not incidentally, Roxas had a strong following among the legislators.

In June, the Congress elected Roxas president of the Senate and head of the Committee on Appointments that reviewed all the permanent posts filled by Osmeña. Roxas forced Osmeña to drop from the cabinet advocates of a purge. He also demanded an early presidential election, presumably before Osmeña secured trade concessions or economic assistance from Washington.

Roosevelt's death, in mid-April, left Philippine policy in more of a muddle than ever. Late that month Ickes urged the new president, Harry S. Truman, to name moderate Republican Charles P. Taft high commissioner. (A nasty move, given Arthur MacArthur's struggle with Charles's father.) When Truman indicated that he understood the importance of finding someone "strong enough to keep General MacArthur in line," Ickes "held out [his] hand and congratulated him upon his attitude toward MacArthur." Hope that Truman would put the general in his place convinced the contentious interior secretary "to overlook a good many things" he disliked about the new president.

A few weeks later, Ickes told Truman that the economic situation in the

Philippines had become desperate, but that MacArthur's censorship prevented reports from leaking out. The president then "indulged in some pretty vigorous Missouri expletives in expressing his opinion of MacArthur" as nothing but a "stuffed shirt" lacking in "military ability." Despite his indignation, Ickes discovered, Truman, like FDR, seemed "afraid to put MacArthur in his place." He waited until July to appoint Paul McNutt high commissioner, with the understanding that the position would not actually be filled until MacArthur assumed his new post in Tokyo.[32]

When MacArthur learned he would probably be selected to command occupation forces in Japan, and especially after Tokyo's sudden surrender in mid-August, he hastened to unburden himself of responsibilities in Manila. Although still determined to help his friends, Philippine politics lost its intense, personal significance to him. On August 22, MacArthur's headquarters announced that by the end of the month military control of the Philippines would cease, including "military help in the public services of health, education, welfare and reconstruction." The next day MacArthur handed over all accused collaborators to the commonwealth government. On August 30, the general flew to Japan, leaving behind a political void and a bankrupt regime.

During the following six months, Osmeña and Roxas argued over the fate of collaborators. Despite requests from the commonwealth, MacArthur refused to turn over evidence his staff had gathered on the suspects, effectively blocking the investigation. In September 1945, High Commissioner Paul McNutt reported to Truman that "enemy collaborators" controlled the Philippine legislature and within a short time would dominate the entire government.

By the end of the year, Roxas formally challenged Osmeña for the presidency. The former appealed to the old oligarchy while the latter, though an oligarch himself, scrambled for support among former guerrillas and the moderate wing of the Huks. A Justice Department investigator (Walter Hutchinson) informed Truman in January 1946 that only direct American intervention would assure punishment of collaborators and prevent their control of an independent Philippines. In March, from Tokyo, MacArthur issued a "strong" protest against this recommendation and urged the president to avoid further involvement in Manila's politics.

Within the Philippines, special "peoples' courts" convicted a few minor collaborators but delayed action against prominent suspects. Legal judgment awaited the outcome of the presidential election in April. Few doubted that if Roxas won the upcoming contest, he would pardon most collaborators. Following Ickes's departure from the administration, none of Truman's cabinet or close advisers voiced much interest in these developments. Instead, growing tensions with the Soviet Union and civil war in China caused many American officials to conclude the Philippine oligarchy, whatever its flaws represented the only barrier to communism.

Like Ickes, MacArthur's old enemy, Drew Pearson, charged the general with manipulating Philippine politics for both ideological and selfish reasons. The general, Courtney Whitney, and Filipinos Mike Elizalde and Andreas Soriano, Pearson believed, led a secret business clique that supported Roxas to further their own "economic position." For years rumors had circulated regarding MacArthur's secret investment in a Philippine gold mine—an investment this group had allegedly promised to protect.[33]

Campaigning as the favored candidate of both MacArthur and Washington, Roxas won the April 1946 presidential election. During the new president's subsequent visit to Tokyo, MacArthur embraced his protégé and described his election as the final refutation of the "irresponsible charges of collaboration." Roxas promptly pardoned most accused collaborators, but barred Huk leader Luis Taruc from taking the legislative seat he had just won. Even moderate guerrilla leaders were frozen out of power.

Independence came on July 4, 1946, but social injustice, corruption, and strong-arm political tactics made a mockery of the event. Forced out of electoral politics, the Huks soon began a military insurgency that, during the next five years, came close to toppling the Manila government.[34]

The general returned to Manila in July 1946 to celebrate the nation's independence. Inadequate American assistance and dreadful mismanagement by local officials had left the country worse off than a year before. The ceremony took place on a temporary site, since the former seat of government remained, in the words of historian Theodore Friend, "an obscene pyramid of brownish dust":

> Manila was a quagmire in the rainy season; and in the hot season the red dust above the city, thrown up by the heavy army traffic, made it look, from a distance, afire. The city was full of jerry-built shelters, and its hasty bazaars were full of gimcrack goods. Soldiers, sailors, and peddlers jammed its sidewalks; whores and pimps and pickpockets, confidence men and influence mongers; ex-guerrillas still in jungle uniform, and throngs of common men and women, tired and unemployed. To one observer who had loved the old city, the new Manila looked like a carnival in hell.

Choked with emotion, MacArthur told those assembled that "despite racial, cultural and language differences" Americans and Filipinos had "forged an affinity of understanding which survived both the vagaries of peace and the shock of war." Philippine independence signaled the "end of mastery over peoples by force alone, the end of empire as the political chain which binds the unwilling weak to the unyielding strong." As the ceremonies ended, MacArthur turned to his wartime aide, Carlos Romulo, and declared "Carlos, America buried imperialism here today!"

Actually, the terms of independence continued to bind the Philippines to the United States. America retained extensive military base rights and its nationals were constitutionally granted "business citizenship." Through both trade and aid, the economic links between the two countries also retained a colonial aspect. Moreover, the American style of decolonization had almost no impact on the tortured fate of the French, Dutch, and British empires in Asia.[35]

For MacArthur personally, the descent of the Philippines into near chaos seemed a sorry end to one of his life's dreams. However, this disappointment quickly gave way to an even greater opportunity: the chance to remake Japan.

8
Destination: Tokyo

MacArthur experienced both exhilaration and disappointment as his forces liberated the Philippines. Despite success in outwitting his rivals and redeeming his pledge to return, liberation had destroyed much of the country he loved and brought political chaos and class warfare in its wake. Nor did his victory break Tokyo's will to continue the war. As operations in the archipelago dragged on (tens of thousands of Japanese troops held out until after the formal surrender), MacArthur focused his attention on the looming assault upon Japan.

The theater divisions of 1942 placed the enemy's home territory in Admiral Nimitz's zone of operations. Many army officers besides MacArthur resented this monopoly. Yet neither service could agree on who should lead an invasion of the home islands. In April 1945, the Joint Chiefs appointed MacArthur, by now a five-star general, Commander-in-Chief, United States Army Forces, Pacific (AFPAC), assuring him command of all army units participating in the operation. (Except for the 20th Air Force in the Marianas, he now controlled almost all army ground and air units in the Pacific.)[1]

Following Germany's surrender in May, MacArthur still voiced fears that Marshall, Eisenhower, or other European commanders might be sent to the Pacific to usurp his position. He also complained that too many black soldiers were assigned to SWPA in order to embarrass him. General Marshall, he suspected, hoped to undermine his popularity by insisting that he publicly endorse an unpopular proposal for a postwar military draft. MacArthur described Eisenhower's celebrated return to America in July as a "depressing spectacle." The "unfriendly press" honored European commanders, while no one in his theater had "had their faces on *Time* and *Life* for a long period of time."[2]

The general's egotism occasionally forced his superiors to question his ability to command the final assault against Japan. Secretary of War Stimson probably echoed Marshall's qualms when he confided to his diary that:

MacArthur stands out as the manifest personality who has won the right to command the final land attack on Japan by virtue of his skillful work in the Southwest Pacific and the Philippines, but his personality is so unpleasant and he has so affronted all the men of the Army and Navy with whom he has to work that it is difficult to get combined assent on the proposition. Furthermore, he has a staff which has been so influenced by his own personality as to be a source of danger to harmony in what ought to be a most harmonious operation.

Yet, like Roosevelt, the military chiefs saw no realistic alternative to giving MacArthur a (if not the) major role. Having built the SWPA commander up to nearly mythic proportions, they would be hard pressed to justify passing him over. However, few military or civilian planners felt disposed to consult with the general about the political, diplomatic, and military developments about to alter radically the course of the war in the Pacific.[3]

Army and navy strategists argued at length about the allocation of resources and the coordination of plans to assure Japan's defeat. In April, MacArthur called for an early landing on Kyushu and Honshu. The navy, he complained, cared more about denying him command of these operations than with winning the war. Nimitz and King wanted to "piddl[e] around for a long time before doing anything against the Japanese homeland." The navy liked any strategy so long as it cut the army out. This explained their proposal to form a "ring around Japan and conduc[t] bombing by the air corps under their control." Strategic bombing "could do a lot to end the war" and, he told General Hap Arnold, he "liked" the plan to destroy "30 Jap cities." But, "in the final analysis," the "doughboys w[ould] have to march into Tokyo."[4]

Although MacArthur respected the contribution tactical air power played in SWPA's progress, he doubted strategic bombing would destroy the enemy's will or ability to fight. He knew that the air war over Germany had not been decisive and that high altitude bombing remained notoriously unreliable. While reluctant to admit this, air commanders in Europe gave up efforts to attack solely military and economic targets. Instead, they obliterated whole cities.

Still, plans went forward to begin air attacks upon Japan by a powerful new weapon, the B-29 bomber. Based in the Marianas and nearby Pacific islands, the 20th Air Force came under the operational control of General Curtis LeMay. These fliers developed techniques for low level, night incendiary bombing which, they argued, would bring Japan to its knees.[5]

On March 9, 1945, over three hundred B-29s raided Tokyo. The attack burned out sixteen square miles of the capital, killed about 83,000 civilians, and injured 41,000 others. During the next few months LeMay's B-29s roamed over Japan practically at will, gutting nearly 40 percent of major urban areas and rendering millions of Japanese refugees in their own country. Meanwhile, the navy and the air forces effectively isolated the entire nation by mining shipping lanes and sinking most of the merchant fleet.[6]

As death rained from the sky in Japan, SWPA units carried out endless operations in the Philippines. When forces under Nimitz captured Iwo Jima and Okinawa, MacArthur criticized the assaults as too costly. Neither a naval blockade nor an air offensive could supplant an invasion—presumably led by himself.[7]

By mid-May, after much haggling, MacArthur and Nimitz reached an accord on invading Japan. They recommended a ground attack upon Kyushu about November 1 and Honshu three months later. The Joint Chiefs soon approved this plan (code-named OLYMPIC) and President Truman assented on June 18. Even though MacArthur's role as ground commander seemed assured, secret developments rendered all this planning obsolete.

Preparations for the Defeat of Japan

During the critical months before Japan surrendered, MacArthur immersed himself in preparing for the grand, final battle of the war. Although slated to play a major part in the enemy's defeat (and already discussed as a possible occupation commander), his views on negotiating surrender, coordinating policy with the Allies, or drafting postwar plans were not solicited by higher authorities in Washington. As in Europe, where the Nazi surrender of early May and establishment of occupation zones had raised complicated policy issues, Japan's defeat was certain to unleash powerful, suppressed antagonisms. Yet neither President Roosevelt nor his successor encouraged MacArthur's contribution to the planning process.

Although there is no evidence that he thought very much about postwar Japan before mid-August, the general broached the subject with Robert Sherwood the previous March. Speaking phrases that resembled closely Roosevelt's idealistic anticolonialism, MacArthur predicted that defeat would eliminate the "sanctity" of the emperor and military, creating a "spiritual vacuum" to be filled with American values. Through liberating the Philippines and defeating Japan, he continued, the United States could exercise the "greatest influence on the future development of Asia." Instead of wasting this opportunity in an "imperialistic" search for "commercial advantage," the United States should exercise "essential liberalism" in order to win the "friendship and cooperation of the Asiatic peoples far into the future."[8]

The general betrayed none of the overt racism and hatred of Japan so common in wartime America. For example (besides the internment of over 100,000 Japanese-Americans), during 1942–43 the White House worried that popular resentment of Japan as enemy #1 might undermine the Europe First strategy. A Gallup poll taken during 1944 found that 13 percent of Americans endorsed the extermination of all Japanese. One third of respondents to a June 1945 poll called for the summary execution of Emperor Hirohito, while most others favored condemning him after a trial.

Influential Americans spoke glibly of ravaging the Japanese population. Senator Lister Hill (D–Alabama) called on the armed forces to "gut the heart of Japan with fire." Senator Ernest McFarland (D–Arizona) demanded that the "Japs . . . pay dearly through their blood and the ashes of their cities" for attacking America. One adviser to the State-War-Navy Planning Committee proposed the "almost total elimination of the Japanese as a race." General Hap Arnold proposed dropping leaflets over Japan addressed "Dear Hirohito" and threatening "completely [to] annihilate the Japanese people" unless Tokyo surrendered unconditionally. Elliott Roosevelt, the president's son, told Commerce Secretary Henry Wallace he thought the air force should continue attacking "until we have destroyed about half the Japanese civilian population." Franklin D. Roosevelt, revealed his own racial antipathy in a lively correspondence with an anthropolgist from the Smithsonian Institution, Dr. Ales Hrdlicka. The two men discussed the possibility for crossbreeding warlike Japanese with more docile Pacific islanders. Compulsory eugenics, they hoped, might eliminate the "primitive" brains and "barbarism" of the enemy race.

Few officials could match the vitriol of Senator Theodore Bilbo (D–Mississippi). After careful consideration, he decided how to visit "exact justice" upon the enemy. "Personally," Bilbo wrote to MacArthur at the end of the war, "if I had my way about it, I would sterilize every damn one of them so in one generation there would be no more Japs."[9]

Although these horrors did not become part of America's formal strategy, the cumulative affects of the air war, naval blockade, and encirclement of Japan took a staggering toll. By August, about one and a half million Japanese soldiers and sailors had been killed. Civilian losses approached a staggering half million.

Amidst this destruction, officials both in Japan and the United States discussed ways of ending the war through negotiation. In Tokyo, many bureaucrats who had been swept along by the military's drive for war now recognized the importance of avoiding a fight to the finish. If the Allies offered a "soft" peace, these "moderates" (sometimes called "conservatives," sometimes "liberals") believed, the emperor could be persuaded to throw his influence behind a drive for compromise settlement. Unfortunately, this "peace faction" lacked the power to depose their hardline rivals and found it difficult to get their ideas taken seriously in Washington.[10]

Yoshida Shigeru, a well-connected diplomat, exemplified the role placed by these moderates in the last stages of the war. The diplomat joined members of the nobility, business elite, and bureaucracy in forming the secret YOHANSEN group. These representatives of the traditional elite resented the influence of the military and feared that a prolonged war would bring defeat and domestic revolution followed by a harsh occupation.

In February 1945, the group's titular head, Prince Konoe Fumimaro presented a grim portrait to the emperor. He predicted defeat coupled with social upheaval and the destruction of the *kokutai*, or traditional imperial political system. The militarists had blundered into a losing war, partly because of their own radical notion of remaking Japan on a Soviet model,

Konoe charged. He elaborated on an earlier warning that "leftist revolution" was more frightening than surrender. Japan could withstand an American occupation more easily, the prince asserted, than a revolutionary upheaval, which might occur if the nation fought on to the bitter end. To "save Japan," the military fanatics had to be "wiped out" and peace feelers extended before the Soviet Union joined the war or America began an invasion.[11]

Unfortunately, the emperor failed to act on this advice until the climactic events of early August. Only then, as the nation reeled under the twin blows of the atomic bomb and Soviet entry into the war, did the militarists falter and the emperor intervene decisively.

As Konoe and Yoshida suspected, some influential Americans agreed that the war could be ended through negotiations. Since 1941 a small "special research division" within the Department of State discussed possibilities for a postwar settlement. But diplomatic specialists like Hugh Borton, George Blakeslee, and their counterparts in the War Department's Civil Affairs Division remained on the periphery until late in the war. They learned of developments like the December 1943 Cairo Declaration (which pledged to strip Japan of its conquests) by reading the newspapers. Things improved a bit in 1944 when the State Department formed a Postwar Programs Committee to meet with the Army's Civil Affairs Division.[12]

By the end of that year State Department and army experts drafted plans for Japan that resembled closely those later implemented. They called for an exclusive American occupation with only token Allied representation, utilization of the existing Japanese government, and provisional retention of the emperor. Following demilitarization and political reform, Japan would regain membership in the Asian-Pacific community.[13]

Advocates of strategic planning received a boost in December 1944 with the creation of a high-level coordinating group. For some time Henry Stimson had fretted over the "present chaotic situation" in Washington where cabinet agencies and the White House often worked in ignorance of or at cross purposes with one another. Stimson, Navy Secretary Forrestal, and Secretary of State Edward R. Stettinius organized the State-War-Navy Coordinating Committee (SWNCC, called "swink") as a forum for assistant secretaries and technical specialists to draft policy without first going to the president. SWNCC's Subcommittee for the Far East formulated a program for Japan in light of broad presidential guidelines.

Roosevelt made it hard for anyone to know his mind. At the Yalta conference in February 1945, for example, FDR reached agreements with Joseph Stalin and Winston Churchill regarding Japan, China, Korea, and Southeast Asia. Yet, the president declined to share the substance of the accords with the rest of the government. Thus, planners lacked precise information about the Soviet pledge to enter the war, decisions reached on the reshuffling of Japanese and Chinese territory, and related topics. Officials of SWNCC filled in the gaps through reliance on rumor and snippets of information supplied by Charles E. Bohlen, Roosevelt's Russian interpreter at Yalta.[14]

During the last few months of his life Roosevelt played havoc with the postwar planning process by encouraging his friend Treasury Secretary Henry Morgenthau to develop programs for postwar Germany. The State and War departments learned only belatedly that Morgenthau's Informal Policy Committee on Germany (IPCOG) recommended the division of Germany into zones, the elimination of central political authority, and permanent restrictions on heavy industry. Early in April 1945, SWNCC planners heard that the treasury secretary hoped to apply this formula to Japan as well. The Subcommittee on the Far East rushed completion of its own draft, "U.S. Initial Post-Surrender Policy." It confirmed the call for an American occupation relying heavily on the existing Japanese government and rejected the kinds of punitive reform Morgenthau advocated.

The SWNCC staff did not excuse Japan from responsibility for the causing of the war nor agree that Tokyo had merely stumbled into the errors of fascism and aggression during the 1930s. They recognized that at its roots Japanese society required major political, economic, and social reform. At the same time, they envisioned an occupation that prevented revolution by encouraging controlled change.

Roosevelt's death on April 12 thrust a new, largely unknown figure into the White House and eliminated Morgenthau's patron. As the treasury secretary lost influence, other cabinet officials and advisers filled his place. Stimson and Forrestal of the War and Navy departments, along with Under Secretary of State Joseph C. Grew, Ambassador W. Averell Harriman, and presidential chief-of-staff Admiral William Leahy all gained a measure of influence over the inexperienced Harry Truman. Although this group had also served FDR, he often played them off against one another and seldom followed their advice to the letter. FDR subordinated their recommendations to what he believed were his own, superior instincts. In contrast, Truman assumed office with little preparation, few bureaucratic allies, and a sense of inadequacy.

Between Truman's assumption of power in mid-April and the beginning of August, American policies toward the Pacific war and the major nations of the region—Japan, China, and the Soviet Union—underwent a major reevaluation. Officials agreed on the need to defeat Japan quickly and with a minimal loss of American lives. Consensus, however, ended there. The president and his advisers disagreed over how to most speedily compel a surrender (through diplomacy or force), whether the Soviets should be encouraged to enter the Pacific war and at what price, how to buttress the Chinese Nationalists against domestic rivals and Russian pressure, and finally, how to insure the protection of American interests throughout Asia and the Pacific.

Recognizing China's failure as a wartime ally, Roosevelt and his military advisers at Yalta encouraged Soviet entry into the Pacific war by offering

Stalin territorial concessions in northeast Asia. In return, the president sought Stalin's pledge not to meddle in China's internal politics. As noted earlier, MacArthur, like most senior officers, applauded this bargain.

However, even before Roosevelt's death, the Western Allies came to fear the extension of Soviet power at the same time as they hoped to use that power against Japan. During March and April, Roosevelt and Churchill engaged in a series of increasingly vitriolic disputes with Stalin over the surrender of German forces in Europe. FDR and Churchill also condemned the Soviets for imposing puppet regimes in Poland and the Balkans, despite more liberal promises made at Yalta. The Kremlin felt the western allies were challenging the integrity of its security zone, while London and Washington worried that Stalin had begun to construct a new Soviet empire and to export revolution.[15]

Both ill-informed about the nuances of the wartime accords and far more of a gut-level anti-Communist than his patrician predecessor, Truman pressed for an early showdown with Moscow. Admiral Leahy shuddered with pleasure as the president revealed his intention of lambasting Soviet Foreign Minister Molotov at a meeting in Washington late in April. Leahy wrote in his diary that with the impending defeat of Germany, "no particular harm can now be done to our war prospects even if Russia should slow down or even stop its war effort in Europe and Asia."[16]

Averell Harriman, James Forrestal, Joseph Grew, and Henry Stimson all encouraged the president's harder line. During May, for example, Harriman warned Truman of Stalin's expansionist designs in northeast Asia. He especially feared imposition of Russian control in Manchuria, Korea, and Japan.

The White House inner circle hoped that technology would alleviate American dependency on Soviet assistance in the Pacific War. Since 1941, scientists working on the Manhattan Project had made remarkable progress in developing an atomic bomb. Intended originally for use against Germany, it now held the promise of defeating Japan without prohibitive American casualties or the risks of Soviet entry.[17]

However since a weapon would not be ready for testing until midsummer, its use, not to mention impact, remained an uncertain factor. In the interim, some of Truman's advisers suggested he accept Japanese peace feelers, ending the war without use of the terrible new weapon or reliance on the Soviets. Terrified that Stalin had replaced Hitler as the threat to western civilization, Under Secretary of State Grew advocated canceling the Yalta accords, lest Russia enter the Pacific war and become the overlord of Asia. "Mongolia, Manchuria and Korea w[ould] gradually slip into Russia's orbit," followed "in due course by China and eventually Japan." Grew appealed to Truman to inform Tokyo that America would leave the emperor on the throne. This, he believed, would bring about a surrender and the reconstruction of Japan as the "cornerstone" of a "peaceful future."[18]

Secretary of War Henry Stimson feared that using the atomic bomb against

Japan would heighten tension with the Soviets, spur an arms race, and condemn to death tens of thousands of civilians. As one of the few senior members of the government who understood the scientific, military, and diplomatic dimensions of the new weapon, Stimson hoped to channel its awesome power in constructive directions. Even though he hoped to share atomic information with Moscow eventually, Stimson still favored keeping the project secret until after a negotiated surrender or demonstration of the weapon's effectiveness.

On June 6, the secretary of war advised the president that the existing policy of telling Moscow nothing about the new weapon should be continued until "the first bomb had been successfully laid on Japan." The president had already reached this conclusion, telling Stimson he had delayed the impending summit with Stalin at Potsdam to allow more time to test-fire the weapon. Though neither Stimson nor Truman discussed using the bomb to intimidate the Russians, both agreed that its early use (and continued American monopoly) might assist "settlement of the Polish, Rumanian, Yugoslavian and Manchurian problems" with Moscow. Since Russian and Chinese negotiators had deadlocked over key details of what privileges the Soviets would gain in Manchuria, Truman's advisers feared the impasse might provide Stalin with an excuse to intervene in China. The sooner Japan capitulated, the less pretext or opportunity would arise for Soviet intervention. Either the atomic bomb or a new diplomatic strategy might induce an early surrender.[19]

Early in July, Stimson discussed with Truman the possibility of offering Tokyo surrender terms before an invasion or the use of the atomic bomb. "Given the increasingly fanatic resistance of Japanese forces, such as on Iwo Jima and Okinawa," he worried that American "casualties during an invasion of Japan might total one-half million." An assault on the home islands would cause a "far more bitter finish fight than in Germany," leaving Japan "even more thoroughly destroyed." If some other way could be found to "secure for us the equivalent of an unconditional surrender," it was madness to plan an invasion.

The secretary of war maintained that Japan was not led only by mad fanatics with an "entirely different mentality from ours." The country had progressed from "isolated feudalism" to the "position of one of the six or seven great powers" through the ability of its ruling classes. Only the "seizure" of power by the "fanatical military group in 1931," removed Japan from the fold of enlightened nations. Its true "liberal leaders," Stimson argued, hated the militarists and wanted to end the war.

They should be told of the "overwhelming character of the force" about to be unleashed upon Japan. At the same time, Washington should spell out terms of a moderate occupation designed to reintegrate Japan into the ranks of industrial democracy. Stimson also favored continuation of a "constitutional monarchy under her present dynasty." These pledges, if offered before an invasion, use of the atomic bomb, or Russian declaration of war, might allow the peace faction to triumph.[20]

American military leaders tended to discount the likelihood of a negotiated settlement or the influence of Japanese moderates. Like MacArthur, they believed that an invasion, Soviet pressure, and perhaps, the atomic bomb would all be required to force a surrender. Stimson and Marshall rejected Grew's extreme anti-communism, noting that the Yalta accords, while imperfect, placed important constraints on Soviet expansion. A premature break with Stalin would give the Russians "an excellent excuse for immediate aggression in the Asiatic area." Even more than civilian officials, military planners believed the "impact of the Russian entry on the already hopeless Japanese [might] well be the decisive action levering them into capitulation."[21]

Swayed by this argument, on June 18 Truman approved plans for a November invasion of Japan on the assumption that the atomic bomb would not be ready and it would be desirable to have Russian assistance. At the same time, the president expressed a hope that the perfection of the secret weapon might eliminate the need for both an invasion or Russian help. The Joint Chiefs ordered preparations for a swift move into Japan in case of an early collapse of resistance prior to invasion.

As so often occurred, MacArthur and Nimitz drafted rival schemes. Nimitz's Campus Plan had the navy and marines seizing Tokyo Bay and "key positions ashore," with army ground forces arriving later. In contrast, MacArthur's Blacklist Plan called for a massive movement of ground, air, and navy forces to occupy a major portion of Japan.

Right through mid-August, the two theater commanders argued over the logistics of the operations and, although they did not admit it, who would receive the glory of arriving first in Tokyo. Even when the Joint Chiefs settled on a version of Blacklist, the prolonged dispute delayed plans for deploying American forces elsewhere in the wake of Japan's collapse. Preparations for sending troops to Korea, north China, and Manchuria were set back several weeks, with grave, but as yet unseen, implications for the future.[22]

In mid-July, as the Pacific commanders continued to plan for an invasion, the leaders of the Grand Alliance convened the last wartime summit. Gathering in the once opulent Berlin suburb of Potsdam, Truman, Stalin, and Churchill (replaced in midconference by Laborite Clement Attlee, after the general election) argued over German boundaries and reparations, postwar regimes in Poland and the Balkans, and unresolved details concerning Russia's entry into the Pacific war.

The delegates made little progress in resolving any of the European problems during the two-week parlay. The western allies refused Soviet demands for more German reparations, while Moscow balked at establishing broader-based governments in Eastern Europe. Each side seemed certain the other intended to dominate the continent. Unable to reach any firm accords, the three powers deferred an open break by creating a postwar Council of Foreign Ministers (CFM) to meet at future intervals.

While traveling to Potsdam, aides reported, the president seemed to be "losing his interest" in Soviet entry into the Pacific War. Although on one level Truman desired assurances of Russian assistance should Allied forces have to invade Japan, he feared the consequences of such help. The president's anxiety and lack of familiarity with complicated diplomatic issues were magnified by the fact that the new secretary of state, James Byrnes (who knew little more than he), sheltered Truman from more experienced diplomats.

For example, Harriman complained that Byrnes "played his cards very close to his vest" and relied on his own judgment. He left Harriman and Stimson "sit[ting] in the sun together outside [Stimson's] villa talking about when and how the Japanese would surrender and how to deal with the Russians after that." Stimson noted dryly that his "assistance, while generally welcome, was strictly limited in the matters in which it should be given."[23]

On July 16, Truman received a coded message forwarding word of the successful atomic test in New Mexico. The information "greatly cheered up" both the president and Churchill (whom Truman told immediately). The prime minister convinced the president to delay informing Stalin of the test until a time when the news might have a greater impact. The two western allies were further heartened on July 21 when a full report on the atomic test revealed the full power of the weapon. The information "tremendously pepped up" the entire delegation.

Byrnes and Truman hoped the bomb could be used to compel a Japanese surrender before Soviet forces were prepared to attack (about August 15, by their estimate). The secretary of state planned to issue a unilateral ultimatum "giving the Japs two weeks to surrender or face destruction" since the "secret weapon" would be ready by then.[24]

At dinner on July 23, Stalin proposed that the Allies drink a toast to "their next meeting in Tokyo." Truman and Byrnes went through the motions but kept their fingers crossed. Earlier, Stalin had mentioned that his forces might not attack Japan until the Chinese Nationalists approved a treaty granting Moscow the Manchurian concessions promised at Yalta. The president knew that an atomic weapon would probably be ready for use by the first week in August. Since Sino-Soviet treaty talks had deadlocked (in fact, American officials urged their Chinese allies to delay a settlement), it seemed possible that Japan could be subdued before a Russian offensive.[25]

On July 24 (the second week at Potsdam), during a private discussion with the president over targets, Henry Stimson urged Truman to avoid hitting cultural treasures like the city of Kyoto. If it were bombed, he warned, "the bitterness which would be caused by such a wanton act [of destruction might make it] impossible during the long postwar period to reconcile the Japanese to us in that area rather than to the Russians." Such a grave, misguided act might inadvertently become the "means of preventing what our policy demanded, namely a sympathetic Japan to the United States in case there should be any aggression by Russia in Manchuria." Even though

Truman described the "Japs" as "savages, ruthless, merciless and fanatic," he agreed to spare both Kyoto and Tokyo.[26]

The Joint Chiefs still doubted that the atomic bomb would prove deci-sive either in defeating Japan or containing the Soviets. If the bomb failed to bring Japan to its knees, an invasion would still be necessary. In that case, Soviet help would be vital. Truman, however, seemed convinced that the atomic bomb would fulfill its promise.

On July 25, the Joint Chiefs instructed MacArthur to be prepared to enter Japan in the event of a sudden surrender. They also raised the pos-sibility of deploying American forces to southern Korea, a zone that by Allied agreement, was consigned to the Soviets. When MacArthur re-sponded with a proposal to enter Japan twelve days following a surrender, Marshall requested a plan for more rapid movement.[27]

Secretary of State Byrnes, who complained that an "awful mistake" had been made in allowing Russia to become so powerful, widened the breech among the Allies on July 26 by releasing the Potsdam Declaration without bothering to inform the Soviets. It called upon Japan to surrender at once or face total destruction. The proclamation balanced the threat with prom-ises of moderate treatment after surrender and the eventual restoration of sovereignty. Still, it made no mention of either the atomic bomb or future status of the emperor.

Even though Stimson, Grew, and the new British foreign minister, Ernest Bevin, considered the emperor a "vital instrument" to "effectively control Japan," Byrnes and the American military chiefs dissented. For a variety of reasons, they opposed a public pledge to retain the imperial institution. The politically astute secretary of state seemed especially concerned by the results of a recent Gallup Poll finding that Americans overwhelmingly fa-vored Hirohito's arrest or execution. These arguments, as well as confi-dence in what he called an "entirely new weapon," convinced Truman that he need not compromise or seek the help of the "Russians or any other nation" in defeating Japan.[28]

On August 6, the first atomic bomb fell upon Hiroshima. Two days later (a week sooner than Washington expected though in accord with earlier bilateral agreements), the Soviet Union declared war upon Japan. The next day, the air force attacked Nagasaki with a second atomic weapon. Fi-nally, on August 10, the Japanese asked Swiss intermediaries to inform the Americans of their willingness to accept the Potsdam Declaration.

Tokyo still requested guarantees that the "prerogatives of the Emperor as a sovereign ruler" would remain intact. At a cabinet meeting that day, Truman explained his inclination to accept this offer even though it signi-fied a major concession. He feared the Russians might get "too far into Manchuria" and refuse to sign a treaty with China (they did four days later, however). The president and Byrnes both felt that the only way to halt Soviet expansion was to make a quick peace with Japan. Truman ordered that Tokyo be informed that if the emperor signed and issued surrender orders and pledged obedience to the American occupation commander, he could remain on his throne pending the "freely expressed will of the Japanese people." Commerce Secretary Wallace left the meeting convinced that all policy goals in Asia had been subordinated to the determination to keep the Soviets out of China and Japan.[29]

At no time during these complex deliberations did anyone in Washington consult with or even inform MacArthur. For example, the War Department waited until almost the last minute (the end of July) to notify the SWPA commander about the impending use of the atomic bomb. (In contrast, Eisenhower and several other commanders learned of the weapon much earlier, a fact that enraged MacArthur.) Even when Truman decided (about August 8) to select the general as occupation commander, no one discussed with MacArthur the terms and responsibilities of the post.

In an effort to appear better informed, the general took a group of journalists into his confidence in Manila on August 6, just before news of the attack upon Hiroshima. He declared that imminent "Russian participation" in the Pacific War would hasten Japan's surrender. Each Russian killed meant "one less American had to be." The next day aides quoted the general as convinced Japan would surrender "as the result of the atomic bombing . . . and because of the perspective entry of Russia into the war."

In an interview with *Time* correspondent Theodore White, the general paced about his headquarters declaiming that the days of heroic warfare had ended. The "scholars and scientists" had seized control of future battlefields away from generals and made "men like me obsolete." There would be "no more wars," he intoned, "no more wars."

On the eve of Japan's defeat, MacArthur worried that the navy still hoped to keep him out of Tokyo. He complained to Eichelberger that while he could soon dictate terms to Tokyo, he greatly feared the navy trying "to horn in" and carry out a "landing of their own in Japan."[30]

Administration officials worried more about how to convince the Soviets to halt their offensive in northeast Asia. On August 10–11, Ambassador Harriman spent much of the night arguing with Foreign Minister Molotov over Russian policy. Besides his government's aim of clearing the Japanese from Manchuria, Molotov demanded an occupation zone in Japan and a voice in selecting the allied occupation commander.

Harriman refused any concessions and the Soviets, to his surprise, dropped their demands. A bit self-importantly, Harriman concluded he had "settled that night" the entire postwar role of the Soviets in Japan. Actually, Stalin probably had consigned Japan to the American sphere from the start. Molotov may have demanded concessions in the expectation they would be refused. Washington's rebuff would justify Moscow's refusal to permit the western Allies much say in the Soviet-dominated Balkans.[31]

In Japan, the emperor assured peace by declaring his willingness to serve under the American commander. By August 14, he convinced most of the military leadership to surrender. The next day Hirohito spoke by radio to his subjects, explaining that the "new and most cruel bomb" used by the enemy had inflicted a dreadful toll on "innocent lives." Prolonging the war would only destroy Japan and threaten all "human civilization." Quoting Buddhist texts, he declared Japan would pave the way for peace by "enduring the unendurable and suffering what is insufferable."

In a letter to his son written three weeks later, the emperor provided a personal explanation of the disaster. "Our people," he told Crown Prince Akihito, "believed in the imperial state too much, and despised Britain and the United States." The military had "placed too much significance on spirit, and were oblivious to science." Too many banzais and not enough technology, he suggested, doomed the nation. Business leaders reacted more gleefully. One executive, who later served as foreign minister, Fujiyama Aiichiro, recalled that "when it was learned that the occupying power would be the U.S.," industrialists "uncorked their champagne bottles and toasted the coming of a new industrialist era."[32]

Final Preparations

Douglas MacArthur learned on about August 10 that Truman had selected him as the supreme occupation commander. Excitedly, he told Eichelberger he intended to "live in the Palace in Tokyo." At some visceral level, he interpreted his appointment as a posthumous victory over FDR. He delighted in telling friends that Roosevelt had insisted "he should not command" in Japan. But the president's death, criticism of the "high navy losses at Okinawa," and favorable "public opinion" made him the only choice. The terms of his appointment gave him full power to arrange the formal surrender and "rule the state" of Japan as he saw fit. General Order #1, which the president issued on August 15, further empowered MacArthur to supervise the surrender of Japanese forces throughout East and Southeast Asia.[33]

A Japanese delegation reached Manila on August 19 to arrange for the arrival of occupation forces. Like many in the general's entourage, Eichelberger ridiculed the delegates as a band of "goggle eyed little bucktoothed birds." MacArthur not only abhorred these racial epithets but sided with the Japanese who begged the Americans to permit the emperor to issue a statement using the special imperial pronoun rather than the commonly used "I." The general explained that he did not want to "debase" Hirohito in the eyes of his own people, "as through him it will be possible to maintain a completely orderly government."[34]

Fortunately, a typhoon delayed the arrival of American forces for two days beyond the August 26 target. This permitted the Imperial Army to suppress some last-ditch opposition in the lower ranks. The nervous pilots of the first American planes to reach Atsugi Air Base outside Yokohama landed on the wrong end of the field, terrifying the Japanese who had laid out an elaborate buffet table. When the wary crews finally deplaned, they feared the beer and food might be poisoned. Only after a Japanese officer downed a few glasses did the Americans plunge in. Still, one participant remembered, it all seemed like a "party that wasn't going off very well."

At this point an American interpreter broke the ice by inquiring after the health of a famous Kabuki actor he had known while a student in prewar Tokyo. This sensitivity about Japanese culture so relaxed both groups of soldiers that they began to mingle. Finally, the Americans boarded a public bus and, when it broke down, a fire engine, for the ride into Yokohama. The only upsetting news came later, when the party learned that a naval landing force, under Harold Stassen, had already sailed into Japanese waters to rescue Allied prisoners. Determined not to be upstaged, MacArthur decided that he and his staff would fly to Japan on August 30, in an unarmed plane. (Of course, the airfield and route were secured before his arrival.)[35]

In Washington, President Truman and others questioned the wisdom of appointing MacArthur to his new post. Senator Tom Connally called Truman early in August to warn him that it would be a "big mistake" to appoint "Dugout Doug as Allied Commander in Chief to accept the Jap surrender." The general would use the post, Connally predicted, "to run against [Truman] in 1948."

Harold Ickes also discussed the appointment with Truman, lamenting that, given the wartime hoopla over the general, it seemed "inevitable" that "MacArthur should be cast for this role." Still, he feared this dangerous "man on horseback" would "take every advantage of this dramatic situation to get himself spread all over the papers."

Truman revealed similar doubts about the general's selection, telling Ickes that it was not fair to "blame on him the appointment of MacArthur as the chief commander in the Far Eastern Area." The interior secretary admitted that, politically, the president "couldn't do anything else." The blame was "due to Roosevelt" who made a terrible "mistake in taking MacArthur away from the Philippines." He should have left the general "to clean up his own mess and taken Wainwright out," allowing the Japanese to solve the MacArthur problem. Truman and Ickes touched on the irony that while most "people of the country" considered MacArthur "an outstanding hero," those in the know judged his wartime record a "fiasco." Removing the general from Corregidor, giving him a theater, and approving the return to the Philippines were all examples of how politics had triumphed over sound strategy. Yet even now, Ickes agreed, denying the general a command in Japan would only make a "martyr out of him and a candidate for president." Harry Truman disputed only one point: "MacArthur would probably be a candidate anyway." On that note, the occupation of Japan began.[36]

9
The Supreme Commander

As the Supreme Commander for the Allied Powers, MacArthur entered Japan on August 30, 1945. His entourage found ruined cities, idle factories, and millions of homeless refugees. Approaching Tokyo by road, one American had the eerie sensation of driving away from a great city, since the closer in he went the more everything seemed "completely flat with destruction." The "rubble did not even look like much." Mounting food shortages seemed certain to grow worse as nearly seven million Japanese soldiers and civilians stranded abroad began to trickle home. In evident despair, the fanatic leadership of the "East Asia League" closed shop with a final admonition to its followers. They were urged to obey Americans in hope of speeding an end to the occupation and to "align themselves with world Jewry, which had now proved its invincibility by triumphing over Hitler."[1]

In an impressive ceremony held aboard the battleship *Missouri* in Tokyo Bay on September 2, General Douglas MacArthur accepted Japan's formal surrender. The huge Allied flotilla and the overflights by hundreds of aircraft confirmed, dramatically, the verdict of the war. Shortly afterward, the occupation commander established his headquarters in the Dai Ichi Insurance Building in central Tokyo. For the next seven years, the General Headquarters (GHQ) of the Supreme Commander for the Allied Powers (SCAP) stood at the center of Japanese life.

Finally, it seemed, Douglas MacArthur possessed the authority and responsibility he had spent a lifetime preparing for. Instead of retiring from active duty (he was now sixty-five) and seeking a position in civilian life, he became the virtual ruler of seventy million people. Besides the honor and challenge, the job would keep MacArthur's name in circulation during the three years until the next presidential election.

From its inception, the occupation became synonymous with its supreme commander. Although few Americans could name the man in charge of the German occupation (General Lucius Clay and, later, John J. McCloy) most could readily identify the top man in Tokyo. An American arriving in Japan a few months after the surrender remembered his astonishment at the "degree to which the Occupation had become personalized." Every "occupation action, every policy, every decision was MacArthur's." His name appeared everywhere, those of other officials nowhere. Neither the general nor the censored Japanese press suggested that anyone other than MacArthur contributed to the formulation of policy. In effect, a single individual "displaced the United States Government as far as the Japanese people were concerned."[2]

Nominally allied commander, MacArthur always functioned as an American official. He commanded military occupation forces in Japan and southern Korea and, after 1947, the regional Far Eastern Command. He did not, however, exercise authority over American forces in China.

Within Japan, the existing government retained major responsibility for the normal management of affairs. SCAP (the acronym applied to Mac-Arthur, personally, as well as to the occupation administration) consisted of a dozen staff sections. Except for the few military contingents (such as intelligence), most of the sections paralleled the Japanese cabinet structure. For most of the occupation period, officers long associated with Mac-Arthur, such as Charles Willoughby, Courtney Whitney, and William Marquat, headed the all-important Intelligence, Government, and Economic sections. Under duress, MacArthur tolerated the presence of a State Department representative. This "Political Adviser" to MacArthur—as if anyone could give him political advice—had few responsibilities.

During occupation and after, Courtney Whitney stood closest to MacArthur. A lawyer and acquaintance of the general in prewar Manila, Whitney took command of Philippine guerrilla operations in 1943 and remained at MacArthur's side (almost literally) for the rest of his boss's life. A few weeks into the occupation, the supreme commander selected Whitney to head the critical Government Section, in charge of restructuring Japan's constitution, laws, and political system. Although a staunch conservative, Whitney adopted a pragmatic approach to his job and leaned heavily on his liberal assistant, Colonel Charles Kades, a New Deal lawyer. The latter convinced Whitney to implement many of the changes proposed by reformers in Washington.

Personal rivalry between Whitney and MacArthur's other close subordinate, General Charles Willoughby, may also have resulted in Government Section's pursuit of reform. A German-born immigrant with pretension's of noble birth, Willoughby brought a Prussian demeanor and extremely right-wing views to his intelligence post. MacArthur hit the mark when he once called his aide "my lovable fascist." Willoughby saw Communist and Jewish conspiracies at home, abroad, and especially, in SCAP's ranks. His Counter Intelligence Corps (CIC) spied on Americans and cultivated former members of Japan's secret police and armed forces. Willoughby encouraged many of his chief's worst tendencies, especially after 1949 when MacArthur turned his attention to schemes for "rolling back" Asian communism.

General William F. Marquat, an anti-aircraft officer MacArthur had known in the Philippines, placed third in SCAP's hierarchy. As chief of the Economic and Scientific Section (ESS) for most of the occupation (Colonel Ray Kramer held the position briefly in 1945), he bore responsibility for restructuring Japan's economy and promoting recovery. Although affable and hardworking, Marquat lacked the intelligence and experience to accomplish his huge assignment. Before 1948, most liberal ESS initiatives came from Washington. Later, the administration promoted a "switch in emphasis" that reversed many of the earlier reforms.[3]

At its height, in 1948, about 3,200 Americans served in SCAP. Of these, a few hundred military and civilian experts actually held positions of policy responsibility. Most Americans in Japan were ordinary GIs serving in the Eighth Army. Just as the occupation troops depended largely on the Japanese to maintain order, SCAP officials relied heavily upon the Japanese bureaucracy to gather information and implement reforms.

During the six years he spent in Tokyo, MacArthur followed an incredibly constant routine. Driven from his residence in the prewar American embassy each morning at 10:30, he worked for several hours at the Dai Ichi Building and returned home for a midday meal and nap. In the late afternoon he repeated the journey. The general entertained only at lunch and rarely socialized with Japanese officials. Only twice, before the Korean War, did he even leave Tokyo, to attend independence ceremonies in Manila and Seoul. However, his new headquarters never provided the emotional comfort of prewar Manila. The Japanese, he once remarked to an aide, were a "brooding" people and their country had "an ominous quality" that put him on edge.[4]

The mere presence of allied forces, even in token numbers, made supreme commander extremely uncomfortable. For their part, the Soviets, British, and Chinese all protested bitterly their exclusion from any consultative role in Japan. At the foreign minister's meetings in London (September 1945) and in Moscow (December 1945), Secretary of State Byrnes found it hard to deflect their collective anger. Many observers questioned how the Americans could demand that the Kremlin loosen its grip on Eastern Europe while refusing to create any kind of allied control commission in Japan.

After prolonged haggling, the foreign ministers struck a deal in December creating two oversight bodies. An eleven-member Far Eastern Commission (FEC) would sit in Washington, while in Tokyo, a smaller Allied Council for Japan (ACJ) could observe policy implementation. However, the United States retained an absolute veto over the recommendations of both groups.

Even symbolic Soviet representation in Tokyo, MacArthur warned, risked a "communist revolution." If the Allied Council got a "foot in the door," it would "destroy us." The general's representatives on the Allied Council, a member recalled, spent most of their energies "humiliating" the Soviet delegate. Eventually, mandatory meetings were adjourned seconds after the sessions began. The equally impotent Far Eastern Commission finally got around to sending a proposed reform agenda to Tokyo about two years after the occupation began.[5]

Most Japanese and American accounts of the occupation emphasize what one historian has called the "heroic myth." This asserts that in August 1945 MacArthur began personally to build a new Japan. The general's own account, like those of his closest subordinates, portrays the occupation as the inspired, individual creation of its military chief. Two of MacArthur's most bitter rivals, President Harry Truman and Secretary of State Dean Acheson, enhanced this legend in their memoirs. They attacked the general unmercifully for his conduct during the Korean War and contrasted the debacle to his great achievement in Japan.

General Courtney Whitney, one of the two men closest to MacArthur in this period, recalled that during the flight to Tokyo late in August, the occupation commander spoke like an oracle. Pacing the aisle of the plane, he casually conceived the "policy under which we would work and live for the next six years." Revealing his much celebrated mastery of the oriental mind, MacArthur decreed the transformation of Japan. First, he would "destroy the military power," then build representative government, enfranchise women, free political prisoners, liberate farmers, establish free trade unions, destroy monopolies, abolish police repression, liberate the press, liberalize education, and decentralize political power.[6]

Whitney neglected to mention that, a few hours earlier, MacArthur received an advance copy of the "Initial Post-Surrender Policy for Japan," a document drafted during the preceding months and known officially as SWNCC 150/4/A. (President Truman approved it on September 6.) MacArthur also knew the substance of an even more detailed plan under final review by the Joint Chiefs of Staff. Transmitted formally to him on November 3, this "Basic Directive for Post-Surrender Military Government in Japan Proper" (JCS 1380/5), like its SWNCC counterpart, outlined virtually the entire reform agenda in the order of MacArthur's recital.[7]

The Japanese had many reasons to cooperate with the occupation force. Most obvious, the end of the war brought collective relief to most people. The killing had ceased and virtually all Japanese preferred an American to an allied or Russian occupation. Traditional conservatives in the government and in business supported a controlled purge of the ultranationalists and militarists who had usurped power and led the nation into disaster. Moreover, as the new foreign minister, Yoshida Shigeru, noted, as "good losers" the Japanese might restore in peace much of what the generals lost in war.

The Americans who drafted the occupation program in 1944–45 fell, broadly speaking, into two ideological camps. A group centered around

Joseph C. Grew (prewar ambassador to Japan and, in 1945, under secretary of state) and his assistant, Eugene Dooman, blamed military fanatics, driven to desperation by the depression for hijacking Japanese democracy during the 1930s. Tokyo had stumbled off the path of progress, but with a few reforms, could easily return to the civilized fold.

Other Asian specialists and especially New Deal lawyers and economists disputed this analysis. Militarism, imperialism, and domestic economic exploitation, they argued, lay at the root of the modern Japanese state. The giant industrial combines, the *zaibatsu* (literally, "money clique"), bore responsibility for internal oppression and foreign war. Only a structural political and economic reform could assure the development of a peaceful and democratic Japan. This group did not favor stripping Japan of all its industry or imposing a harsh peace. Instead, they believed that Japan must be restructured along liberal, democratic lines in order to become a peaceful member of the world community.

As a result of both political and personality struggles, Grew and his circle lost influence during 1945, and he retired as soon as the war ended. Such rising stars as Dean Acheson, who replaced Grew as under secretary, and John Carter Vincent, who took over the Office of Far Eastern Affairs, felt Japan deserved more than a rap on the knuckles to atone for the events of the past fifteen years.

The initial reform plan sent to MacArthur struck a compromise between the "stumble" and "root" camps. The Japanese government, minus its most notorious members, was permitted to function. However, the occupation would impose firm guidelines assuring civil liberties and democracy. At the recommendation of New Deal economists, the Joint Chiefs and SWNCC instructed MacArthur to promote a wider "distribution of income and ownership of the means of production and trade," encourage labor unions, and implement a rigorous antimonopoly program.[8]

The supreme commander also straddled the "stumble" and "root" camps. He often spoke of leading a "revolution" against the existing "feudal" order. He meant, of course, a controlled reform program with essentially liberal goals. When challenged by a conservative friend to explain his support for "socialistic reforms," MacArthur defended his actions as "designed to clear the way for the ultimate development in Japan of a healthy economy based upon free, competitive private enterprise." He sought only to bury a "decadent past" and create a "buttress against the spread of any conflicting philosophies of life" like communism.[9]

Theodore Cohen, a liberal Democrat and SCAP labor adviser, observed the general at close range from 1946–50. Like many New Dealers, he first thought MacArthur might be "faking" his support for the reform program. How, he wondered, could a "reactionary" at home turn "crusading reformer in Japan?" Cohen concluded that MacArthur's behavior defied contemporary ideology. He was actually a "political primitive with wide open spaces where his reactionary principles were supposed to be." Raised on the frontier and educated in myths about American greatness, he had little "urban and no industrial experience that might have prepared him for the great American and European social conflicts after World War I." Uncomfortable with the politics and personalities of the New Deal, he shared many nineteenth century populist beliefs about selfish bankers and corporations. Marxism and revolution were confusing abstractions, while, he told Cohen, in times of stress he found comfort by "turning his thoughts to George Washington." [10]

As before, MacArthur captured the attention of the Japanese and American public with his unique rhetoric. His pronouncements had the ring of a nineteenth century dandy inspired by a mix of social gospel and social darwinism. He spoke of the Japanese as childlike and malleable, as clay given him to mold into model Americans. After visiting Tokyo on an educational mission, the great historian of Puritan New England, Perry Miller, described MacArthur's hope of remaking Japan into a "a new Middle West—not, of course the Middle West as it is, or in fact ever was, but as it perpetually dreams of being."

During the furor following his recall in 1951, curious senators asked him to comment on the Japanese characters. The general responded by comparing the racial and cultural structures of both wartime enemies.

> The German people were a mature race. If the Anglo-Saxon was say 45 years of age in his development, in the sciences, the arts, divinity, culture, the Germans were quite as mature. The Japanese, however, in spite of their antiquity measured by time, were in a very tuitionary condition. Measured by the standards of modern civilization, they would be like a boy of 12 as compared with our development of 45 years. Like any tuitionary period, they were susceptible to following new models, new ideas. You can implant basic concepts there. They were still close enough to origin to be elastic and acceptable to new concepts.

MacArthur's special vocabulary (such as the archaic Latin "tuitionary") echoed yet again his father's advice on enlightening the Filipinos a half-century before. [11]

The Western Outpost of our Defenses

Within weeks of his arrival, MacArthur voiced remarkable certainty about the future of Japan. In October 1945, he told Edwin A. Locke, an envoy from Truman, that because of his unique understanding of oriental psychology, he could make Japan serve American interests. The general and his economic adviser at that time, Colonel Raymond C. Kramer warned Locke that imposing economic reforms might make Japan an economic "albatross around our necks." The loss of empire had already "badly unbalanced" industry, and they thought it wise to enact cosmetic changes while leaving real economic power in "identical hands." Kramer opposed giving industrial equipment as reparations to Japan's wartime victims, all of whom he judged "incapable."

MacArthur said little about eradicating militarism but spoke of the danger of "underground communist agitation" among "so-called liberals" who, on Moscow's orders, targeted Japan for a "successful communist revolution." Only weeks after the formal surrender, he spoke of Japan as an incipient ally. While the general nodded approval, Kramer told Locke that after demobilizing its armed forces, Japan should be given the "green light" to expand economically throughout the Pacific. Because of their superior talents, "discipline, and honesty," the Japanese should be encouraged to serve as "the natural leaders of Asia." Other visitors reported MacArthur's aides telling them that Japan should be developed "as a base for possible future military operations against Russia."[12]

While the formal occupation agenda stressed civil liberties, democratic reform, and economic change, MacArthur often articulated a different set of priorities. In July 1946, for example, he told the visiting navy secretary, James Forrestal, that the administration ought to do more to halt the Communist advance in China. He described Japan as the "western outpost of our defenses" and the key to holding the line against Communist revolution in Asia. Many proposed reforms, he suggested, opened the door to communism.

The general also defended the emperor (whom some Americans still hoped to try as a war criminal) as an innocent hostage of the military fanatics. Hirohito, MacArthur claimed, bore him such personal gratitude for his liberation that the descendent of the sun goddess had requested permission to "become converted to Christianity."

Christianity often seemed a metaphor for MacArthur's vision of postwar Asia. His most important mission, he told James Forrestal, was to secure Asia for Christianity and deny it to Marxism. The general actually compared his own service in Tokyo to biblical accounts of Christ's agony on the cross. He took solace that even though crucified, "Christ nevertheless prevailed."[13]

He condemned all critics as heretics unworthy of the new faith. For example, labor unions that rejected his "spiritual revolution" were characterized as on the "extreme radical left." MacArthur informed Congress that he built a "powerful bulwark for peace in the Pacific" through the "reformation of the Japanese people." He had broken their idols of "legendary ritualism" and taught them the "maturity of enlightened knowledge and truth."

MacArthur estimated that at least 2 million Japanese had "moved to embrace the Christian faith as a means to fill the spiritual vacuum left in Japanese life by the collapse of their past faith." This "spearhead of Christianity," as he called it, would mobilize and transform the "hundreds of millions of backwards peoples" throughout Asia.[14]

MacArthur denied having an "obsession" with the conversion of Japan or of using state power to favor a single faith. He insisted that so long as SCAP did not officially "oppress" any other religion, occupation authorities "had every right to propagate Christianity." Even though the general never attended church, he promoted the Christianity as part of his "mission as a peacemaker" and as an effective antidote to revolutionary dogma.

MacArthur argued that unless he filled Japan's "spiritual vacuum," with Christianity, it would "be filled with Communism." The general sent a public letter to the president of the Southern Baptist Convention in 1946 that noted "an opportunity without counterpart since the birth of Christ for the spread of Christianity among the peoples of the Far East." The occupation commander exempted missionaries from restrictions on foreigners entering Japan. Evangelist Bob Jones wrote excitedly to MacArthur that eleven hundred of his followers were "learning how to load the Gospel Gun and how to shoot it." Journalist John Gunther came away from an interview with the impression that the general thought "of himself and the pope as the two leading representatives of Christianity in the world today."[15]

MacArthur reacted to the election of moderate socialist Katayama Tetsu as prime minister early in 1947 by proclaiming the "spiritual implications" of the vote. For the "first time in history," Japan had a "Christian leader." The occupation commander deemed it a divine sign that "three great Oriental countries [Japan, China, and the Philippines] now have men who embrace the Christian faith as heads of their governments." The advance of this "sacred concept" would bind America and Asia in erecting an "invincible spiritual barrier" against godless ideologies. God, however, moved in mysterious ways since all three Christians soon lost power.[16]

MacArthur spoke this way in private as well as in public. In 1948, for example, he harangued visiting diplomat George F. Kennan for several hours on evangelism. Whenever Kennan tried to discuss strategy, industrial recovery, and containment, MacArthur broke into a monologue on his spiritual mission to the heathen. He told Kennan (who, privately, described his host as a cross between Catherine the Great and Stalin) that his most spectacular achievement had been "bringing to the Japanese two great appreciations which they had never before perceived and which were destined to revolutionize their thinking, namely, democracy and Christianity."

Armchair strategists in Washington, ignorant of his power to stir the Asian masses, failed to realize that the "great events of the next 1000 years would transpire" in the Far East. Through his Japan converts he would "plant the seeds of . . . democracy and Christianity" among a "billion of these Oriental Peoples on the shores of the Pacific." His actions would "fundamentally alter the course of world history."[17]

Either MacArthur did not know or did not care that most Asians, outside the Philippines and Indochina, remained largely indifferent to Christianity. Even his proud claim of winning "two million" new Japanese converts had a fraudulent basis. In preparing a speech on "spiritual revolution," MacArthur requested the SCAP's Civil Information and Education Section furnish details on the numbers of Japanese Christians. The officers handling religious affairs determined that there had been slightly over 200,000

Christians active before the war, but the number had declined since 1941. The head of the section protested that "this wasn't enough for MacArthur" who would be furious. The junior officers returned to their desks and debated how many additional "zeros" must be added to satisfy the general. Predictably, the next week MacArthur spoke of how the Japanese had "heard the Sermon on the Mount" and Christianity was sweeping the land. Through moral example and persuasion, he asserted, the 200,000 Christians had now grown to a community of 2,000,000.[18]

Despite such excesses, most Japanese accepted, and many revered, MacArthur. His haughty style mirrored the authoritarian cultural traditions he had been sent to change. Moreover, Japanese credited him for the generally good behavior of his forces and the vital relief supplies that prevented starvation. Officials of SCAP as well as the Japanese press and government attributed all aspects of the reform program to the general's authority. Also, censorship prevented any open criticism of MacArthur and the occupation.

The general's headquarters encouraged a degree of idolatry that made critics seem petty naysayers. Although rarely discussed, SCAP censored both the Japanese press and dispatches by American correspondents. Early in the occupation, MacArthur's public relations officer, General LeGrande A. Diller, warned the American press corps that he would not "let you give MacArthur's critics in the States any ammunition." Besides overt censorship, he urged his audience to remember that "the Army controls the food here." During the first five years of the occupation, MacArthur ordered or approved the expulsion of seventeen journalists on a variety of charges. Correspondents who wrote unflattering reports while outside Japan were denied reentry permits. The mere threat of expulsion or exclusion discouraged critical reporting.[19]

The American media often assisted these excesses by reprinting as fact sycophantic hyperbola. For example, *Time* magazine on one occasion described the scene in SCAP headquarters where "bleary eyed staff officers looked up from stacks of paper [and] whispered 'God, the man is great.' " The magazine quoted General Ned Almond's description of his chief as "the greatest man alive." Not to be outdone, the occupation commander's top air officer, General George E. Stratemeyer "put it as strongly as it could be put: He's the greatest man in history."[20]

None of these unflattering traits negates the importance of the occupation or the degree to which Japan was transformed during MacArthur's tenure. Especially through 1948, the general and his staff participated actively in what they conceived of as a "controlled revolution." MacArthur relished the opportunity to rule one nation and prove his ability to lead another. But the achievements of the occupation were collective rather than individual. Although he never acknowledged this in public, MacArthur had no qualms about utilizing the talents of American social reformers and Japanese politicians in remaking Japan.

The Controlled Revolution

In October 1945, SCAP issued a civil liberties directive ordering the release of political prisoners, legalization of all political parties (including the Communists), and protection of the rights of free speech and assembly. The cabinet of Prime Minister Higashikuni Toshihiko resigned in protest, warning that the Americans had opened the gates to revolution. Nevertheless, the avalanche of reform continued.

Early in 1946, members of SCAP Government Section presented a startled Japanese delegation with a finished draft of a new, democratic constitution. Unless it were accepted quickly by the cabinet and Diet, SCAP intended to submit it as a referendum to the Japanese people. More terrified by the prospect of a popular vote on the constitution than by its democratic content, the conservative oligarchs accepted the new document.

The constitution, which had to be laboriously translated into Japanese, stripped the emperor of temporal power, strengthened the role of the Diet, broadened voting rights, enhanced the power of local government, provided for legal equality of the sexes, and in Article 9, forbade creation of armed forces or the conduct of war.

Emperor Hirohito eased its adoption by issuing a proclamation repudiating his divinity. This conversion virtually insured he would not be deposed or tried as a war criminal, as some American and many foreign officials still urged. "With him as a figurehead," MacArthur commented, "our job is so much more easy." Congressman Maury Maverick of Texas quipped that the real "reason the Emperor came out and said he wasn't God was because he found that MacArthur was."

By instituting a purge early in 1946, SCAP neutered the most extreme political opposition. This was by no means an unpopular act, as even moderate conservatives were eager to cleanse the political landscape of leading militarists and ultranationalists. However, the broad scope of the purge order, SCAPIN 550, required the removal from national government of politicians, bureaucrats, military officers, and police linked to aggression. Later, business officials with ties to foreign expansion were included, as were local government officials.

In spite of its comprehensive appearance, the purge did not rock the political landscape. About twenty young Americans had the task of investigating some 2.5 million Japanese! Most of the actual work fell to Japanese assistants who exercised subtle control over the program. The purge (which involved removal from office and loss of certain political rights) eventually affected about 200,000 people, 80 percent of whom came from the military. Some party leaders also lost their posts, although this had little affect on their organizations.

When a conservative American friend asked the supreme commander how he could justify this program, MacArthur explained that by going through the motions he prevented a more draconian purge favored by vindictive liberals. His aides targeted mainly army and navy officers, not politicians or business leaders. In any case, a far smaller percentage of Japanese were investigated than were Germans. Ironically, the most vigorous enforcement of the purge began in 1949 and 1950, with leftists as the target.

During 1946, SCAP planners challenged directly the endemic problem of landlordism. Although his instructions left the question of land reform somewhat vague, this issue, like disarmament, evangelism, and trust-busting, captured the general's personal attention. MacArthur endorsed a fairly radical proposal by SCAP civilian experts to eliminate large landholdings with the aim of creating a viable, smallholder rural base. This would transform the rural economy, expand production, enlarge domestic consumption, and most significantly, eliminate the potential for the kind of peasant uprisings then sweeping China and Southeast Asia.

MacArthur justified the programs on both idealistic and pragmatic grounds. "The land reform measures," he wrote, were "designed to tear down the large feudalistic land holdings in order that those who till the soil will have the opportunity to reap the full benefit from their toil." This action would create in Japan a "new class of small capitalistic landowners which in itself will stand firm against the invasion of proponents of socialism or collectivism."

Over the objections of Japanese officials, SCAP pushed land reform through the Diet. Colonel Charles Kades, a ranking member of the Government Section and a leading reform advocate, witnessed a remarkable scene in MacArthur's office that probably explains part of the supreme commander's interest in the question. Nearly half a century before, the general told his staff, his own father had advocated land redistribution in the Philippines. When Kades brought him word that the Diet had passed the land reform law, MacArthur "looked up toward the ceiling, toward a picture of his father which hung on the wall and said 'How am I doing Dad?' "

The new law compelled landlords to sell most of their holdings to tenants, who received credits to purchase the land. The rampaging inflation of the postwar years quickly nullified the payment given the landlords and eased the debt of the tillers. Eventually, nearly a third of Japan's land changed hands. Land reform created a class of small farmers loyal to the conservative parties which first opposed the law. As MacArthur predicted, Japan avoided the rural insurgencies that brought havoc to much of Asia. In 1950, Chiang Kai-shek told the general ruefully that if he "could have done in China what you did in Japan, I would still be there today."

During these first two and a half years, numerous other institutions underwent similarly dramatic change. SCAP reduced the power of the feared police force, reorganizing it on a local basis. Public education was remodeled on Western lines and no longer inculcated extreme nationalist values. New codes swept away the legal sanctions used to suppress the labor movement. By the end of 1946, nearly four million Japanese workers joined

MacArthur greets General Kuzma N. Derevyanko, Soviet delegate to the Allied
Council in Tokyo, late 1945

MacArthur greeting Manuel Roxas at Philippine independence ceremony
in Manila, July 1946

Enthusiastic Japanese declare their political preferences on
the eve of the Wisconsin presidential primary, spring 1948

MacArthur received by Chiang Kai-shek during visit to Taiwan,
July 31, 1950

President Truman awards MacArthur Distinguished Service Medal at the conclusion of the Wake Island meeting, October 15, 1950

A private moment between President Truman and General MacArthur,
Wake Island, October 15, 1950

MacArthur addresses Congress following his recall, April 19, 1951

unions, ten times the prewar total. Soon, half the urban work force belonged to trade unions.[21]

The Limits of Controlled Revolution

Japan's political, economic, and social landscape underwent a thorough redesign during the three years following its surrender. But observers sometimes overstate the degree of transformation and the American role in those changes. MacArthur often seemed more concerned with democratic form than content. Officials of SCAP pointed proudly to early and orderly elections. Yet the political process continued to favor the elite coalitions of the prewar period. At the beginning of the occupation, journalist I. F. Stone predicted that the decision to "leave in power the very elements we are pledged to eradicate" would distort democratic change.

One SCAP veteran, John Maki, confirmed this trend in an analysis written early in 1947. Despite significant political and legal reforms, he asserted, the occupation had only scratched the surface. The powerful career bureaucracy remained intact, altered neither by purge nor free elections. Traditional conservative parties, better organized and financed than liberal and socialist rivals, still dominated the Diet. Maki traced these failures to MacArthur's "decision to act through a Japanese government [which had] the inevitable result of creating both the appearance and the actuality [of supporting] that government on issues of critical internal importance." Whether intended or not, this action "created an identity of interest" between the occupation and the old regime.[22]

Beginning with the first postwar election in April 1946, two conservative parties (the Progressives and the Liberals) dominated the Diet. Personality issues and ties to industry, rather than ideology, distinguished them. The left, still weakened by decades of oppression, remained splintered among several midly socialist groups and a small Communist party. MacArthur's insistence on early parliamentary elections provided the entrenched and well-financed conservatives with an important organizational edge.

The conservative coalition cabinet led by Yoshida Shigeru held power from the spring of 1946 through the spring of 1947. The Socialists gained enough seats by June 1947 to become briefly the largest single party and organized a short-lived minority government under Katayama Tetsu. By early 1948 the Socialist-led coalition collapsed and the pendulum swung right, with a succession of conservative coalition cabinets taking power. Yoshida Shigeru organized a new Democratic-Liberal party (conservative) in the spring of 1948 and assumed the premiership after an electoral victory in October 1948. With its representation increased early in 1949, Yoshida's cabinet remained in power past the end of the occupation. Its successors would retain their grip into the late-1980s.

Initially, MacArthur condemned Yoshida as "monumentally lazy" and "inept." He felt the prime minister often tried to delay or distort reforms mandated by SCAP. But with the conservative turn in American foreign policy by 1948, the two leaders cooperated in imposing renewed controls on labor and in resurrecting large industry. Gradually, in 1949 and 1950, MacArthur reduced the size of SCAP and turned over increasing authority to the Yoshida administration. Despite their initial fears about civil liberties and constitutional reform, Japan's postwar conservatives ultimately found MacArthur a collaborator in the creation of a postwar conservative hegemony.

Besides MacArthur's lack of concern with encouraging a viable multiparty system in Japan, observers noted other disturbing trends. As with his reports of easy advances during the Pacific War, the general frequently proclaimed offensives against "feudal" institutions and then, almost as suddenly, declared victory. These announcments often coincided with political developments in the United States rather than with objective factors in Japan.

MacArthur relished crusades that appealed to a broad political spectrum of Americans. Hence, he pursued with fervor the trials in Manila and Tokyo of notorious generals (especially those who opposed him successfully) and wartime leaders long after the Truman administration judged the tribunals an embarrassment. Most Japanese evinced little interest in the trials, which were often characterized as "victor's justice."

Despite his determination to punish selected offenders, MacArthur cooperated secretly with the decision by army and State Department officials to protect Japanese military doctors who conducted gruesome medical experiments on Chinese civilians and Allied (including American) prisoners of war. In return for information on chemical and germ warfare, SCAP blocked prosecution of the admitted perpetrators and dismissed Soviet protests as mere propaganda.[23]

The American and Japanese public alike responded enthusiastically to land reform and the abolition of the Japanese armed forces. At times, the supreme commander's enthusiasm for labor reform and trust-busting actually surpassed the changes called for in the postsurrender program. As became clear, the degree and timing of this enthusiasm reflected, in part, a political calculation. As occupation commander, MacArthur developed a marvelous ability to finesse thorny ideological problems. The general often spoke as a populist, championing labor rights and condemning monopolies. However, he also made certain that his staff and the Japanese bureaucracy tempered any overly radical program. As a result, MacArthur could (and did) point to a record of both promoting change and preserving tradition.

SCAP's handling of organized labor and Japanese monopolies demonstrated this pattern. Initially, MacArthur's support for union rights won him high praise in international labor circles. But when union militancy threatened the power of the conservative political and economic elite, SCAP intervened forcefully on the side of capital.

Decrees by SCAP and Diet legislation permitted labor organizers nearly free reign to build a following among industrial and government workers beginning early in 1946. The collapse of the economy, high unemployment, and hyperinflation drove workers into the ranks of organized labor. When neither the series of weak, conservative cabinets in 1945–46 nor SCAP proposed a solution to the crisis, the economy practically collapsed. Within a year of surrender the inflation rate approached 1000 percent, production shrank to a fraction of capacity, and wages received by workers fell far below minimum living requirements.

Retaliating against what they saw as an industrialists' plot and SCAP indifference, unions adopted militant tactics. Many workers went on strike while others took control of factories and ran them without the owners. In the spring of 1946, unions led mass demonstrations in Tokyo and other large cities demanding more food, higher wages, and the resignations of first the Shidehara and then the Yoshida cabinets.

MacArthur's blanket support of workers' rights faded quickly when labor demonstrators tried to flex political muscle. In May 1946, he denounced demonstrations against Yoshida's efforts to form a new government. The general warned "against mob disorder or violence," and decried a "growing tendency towards mass violence" and intimidation under organized leadership." Union actions, he declared, threatened "orderly government" and the goals of the occupation. Unless the Japanese government took steps to bring "disorderly minorities" and "minor elements" under control, SCAP would intervene to "control and remedy such a deplorable situation."

Labor and Socialist leaders were stunned when MacArthur compared them to the deposed "feudalistic and military" elements. Conservatives, on the other hand, interpreted the general's remarks as an endorsement of their efforts to retain political and economic power. MacArthur provided a vital prop for conservative power faced by its first real challenge.

A few months later, the supreme commander condemned most "strikes, walkouts, or other work stoppages as inimical to the objectives of the occupation." Privately, SCAP personnel threatened to arrest union leaders. On the anniversary of Japan's surrender, September, 2, 1946, MacArthur warned of the "dread uncertainty arising from impinging ideologies which now stir mankind." He boasted of providing the "spiritual revolution," which filled the "complete vacuum" of defeat and called on the Japanese to reject the "slanted propaganda" of the "extreme radical left."

When workers ignored him and scheduled a general strike for February 1, 1947, MacArthur banned the action. He accused unions of wielding a "deadly social weapon," which risked chaos in order to further the purposes of a "small minority." The next year MacArthur accused public sector unions of "subverting the public interest" and decreed an end of their right to strike or even to bargain collectively.[24]

Japan proved the ideal setting for Douglas MacArthur to exercise his talents and nurture his resentments. Until mid-1947, he functioned with minimal supervision by the hated bureaucrats in Washington. He relished the international status of his rank, while effectively blocking token allied participation. Even his complaints about being victimized by a "Europe-centered" administration must be taken with a grain of salt. He required distance from Washington, independence of action, and the opportunity to monopolize and slant the news.

At various times between 1945 and 1950 MacArthur advocated both preserving and dissolving industrial combines, protecting and suppressing organized labor, banning and building permanent American military bases in Japan, cooperating with and excluding the Soviets from a peace treaty, and ignoring and hiding behind the Allied consulting organs created in 1945. The policies advocated by the Truman Administration also underwent a metamorphoses during these years. However, the reasons for the changes differed markedly.

Officials in Washington responded to events such as the Communist revolution in China, Soviet-American rivalry, the growing cost of supporting and protecting Japan and the idea that the containment of Soviet influence required the reconstruction of Germany and Japan. There is little in the record to suggest that MacArthur's ideas evolved in response to such weighty international developments.

The general set his compass more to the poles of political ambition and personal frustration than to any ideological or geopolitical vision. The speed with which he struck new poses alternately confused and outraged his many critics in Washington. The occupation represented an unprecedented chance to prove that he combined administrative and political talent with military heroism. By rebuilding the vanquished foe, MacArthur would demonstrate his ability to lead America.

The prism of American politics continually refracted the general's actions. As one officer told journalist Mark Gayn, MacArthur knew that, at the age of sixty-six, the upcoming presidential race was "the last election he can enter or influence. We all think he will make a try at one or the other."[25]

10
Occupied Japan and American Politics, 1945-49

In Japan Douglas MacArthur combined dedication to work with a restless drive to achieve wider recognition. During the Pacific War a British colleague observed his tendency to confuse "his emotions and ambitions for principle." Now, along with age (he turned 65 in 1945), came the realization that his military career would soon end and a single opportunity (1948) existed to seek political office. If he hoped to become president, SCAP would have to serve as his catapult.

Within days of his arrival in Tokyo, MacArthur threw down a political gauntlet. On September 17 he issued a statement declaring himself the victor in the most significant military and political contest "in history." The general praised his own decision to permit the Japanese to supervise their own demobilization and lavished compliments on the talent, goodwill, and honor of the defeated enemy. Because of this cooperation, he predicted a "drastic cut" in the size of occupation forces, down to about 200,000 men. He would send the boys home as "rapidly as ships can be made available," perhaps by Christmas. MacArthur spoke of completing the occupation within "three years." Eighth Army commander General Robert Eichelberger boasted that the whole operation might "not last longer than a year."[1]

MacArthur's alleged coddling of Japan and talk of "sending the boys home" evoked howls of protest in foreign capitals as well as in Washington. Senator Richard Russell (D–Georgia) rose to denounce a soft peace and called for resuming atomic and incendiary raids upon Tokyo until the Japanese were "brought grovelling to their knees." Although less dramatic in their response, War Department officials feared MacArthur's remarks would unhinge the effort to stabilize force levels and even contradicted his private request for 400,000 occupation troops. Most military planners considered the general's statements a sop to homesick soldiers and their families anxious to put the Pacific War behind them. They worried, however, that public calls for demobilization would sabotage the president's appeal to a reluctant Congress to extend the draft.

Before speaking, MacArthur had considered carefully the impact of his statement, relying on the advice of two political supporters, Robert E. Wood and Philip LaFollette. Wood had told him of a plot hatched by liberals anxious to ruin MacArthur by blaming him for an extended occupation. The administration wanted to maintain a large conscript army and its call for an extensive reform program really was a "very clever ruse on the part of politicians" who planned to put "the burden of the blame on Mac-Arthur." Wood urged that he announce minimal needs and block the attempt to "pillor[y]" him "in the eyes of the public."

LaFollette, a MacArthur booster since serving in SWPA, had already begun organizing a political movement on the general's behalf in Wisconsin. He, too, wrote his friend about the storm cloud "on the horizon that touches you." Even though the administration intended to use troops in Europe, Truman schemed to "shift onto your shoulders responsibility in the public mind for retaining men in service." By pretending that "MacArthur's needs" determined the pace of demobilization, Democrats hoped to take the "heat" off themselves by "transferring it to you."[2]

These communications shaped MacArthur's pronouncements and also convinced him to reject invitations from President Truman and General Marshall to return to the United States. When they asked that he attend a national victory celebration for ranking commanders, MacArthur declined, citing the "extraordinarily dangerous situation in Japan." Also, he wanted to avoid "complications" should congressional committees seek his testimony on any "controversial issues," such as the draft.[3]

MacArthur still feared collusion by Marshall and the "navy crowd." The new president, he claimed, wanted to appoint Admiral Nimitz as SCAP until someone warned him that slighting MacArthur would cost "5,000,000 votes" in 1948. Truman, he complained, had started down the same path as FDR, "trying to fool the people" by casting him as a villain. The general suspected his enemies hoped to lure him home with the promise of honors and then force his retirement. MacArthur told Eichelberger that he almost gave in. But now that the "President, the State Department and Marshall have all beeen attacking me" and since the "communists booed me" (the Russians, like the British and Chinese, protested unilateral American control of Japan), he would stay on. Soviet denunciations of his policies might even have compelled Washington to keep him in place. Thanks to the Bolsheviks, he chortled, "I am on top. I would like to pin a medal on their-ass." Through his secretary, Bonner Fellers, MacArthur informed Robert Wood that the call for rapid troop reductions was a preemptive strike against the "State, Navy and War departments."[4]

Embarrassed by MacArthur's public remarks, the president, Marshall, and Dean Acheson contradicted his assertions and instructed him to avoid public speculation on policy. Privately, Truman characterized the episode as a purely "political statement" designed to boost MacArtur's following.

The president fumed he was "tired of fooling around" and would soon "do something with that fellow [MacArthur] who had balled things up."[5]

General Marshall (about to retire) and Dean Acheson (whom Truman nominated as under secretary of state on September 19) dispatched Assistant Secretary of War John J. McCloy on a troubleshooting mission to "muzzle the talkative Generals." He would first visit Frankfurt to confer with Eisenhower over the removal of General George S. Patton for his pro-Nazi remarks. "However," McCloy told Harold Ickes, "the real objective was MacArthur in Tokyo." The interior secretary immediately told Dean Acheson "how much I thought of him for setting Gen. MacArthur back on his heels." For Acheson, the incident proved the first of many with MacArthur.[6]

Neither McCloy's visit nor the administration's rebuke had much impact on the supreme commander's efforts to recast the occupation's reform program more to his liking. For example, MacArthur resisted pressure to begin a reparations and antimonopoly program, both of which were called for in the initial surrender plans. Uneasy with such "radical" ideas, he preferred making a compromise deal with Japanese leaders. Immediately after the surrender, Japanese political and business officials urged that SCAP soften proposed economic reforms. In October, *zaibatsu* representatives met with Colonel Ray C. Kramer, then SCAP economic chief, to discuss a plan devised by the Yasuda family and later endorsed by the Mitsui, Mitsubishi, and Sumitomo clans as well. It called for a voluntary dissolution of holding companies and the resignation of family members from major combine subsidiaries. Family shares in the holding companies would be sold by a public commission with the proceeds placed in ten-year government bonds.

While reducing direct family control of the combines, the Yasuda plan made no provision for dissolving the major operating subsidiaries (the industrial heart of the operation) or for insuring a wide distribution of ownership and control. Still, MacArthur and Kramer liked the proposal.

The general defended his position on the grounds of oriental psychology and anti-communism. As he and Kramer put the case to a skeptical White House envoy, many of the reforms proposed by Washington would permit "so-called liberals," really "Soviet agents," to carry out a "communist revolution." Purges, reparations, antimonopoly measures, and the like would blow the lid off the cauldron. Better to go through the motions of reform while leaving real power in "identical hands" and encouraging Japan's reemergence as the "natural leader of Asia."

MacArthur proved as good as his word. On November 6, 1945, his headquarters issued SCAPIN 244, dissolving the family holding companies, an order that resembled the Yasuda plan. The general also undercut the reparations program by permitting the reclassification of former military facilities as industry vital for civilian production and therefore exempt from seizure.[7]

The general's sabotage of plans for fundamental economic reform stunned numerous planners in Washington. Assistant Secretary of State for Economic Affairs Will Clayton convinced the Justice Department to organize a "Special Mission on Japanese Combines" to redress MacArthur's actions. Led by Northwestern University economist Corwin Edwards, the group visited Japan early in 1946. It complemented the work of a reparations mission (created earlier under Edwin Pauley) in seeking "ways and means that would effectively destroy the power of the *zaibatsu*."[8]

The members of the two economic missions devised a comprehensive reparations and antimonopoly program more in line with the spirit of the postsurrender plan. Edwin Pauley informed MacArthur that reparations should be used to propel all "East Asia [toward] political stability and peaceful progress." Although he opposed "pauperizing" Japan, he insisted that Tokyo should not again "gain control" or "secure an advantage over her neighbors." Japan must never be allowed a "place of leadership and control" in Asia.

By mid-December, Pauley and his staff forwarded a tough report to the president. Rebuking MacArthur for stalling plans to restructure Japanese industry, it took aim at the "excess capacity" of the giant combines— essentially, heavy industrial plants, which had grown in response to war orders or depended on raw materials imported from conquered Asian states. Pauley called for an immediate interim program of stripping excess capacity followed by more comprehensive removals after war victims filed claims. His staff also traveled widely in North and Southeast Asia to study the regional development needs of claimants.[9]

The Edwards mission attacked the *zaibatsu* as among the "groups principally responsible for the war and . . . a principal factor in the Japanese war potential." Democracy and peace, Edwards insisted, required breaking up the combines. His group charged the giant monopolies with enforcing "semi-feudal relations between employer and employee, holding down wages, and blocking the development of labor unions." They stifled small business, inhibited the rise of a middle class, held down domestic consumption and gave "incentive to Japanese imperialism." Breaking the *zaibatsu* operating companies into small, competitive units would "lay the foundation for a Japanese middle class and competitive capitalism."[10]

President Truman supported these findings, ordering that the reparations program "should be implemented as soon as the details are worked out." The Joint Chiefs forwarded the Edwards antimonopoly plan to MacArthur in October 1946 as an "interim directive." MacArthur considered the program "outside meddling." Officials at SCAP praised the proposals as "admirable" but impossible to implement. The head of the Economic Section dismissed the reforms as "idealistic" as well as "too sweeping," "unworkable," "too liberal," and "unwise."

Despite the potential for a confrontation, MacArthur and the Truman administration both shelved the dispute. The president and his top political assistants, as well as the army, actually had little interest in pressing the reforms advocated by the technical experts. This indifference, which persisted during 1946 and early 1947, rendered the dispute moot. The result, however, benefitted neither the occupation administration nor the Japanese, as the unsettled agenda contributed to a downward economic spiral. Uncertain what level of industry would be permitted or what might be taken as reparations, industrialists had little incentive to shore up a deteriorating economy. Some business leaders actually contributed to economic decline, hoping that the United States would soon walk away from insurmountable problems and let the *zaibatsu* go their own way.[11]

The protracted impasse resulted in falling production, rising unemployment, soaring inflation, and widening trade deficits. The army's government and Relief in Occupied Areas Program (GARIOA) provided a vital margin of food, medicine, and fuel to prevent mass starvation. By law, however, this assistance (totaling about $400 million per year) could not be used for economic recovery. MacArthur and his top economic adviser, General Marquat, still attributed economic problems to the legacy of war. Time and emergency relief, they believed, would assure recovery. The supreme commander opposed most structural reforms fearing they would prove slow, expensive, and controversial.

In meetings with Dwight Eisenhower, Herbert Hoover, and James Forrestal during 1946, MacArthur said little about Japan's economic muddle. He spoke, instead, about oriental psychology and American politics. Eisenhower, he charged, was a "slave to the press," a poor chief of staff, and an "enemy" of West Point. Instead of wasting time on a "2000 to 1" shot mediating China's civil war (and generating publicity for a presidential campaign!), Marshall ought to send military aid to Chiang. MacArthur charged that Roosevelt had provoked the Pearl Harbor attack because a foreign war would help his domestic program. Later, FDR spurned Japanese peace feelers, extended the war, and encouraged Soviet expansion. Under Truman, "vindictive liberals were attempting to destroy" Japan's economy. By undermining Chiang Kai-shek and pushing radical nostrums on SCAP, the administration would force Tokyo "to go Communist both to get free and secure Russian protection." Truman's main interest seemed sabotaging MacArthur's effort to construct an "ideological dam in the Pacific . . . against the Asiatic tide of Communism".[12]

After visiting Tokyo, Eisenhower told Truman he "would have to face the prospect of MacArthur's returning" in 1948 "to launch a campaign for" the White House. On the other hand, Truman added wryly to an aide, MacArthur had sent him a message warning that "Eisenhower would be a candidate for the presidency!" Both generals showed signs of "Potomac fever" and "brass infection."[13]

As in 1942, Republican congressional victories in 1946 aroused hope for a GOP triumph in the coming presidential election. By early 1947, MacArthur's actions suggested his incipient candidacy. Simultaneously, the adminstration reorganized the foreign policy apparatus in Washington and proclaimed Germany and Japan as pillars of the new "containment program." This unwelcome intrusion threatened the general's conduct of the occupation and undermined his plans for using Japan as a campaign platform.

The Truman Doctrine of March 1947 (pledging military and economic assistance to governments battling internal or foreign Communist threats) and, shortly afterward, the proposed European Recovery Program established Germany and Japan as the pivots of American foreign policy. While no one questioned the depth of MacArthur's anti-communism, his idiosyncratic response to global developments frightened, then outraged the Truman administration.

MacArthur evaluated Japanese security largely in terms of conventional military threats. Since he like other American officials did not anticipate a Soviet invasion of the islands, the general downplayed the danger of Tokyo falling behind the iron curtain. But when Truman's advisers began to speak about the critical importance of restoring regional economies and world trade, MacArthur prattled on about oriental psychology, spiritual uplift, and religious conversion. As the tide of battle in China turned in the Communists' favor, most Asia specialists in Washington opposed further American involvement. MacArthur, by contrast, argued for increased assistance to the crumbling Nationalist regime. And, when the administration proposed that Congress approve an emergency recovery package for Japan, the supreme commander dismissed the notion as foolish, calling instead for an immediate peace treaty ending the occupation.

Washington Rethinks the Occupation

Before 1947, only a handful of liberal publications, such as the *Nation* and *The New Republic,* criticized MacArthur's performance. They faulted him for treating the enemy too leniently. Then, in January, *Newsweek* charged that the occupation had swerved dangerously "left." Its Tokyo correspondent, Compton Packenham, defied SCAP censorship by reporting critically about the purge of business executives. The occupation administration had destroyed the "brains of the entire Japanese economic structure" and played into the hands of the "ever watchful" Soviets. The story quoted a bewildered Japanese as asking why the Americans were "wrecking the country so as to leave it as an eventual prize for the Russians."

Outraged by these assertions, MacArthur published a rejoinder two weeks later assuming personal responsibility for the purge. He considered it "fantastic" that anyone would defend what he described as warmongers in business suits who opposed democratic capitalism. In fact, both the general and his critics overstated the case. The purge affected fewer than 500 executives, not the 30,000 claimed by *Newsweek*.

MacArthur was correct, however, in recognizing the political animus behind the story. Commerce Secretary Averell Harriman and his brother were major shareholders in *Newsweek*. The magazine's foreign editor, Harry Kern, and Tokyo correspondent, Compton Packenham, had close links to

Japanese business and political leaders stretching back to before the war. Both men detested MacArthur and later organized a lobbying group, the American Council for Japan, to challenge SCAP policy. In retaliation against his article, when Packenham tried to return to Tokyo from home leave, MacArthur barred his reentry.[14]

Navy Secretary Forrestal shared Harriman's concern about the need to alter radically occupation policy. He took the lead in bringing several cabinet members together for a discussion of how to "have a run for our side in the competition with the Soviet Union." Containing the Soviets, he argued, required putting "Japan, Germany and other affiliates of the Axis . . . back to work."

Agriculture Secretary Clinton Anderson, Secretary of War Robert Patterson, Under Secretary of State Acheson, and former President Herbert Hoover joined Harriman and Forrestal in discussing how to reverse occupation policies in Germany and Japan. (Hoover blamed New Dealers, not MacArthur for SCAP's mistakes.) All agreed that European recovery depended on the revival of German industry, and as Forrestal put it, everything said about Germany "applied with equal force to Japan." Still, the navy secretary complained, MacArthur went on "wrecking" the Japanese economy while the army seemed unable now, as in the war, to "issue a single order" to the general. Without a reversal of SCAP policy, he predicted a "complete economic collapse." When the War Department declined to offend MacArthur by dispatching new economic specialists to Tokyo, Acheson informed the White House that a "super diplomat" must be sent to "break the grip of General MacArthur, his chief of staff, and the Bataan Group out there." If the administration "saved the face of SCAP," it would "wreck its whole economic policy for the Far East."[15]

Preoccupied with domestic political concerns and the reorganization of the diplomatic and military services, Truman shied away from confronting the occupation commander. The president had recently selected George C. Marshall to replace James F. Byrnes as Secretary of State. The War and Navy departments remained so divided by interservice rivalries and budget disputes that neither service secretary had the inclination or power to challenge MacArthur. Only after James Forrestal became head of a new semiunified Defense Department in the fall did the administration coordinate its efforts. Policy uncertainties in Washington allowed MacArthur to defy his nominal superiors with relative impunity.

Meanwhile, critics within the administration feared that the worsening economic crisis in Europe and Japan threatened American security as much or more than Soviet military power. The emergency postwar relief programs had proved inadequate to the task of industrial reconstruction. By early 1947, Germany and Western Europe, like Japan, had sunk further into debt and decline. The loss of colonies and colonial rebellions in Southeast Asia, the legacy of wartime hatred, the division of Europe, and wartime destruction of industry impeded recovery. Ironically, America's vast comparative wealth made it hard for any nation to compete with it in an open market. The resulting trade imbalance—or dollar gap—so favored the United States that it threatened to swamp its trading partners. Without restored production and trade, they would run out of dollars and raw materials, forcing a turn to protectionism or Soviet aid. Most officials believed another depression would clear the path for wider Communist influence or direct Soviet expansion.

Planners such as Dean Acheson, James Forrestal, George Kennan, and Army Under Secretary William H. Draper, Jr., believed that American prosperity and military security required a massive industrial recovery program for Europe (especially western Germany) and Japan. The United States could stimulate production and trade by providing vital capital and raw materials. Later, the Europeans and Japanese could save their scarce dollars by securing raw materials developing nations and cultivating markets there as well. The integration of regional markets would create efficient economies of scale, soothe old rivalries, insure stability in less developed countries, and blunt Soviet influence.

The new secretary of state, George C. Marshall, assisted the birth of a new policy by reorganizing the State Department. He gave enhanced authority to his talented under secretary, Dean Acheson, and established a powerful Policy Planning Staff (PPS) under the direction of George Kennan. In July 1947, after prolonged debate, Congress approved the National Security Act, creating a semi-unified Defense Department, a formal Joint Chiefs of Staff, a Central Intelligence Agency, and a National Security Council to coordinate policy.

Along with Kennan, Marshall, Acheson, Forrestal, and many civilian and military specialists contributed to the evolving containment program. Collectively they stressed the need to assure the recovery of Europe and Japan. Prosperous industrial democracies, they argued, would not succumb easily to Soviet pressure. Once Russia found its expansion blocked, it would have no choice but to alter its behavior in ways favoring the West.[16]

MacArthur had no real quarrel with the ideological tenets of containment. While he did question the European focus and disinclination to assist China, his opposition centered on the fact that any comprehensive, centrally directed recovery program would impinge upon SCAP independence. Moreover, the general considered Secretary of State Marshall (whose name the recovery program soon bore) a sworn enemy. James Forrestal, formerly an ally, now appeared a turncoat. Allowing outsiders from Washington into Tokyo, under any pretext, risked accepting a Trojan horse; even the suggestion that Japan needed a special recovery program implied SCAP's lack of perfection. Finally, MacArthur feared that linking the occupation to economic recovery might complicate his plans to wrap up the exercise in advance of the 1948 election. Determined to prevent the administration from gaining a foothold in Tokyo, MacArthur pronounced the occupation complete, Japan ready for a peace treaty, and himself available to host a peace conference. As Truman begged Congress to fund Europe's recovery, the general declared "mission accomplished."

First privately, then in a public statement, MacArthur described a three-phase program, the first two parts of which (demilitarization and democratization) he had completed. The final stage, economic recovery, would not occur until the occupation ended and other countries ceased discriminating against trade with Japan. Secretary Marshall and his ilk, the supreme commander complained, "paid heed" only to "Wall Street whose main holdings were in Europe." Truman, like FDR, would scuttle the Pacific in a moment.

MacArthur dismissed the thorny issues of reparations, levels of industry, and trust-busting as mere "technical problems" that experts could tackle after a settlement. Nor did he see any difficulty in protecting Japan. Since the Soviets would honor a neutral Japan, Washington need only place a "simple article" in a treaty "providing UN responsibility for future protection." Truman and Marshall delayed a settlement to make him seem a failure. Unless Marshall struck an agreement with the Soviets for a treaty at the upcoming foreign minister's meeting in Moscow, MacArthur told a confidant, he would shatter the "Pacific air" with "cries" to "conclude a settlement with Japan." [17]

While MacArthur tried to orchestrate a quick treaty, President Truman initiated a dramatic new cold war policy. On March 12, 1947, the chief executive responded to the British withdrawal of assistance to the Greek government (then fighting a leftist insurgency) by announcing what soon became called the Truman Doctrine. He identified Moscow as the source of world disorder and pledged aid to any government fighting a Communist movement. Washington abandoned plans to seek Soviet cooperation in Germany or Japan, opting, instead, to continue these occupations until their recovery seemed assured.

MacArthur convened a rare, on-the-record press briefing on March 17 to challenge Truman's assertions. He called for a quick end to the occupation and the cessation of "economic warfare" against Tokyo. He rejected responsibility for Japan's downward economic spiral and predicted that a new recovery program would do nothing but squander American wealth.

Ignoring the dismal economic situation within Japan, MacArthur spoke of "spiritual revolution." The former enemy had atoned for its sins, owed no reparations to its wartime victims, and could depend upon its own "advanced spirituality" for protection against aggression. Paraphrasing Winston Churchill's remark about how the West's major European problem was "not to keep Germany down, but to keep it up," he took credit for solving the greatest "problem" in Asia, "keep[ing] Japan up." While Truman prepared to waste billions of dollars in Europe, SCAP had already rebuilt Japan at bargain rates. [18]

Nearly all civilian and military experts in Washington disputed Mac-

Arthur's assertions. Marshall, Forrestal, Acheson, Kennan, and others all predicted a collapse in Japan if the occupation ended before a drastic reversal of economic policy. Some suspected that even MacArthur shared this concern but hoped to get out of Tokyo before a disaster tarnished his presidential ambitions. Eighth Army commander General Robert Eichelberger, who served under MacArthur, believed that his boss's push for a quick peace treaty was his way of "placing his hat in the ring for the presidency." Understandably, few in the administration felt disposed to assist him.[19]

Speaking in Cleveland, Mississippi, on May 8, Dean Acheson unveiled the administration's response to MacArthur's challenge. The under secretary of state blamed the growing "dollar gap" trade imbalance on the "grim fact of life" that the "greatest workshops of Europe and Asia, Germany and Japan" remained idle. Since world stability depended on their prosperity, the United States would rebuild the "two workshops" upon which the "ultimate recovery of the two continents so largely depends." The occupations would continue until Germany and Japan became economically secure.[20]

Instead of following the administration's new lead, MacArthur suddenly proposed a spectacular campaign against the *zaibatsu* monopolies within Japan. As planners in Washington resolved to revive centralized, heavy industry, the general proclaimed *zaibatsu* dissolution a litmus test of the occupation. Embracing the formerly despised Edwards plan of 1946 (now called FEC 230), he called for a "controlled revolution" on behalf of genuine private enterprise. In July he ordered the Socialist cabinet of Prime Minister Katayama Tetsu to introduce a bill to break up the combines.[21]

George Kennan and army planners condemned MacArthur for practically inviting "Sovietized totalitarianism" in Japan. SCAP ignored the fact that Japan could "only gravitate into the orbit of one or the other of the superpowers." An immediate settlement and the proposed economic program, Kennan declared, did "not seem to be related to any realistic pattern" of foreign policy goals. The army's Plans and Operations Division feared SCAP's latest action would "breed communism" and require spending vast sums in Japan "to bring order out of chaos."[22]

During the autumn, Kennan and his staff drafted plans to halt or reverse many of the occupation reforms MacArthur came belatedly to embrace. The "radically changed world situation," they argued, required that Japan be made "internally stable," "amenable to American leadership" and "industrially revived" in order to assist the development of all "non-communist Asia." To prevent the possibility of any Soviet influence, Washington should undertake "unilaterally" a program to "crank-up" the Japanese economy and should bind Tokyo to the West through a bilateral defense pact.

The idea of "eliminating Japan as a military power for all time" had changed. Because of "Russia's conduct," the Policy Planning Staff urged that "Hirohito's islands" be made a "buffer state" against the Soviet Union.

While Kennan did not anticipate a Soviet attack upon Japan, he feared that Communist control of Manchuria, northern China, and Korea would create a "lever for Soviet political pressure" unless Japan obtained vital raw materials and markets elsewhere," particularly in Southeast Asia. Ultimately, Tokyo's survival as a non-Communist ally depended on the "willingness of the [American] government to prime the Japanese economic pump."[23]

This rejection of the 1945 reform agenda whipsawed Douglas MacArthur. In the early occupation period New Deal planners tried to push the conservative general to the left. Belatedly, he adopted many of their arguments. Now the Democratic cold warriors around Truman who displaced the New Dealers lambasted MacArthur's adherence to a "leftist" reform program. Caught between the deposed and ascendent factions in Washington, and trying desperately to preserve his political options, he had little room to maneuver. Even accounting for his ambition and monumental ego, his futile effort to stay on the "winning side" of the policy debate rendered him somewhat pathetic.

Criticism came from policymakers and private Americans whom be began allowing into Japan in mid-1947. *Newsweek* editor Harry Kern took advantage of the relaxed travel policy to send James Lee Kauffman to Tokyo. A prominent business attorney with many corporate clients in prewar Japan, Kauffman detested SCAP's economic agenda. In a report he circulated in Washington during the autumn (*Newsweek* printed selections in December) and in discussions with top officials, the attorney denounced FEC 230 (the *zaibatsu* dissolution proposal) as a "socialist idea" certain to destroy Japan. MacArthur seemed eager to put Japan under the "knife of the economic quack" wielded by "radical reformers" and "crackpots" who trusted Japan's "childlike" labor unions. Kauffman urged the administration to protect Japan's industry and industrialists in order to create "the buffer against Soviet Russia" so desperately needed in the Pacific.[24]

Influenced by Kauffman and other informants, Kennan told Defense Secretary Forrestal, Army Secretary Kenneth Royall, Army Under Secretary William H. Draper, and CIA Director Admiral Sidney Souers that out of ignorance or duplicity MacArthur had opened Japan to Communist influence. He claimed that SCAP pushed "socialization," attacked business, and encouraged labor radicalism. Pursuit of MacArthur's program, Kennan predicted, would cause "economic disaster, inflation . . . near anarchy which would be precisely what the communists want." Attacking the *zaibatsu* was a "vicious" scheme to destroy the major barrier to Soviet penetration.

William H. Draper agreed that SCAP had turned Japan into an economic "morgue." The general's entire program, Kenneth Royall charged, "really amounted to socialism, pure and simple, if not near communism." James Forrestal took this argument directly to Truman and the cabinet in November. All occupation policies, he demanded, should focus on enhancing the industrial strength of "both Germany and Japan." The survival of the free world required rebuilding the "two nations we have just destroyed."[25]

MacArthur responded to criticism by rushing to enact the program under attack. He pressed the Japanese government to legislate the dissolution program immediately, threatening the prime minister with retribution if his cabinet or the Diet delayed the anti-*zaibatsu* bill. The general did not care that both the left and right in Japan opposed his plan. (The former preferred nationalization to splitting up the combines while the right and business groups feared that small companies could not compete and would be swallowed up by American corporations.)

When Prime Minister Katayama appealed to MacArthur for a delay, the general not only refused but warned him to cease such "insulting behavior." Unless the cabinet and Diet acted quickly, he warned through a spokesman, SCAP "would not be so kind in the future." When the upper chamber of the Diet almost adjourned without passing the bill, one of MacArthur's deputies informed Katayama that the law must "be passed so as not to embarrass" the supreme commander, who "expected to be nominated for President." There must be "no sign to the world of dissension" in Tokyo. If the Japanese did anything to tarnish MacArthur's image, it would "prejudice the future of Japan when the Supreme Commander became President." The SCAP officer stated that MacArthur did not really care how vigorously Tokyo enforced the law, only that it approve the measure immediately as a symbol of reform.

The occupation chief had no qualms about working both sides of the street. In fact, he tried hard to dispel the fears of American conservatives shocked by the implications of the dissolution program. The general told key Republican party officials visiting Tokyo that "he had nothing to do with the reforms." This leftist mischief was the work of men under him sent by the "Washington Democratic Reds."[26]

MacArthur and the 1948 Election

MacArthur's abortive effort to orchestrate a peace conference in Japan at the end of 1947, like the crusade against the *zaibatsu*, reflected his quest for the Republican presidential nomination. He guessed the public would adore the first treaty with a wartime enemy and would applaud his blow against Japanese monopolies. What better proof could he offer of his enlightened conservatism? In September 1947, as Robert Eichelberger prepared to visit the United States, MacArthur discussed with him possible rivals and sources of support.

By now, Eisenhower had replaced George C. Marshall as MacArthur's chief nemesis. He described his former aide and the current Chief of Staff as a "good time Charlie and a New Dealer." The "Ike for President" movement, MacArthur asserted, was "directed against me and will backfire." On the other hand, he thought a deal might be struck whereby Ike would "come out and say he is not interested in 1948 but might be in 1952 . . . then say I am for my old chief, MacArthur." If Eisenhower agreed, MacArthur would pledge to serve but a single term, opening 1952 for Ike. If his rival ran this time, the occupation commander suggested bringing to light a past scandal. (Later, he told Eichelberger he meant Ike's rumored wartime affair with Kay Sommersby. He charged that Eisenhower was writing his memoirs to raise cash to "go as blackmail . . . to keep her from writing a book" about their involvement. Perhaps Isabel remained on his mind.)

MacArthur admitted his own reluctance to return home to seek nomination. Enemies would accuse him of deserting his post and call him a "politician." He might even be "heckled by the Communists." As in 1944, a stalemate between the two front runners (Dewey and Taft) would provide his best opportunity for the nomination. The general told Eichelberger he would not leave Tokyo until nominated. Once selected, he would go at once to America and, with assurances of "protection by the Republican papers and organization," would campaign from "coast to coast."[27]

MacArthur's leading political backer, Robert E. Wood, thought the general could bridge the isolationist and internationalist camps, represented by Taft on one side and Dewey, Eisenhower, and Stassen on the other. He opposed deeper involvement in Europe while advocating a greater commitment to Asia. Many midwestern Republicans found this an attractive combination. Also, MacArthur's reform record in Japan impressed many liberals. If the voters who admired the man simply as a military hero were combined with these groups, a viable coalition might emerge.

Wood and Hanford MacNider, a SWPA veteran and Iowa businessman influential in the American Legion, urged the general to return home early in 1948 to launch a campaign. Again, he declined. Only when chosen as the Republican nominee, he declared, would the call of the people leave him "no alternative but to consider it a mandate" to return. An earlier departure would make it seem he was "indulg[ing] in the cheapest form of theatricals." MacArthur also admitted that, after a lifetime in the army, he did not relish resigning his commission with no certainty of political victory or subsequent employment. As he told Eichelberger, while remaining "out here . . . I have a certain mysticism." The moment he landed in America, he would become just an ordinary "political target."[28]

The general designated a staff officer, Colonel Harold Eastwood, to coordinate plans with his stateside boosters. Although MacArthur refused to return (despite MacNider's complaints about having trouble "selling the product" from abroad), he permitted his supporters to enter a MacArthur slate in the upcoming Wisconsin primary. He claimed, of course, to be the state's "native son," spoke of retiring there, and hoped his Asia First and antimonopoly strategy would appeal to the curious mix of isolationist and progressive strains in Wisconsin politics. Personnel at SCAP, certain of their chief's winning, rushed to claim staff positions in a MacArthur White House.[29]

Meanwhile, in October, the State and Defense departments reached agreement on a recovery package for submission to Congress. The Economic Recovery in Occupied Areas program (EROA), which resembled the European Recovery Program (ERP) or Marshall Plan, earmarked about $180 million in additional assistance for the period April 1948 to June 1949. But the administration feared that the budget-cutting, Republican-controlled Congress would balk at appropriating money while MacArthur expanded the purge, dissolved major industries, disputed the need for recovery aid, and called for ending the occupation.

When Army Under Secretary Draper appealed for MacArthur to delay Diet consideration of the SCAP-sponsored deconcentration bill, the general repeated his demand that the Diet approve the law promptly. Enraged by the charge that he had opened Japan to communism, the occupation commander described the zaibatsu as a form of "socialism in private hands." While he led a struggle to build "a system of competitive enterprise" the administration seemed to favor a "Socialism of one kind or another."

After the lower house of the Diet passed the antimonopoly law, Army Secretary Royall accused MacArthur of promoting a redistribution scheme that went "beyond American philosophies" and threatened the basic "concept of private property." He demanded that MacArthur pull the legislation from the upper house or, if this proved impossible, block its implementation until Washington appointed a review board to oversee its application. Secretary of Defense Forrestal fumed that no one in Washington could make "MacArthur realize what [is] involved." The whole economic program had been "scrambled" terribly.

When Royall and Forrestal attempted to send a direct order that MacArthur withdraw the "near communist" legislation, a Pentagon officer balked, referring to an informal policy against "issuing an order to General MacArthur." Royall, joined by Eisenhower, brought this astounding claim directly to the president. Truman insisted that an order go to Tokyo and that "if the so-and-so hesitates one minute to veto that bill, I'll bust him to a corporal." Later, Royall claimed mistakenly that he and the president sent the message and forced MacArthur to back down. In fact, they dispatched no direct order. However, in April 1951, when the president relieved MacArthur from command, he and his staff referred to this incident as an early example of MacArthur's defiance of civilian authority.[30]

Army Under Secretary William Draper encouraged Newsweek to attack SCAP in its December 1 issue by printing part of a critical report written months before by lawyer James Lee Kauffman. The magazine accused MacArthur of promoting reforms "far to the left of anything tolerated in America" and of embracing the "lethal weapons" of socialism. Draper then provided conservative Republicans, like Senator William R. Knowland (R–California, who was told that Japan might receive credits to buy California cotton), with critical information about the deconcentration program. Speaking in the Senate, Knowland condemned the antimonopoly plan as the kind of attack on property one would expect from the "government of the U.S.S.R. or even by the Labour government of Britain." He threatened to investigate those responsible for promoting the "most socialist" ideas ever attempted outside Russia. MacArthur responded by steering the deconcentration bill to final Diet passage.[31]

Perhaps to blunt the attack upon him, the general spread unflattering tales about his Democratic and Republican rivals. He told British diplomat Sir Alvary Gascoigne that Truman and Marshall were cowards who had lost China to communism. He dismissed Thomas E. Dewey and Robert A. Taft as, respectively, "very shopworn" and too "provincial." When the general predicted that a "dark horse" would sweep the Republican convention, Gascoigne asked if he meant Eisenhower. Outraged, MacArthur ridiculed Ike's character, politics, and heritage. Eisenhower should not be counted as presidential timber or even a "good Republican" as he carried "Jewish blood in his veins." (Attacking a rival as a Jew was a frequent MacArthur tactic.) But, if, by chance, supporters put MacArthur's own name forward, "he as a good citizen would accept the nomination."[32]

When American conservatives like Robert E. Wood asked MacArthur to explain his apparent conversion to economic liberalism, the general defended himself vigorously. He maintained that the dissolution program targeted only "56 families . . . which controlled about 80% of Japan's prewar commerce and industry and raw materials." These families were linked to the military's program of "world conquest." (This move against a few families was part of the SCAP plan of 1945, but *not* of the broader 1947 program.) The Japanese people favored dissolution so intensely that without reform, a "bloody revolution" would follow the occupation. Preserving the "existing pyramid" would only provide Communists a propaganda coup and a marvelous target for conquest.[33]

Army Secretary Royall took this battle public in a January 1948 speech declaring *zaibatsu* power already "virtually abolished." The time had come to rebuild, not destroy, industry. "Extreme deconcentration" would prevent Japan from becoming a "self-supporting nation" and leave it vulnerable to aggression. Even "the men who were most active in building up and running Japan's war machine," Royall declared, must be pardoned and encouraged to assist with recovery.

MacArthur responded by accusing the administration of defending a "few Japanese families" who enjoyed direct or indirect control "over all commerce and industry, all raw materials." Why should Americans protect those who inflicted "economic exploitation at home and aggression and spoilation abroad." His reforms would prevent "a bloodbath of revolutionary violence." In a further sign of defiance, the general established a Holding Company Liquidation Commission (HCLC), which promptly designated over three hundred firms, possessing more than half of the nation's industrial and commercial strength, as "excessive concentrations of economic power" and targets for dissolution.[34]

Despite these provocations, Truman feared moving against the occupation commander before the spring primaries. Early in 1948 he told Kenneth Royall that he "didn't think . . . he could beat MacArthur" but, he thought, "Eisenhower could do so." Ike confirmed that Truman (with Royall the apparent intermediary) "offered him the opportunity to run on the Democratic ticket" and promised to "step down and run for vice-president." Even though Eisenhower soon took himself out of consideration, the incident reveals the connection between occupation policies and American politics.[35]

Truman probably worried that if he imposed a new program in Japan, MacArthur's supporters (and most Republicans) would accuse him of denigrating the general's accomplishments out of political jealousy. Moreover, the public thought MacArthur served as occupation commander because of his wartime accomplishments and unsurpassed knowledge of the Orient. What, besides personal rivalry, could explain the president's change of heart? Reluctant to make MacArthur a martyr and enhance his candidacy, the president delayed any change in Japan policy.

In the interim, Marshall and Forrestal sent two of MacArthur's leading critics on a mission to Tokyo. During March, Policy Planning Staff director George Kennan and Army Under Secretary William H. Draper visited Japan in a final effort to win cooperation from the occupation commander. If, as expected, he remained inflexible, they would gather additional evidence needed to impose a "switch in emphasis."

The two emissaries traveled separately, both anticipating that MacArthur would receive them as supplicants from a vassal state. For his part, the general believed they would try to embarrass him on the eve of the Republican primaries. He "spoke most bitterly" to British diplomat Sir Alvary Gascoigne about "American tycoons" (such as Forrestal, Harriman, Royall, and Draper) who feared his political following in America and opposed his economic reforms "because they thought they would conflict with their own business interests."

Although the general had always disparaged his status as *Allied* occupation commander, he found new virtue in the title. If Washington ordered him to do things he opposed, he would simply switch hats. Describing himself as an "international official," MacArthur swore that he would not permit Washington to command him about as if he were a "purely American official."[36]

Early in March, as soon as Kennan stepped off the plane after a grueling transpacific flight, MacArthur summoned him for a ritual humiliation. During a late night meeting MacArthur turned his back to Kennan and delivered a two-hour lecture to the diplomat's traveling companion, Major General C. V. R. Schuyler. "Thumping the table for emphasis with a single vertically held finger," the occupation commander delivered a monologue on the history of military occupations since Caesar and his own crusade to bring democracy and Christ to Asia. He then dismissed the meddlesome visitors.

Refusing to be baited (George Marshall warned him at length of how to avoid provoking the general), Kennan secured a second audience. Despite their mutual suspicions, Kennan and MacArthur actually shared some common strategic views. Like Kennan, the general favored a demilitarized Japan in a Pacific Ocean dominated by American air and naval bases. From such islands as the Aleutians, the former Japanese Mandates, the Philippines, and Okinawa, the United States could control "every one of the ports of northern Asia from which an amphibious operation could conceivably be launched." Kennan's opposition to taking permanent bases in Japan (assuming economic stability), pleased MacArthur. Yet, he vigorously objected to the envoy's advocacy of a new occupation agenda.

MacArthur defended everything he had done in Tokyo and objected to Washington's efforts to incorporate Japan into a regional and global containment program. To prosper, Japan needed nothing more than the restoration of sovereignty and a UN security guarantee. He maintained that the business purge affected only "elderly incompetents" who resembled "the most effete New York club men." Similarly, the deconcentration program was not so "extreme" as many feared. Any taint of socialism came via the influence of "academic theorizers of a left-wing variety" forced on him by Washington. The State Department, he charged, had its own share of leftists and should not cast stones at SCAP. The general assumed, naively, that his harangue had altered Kennan's outlook. He then instructed SCAP staff to facilitate the diplomat's travels in Japan.[37]

In fact, MacArthur's posturing, geopolitical nostrums, and blathering about converting a billion Asians to Christianity repelled the diplomat. The "stuffiness" and "degree of internal intrigue" around MacArthur, he wrote to a colleague, resembled "nothing more than the latter days of the court of the Empress Catherine II, or possibly the final stages of the regime of Belisarius in Italy." The "fragile psychic quality" in the general's headquarters resembled the Kremlin under Joseph Stalin. SCAP functioned like a "parasite," feasting on Japan's vitals to insure its own luxury.

MacArthur's underlings and their "shrill cackling" wives, Kennan opined, behaved as if the war had been fought so they might have "six Japanese butlers with the divisional insignia on their jackets." While "many of the Japs deserve a worse fate than to have the tastes and habits of American suburbia imposed upon them," SCAP's program was simply not "intelligible to most Japanese." Liberal social engineering, Kennan predicted, would be rejected like an infection once the occupation ended. Unless immediate steps were taken to resurrect conservative forces, communism would fill the void as Americans departed.[38]

Before leaving Tokyo, Kennan began drafting a comprehensive attack on MacArthur's program which, by year's end, became national policy. The report condemned SCAP for sewing the seeds of a disaster. Occupation costs consumed nearly a third of Japan's budget while reforms endangered the surviving economy. With industry crippled by purges and threats of dissolution, Japan teetered on the edge of an abyss. Given these problems, Kennan condemned talk of an early end to the occupation and insisted that SCAP must cease purges, end the deconcentration program, clamp down on labor unions, and bolster conservative authority. Nothing should be permitted that "operated against the stability of Japanese society" or impeded the "prime objective" of industrial recovery.

William Draper's views parallelled those of his State Department ally. He brought to Tokyo a group of prominent business executives prepared to spearhead an attack upon the "radical policies" of SCAP that, he felt, had turned Japan into an "economic morgue." The delegation, nominally led by Chemical Bank chairman Percy H. Johnston, conferred with many *zaibatsu* representatives whose operations were slated for reorganization. Even though the group's formal report was not released until late April, the delegation told journalists they favored a "businessman's solution to Japan's economic ills." This meant scrapping most of the current program, providing half a billion dollars in new aid, and assisting the former enemy to again become the "workshop of the Far East." The "bad times were over," Draper reportedly told Japanese business leaders.[39]

Political Defeat

The Wisconsin primary held on April 6 would settle the direction of occupation policy by signaling MacArthur's presidential prospects. Since none of the other Republican hopefuls expressed much sympathy with his latest agenda for Japan, only his own victory could protect SCAP. Few midwestern political races had so great a potential impact on American–East Asian relations.

As 1948 began, the political omens seemed mixed. Eisenhower threw a monkey wrench into the MacArthur campaign as early as January 22 when he withdrew from the race, declaring that, except in extreme circumstances, military figures should not seek civilian office. Although Ike mentioned no names, a cartoon in the *Chicago Sun Times* captured his meaning. It showed a boot labeled "Ike's Message" trampling on MacArthur's distinctive cap. The occupation chief "exploded" to General Marquat that "Eisenhower had really aimed the statement at him."

The occupation commander's refusal to declare himself a candidate worried his supporters. On March 9, at their insistence, he finally announced his willingness to accept the Republican nomination. With "due humility" he revealed his desire to save the country in its "hour of momentous import, national and international, spiritual and temporal."

On the surface, the general's prospects in Wisconsin looked good. His campaign organization played up his professed "native son" status as well as his reform record in Japan. Philip LaFollette stressed his support for unions and crusade against monopoly. Simultaneously, the conservative Hearst, McCormick, and Gannett newspaper chains emphasized his fierce opposition to Asian communism, reluctance to commit money or troops to Europe, and personal embodiment of "traditional values." The ailing William Randolph Hearst reportedly ordered his editors to print a picture of the general "on page one in the daily newspapers once a week and every Sunday as well." It was especially important to show "flattering pictures" of his "small son," presumably to refute voter concern about his age. Hearst's paper in Milwaukee, the *Sentinel*, endorsed MacArthur as the one prominent American who had responded to "the challenge of Red Fascism with courage and success. His destined place is no longer in Japan . . . but in Washington as President."

Despite these appeals, the general had grave problems. Wisconsin's Republican party was riven by intense factionalism, involving hostility between LaFollette Progressives (who had dissolved their own party) and the traditional Republicans who resented the efforts of the old Progressives to merge with, or take over, the Republican organization. Many of MacArthur's local supporters joined his campaign solely because they hoped his high profile would facilitate their election as convention delegates, an important step in capturing the state apparatus. Entrenched leaders resented the general's intrusion into local politics.

The predominant Republican faction and most state officeholders supported either Harold Stassen or Thomas Dewey, as their delegate slates posed no challenge to the status quo. For example, Wisconsin's junior senator, Joe McCarthy (later one of the general's most voluble supporters) used his considerable gutter skills against MacArthur.

On the eve of the primary, the senator challenged the general's assertion of native son status by putting a special spin on the truth. In a "Dear Folks" letter, Joe McCarthy informed his constituents that:

> General MacArthur has been a great General. But he is now ready for retirement. He would be 72 years old before a term as President ended. Twice before we have had Presidents who became physically weakened during their term of office and both times it had very sad results for our country. . . . The General was born in Little Rock, Arkansas, on January 26, 1880, and not in Wisconsin. . . . Neither his first nor his second marriage, nor his divorce, took place in Wisconsin. He was first married in Florida to Mrs. Walter Brooks of Baltimore. . . . After she divorced him in Reno, Nevada, he was remarried in New York City. Neither wife ever resided in Wisconsin.

As usual, the senator left little to the imagination.

Although even the sedate *New York Times* predicted a MacArthur victory, on April 6, Harold Stassen (of nearby Minnesota) won nineteen convention delegates, MacArthur eight, and Dewey none at all. The unimpressive result under supposedly ideal conditions shook the general's supporters and virtually halted his campaign. After an equally weak showing in Nebraska the small organization dissolved. (MacArthur won a nonbinding race in Illinois against token opposition.) The general urged his main backer, Robert E. Wood, to maintain the effort by "developing and maintaining a friendly beachhead" among each state delegation in case of a convention deadlock. Despite this martial spunk, there would be no return from this defeat. At the Republican convention in June, MacArthur received only eight first-ballot votes, all of which soon switched to the nominee, Thomas Dewey.[40]

Douglas MacArthur, in the words of one observer in Tokyo, seemed "low as a rug" when he learned of the Wisconsin debacle. The same public that adored him as a war hero and national symbol rejected him as a flesh-and-blood politician. In 1952, a Republican party veteran summed up in a telling anecdote his opinion of the general's abortive presidential campaigns. MacArthur's candidacy reminded him of the "guy who walks up to you with a new suit on. [He asks you if you admire it.] You say 'its beautiful.' Then he says, 'you want to buy it?' Now that's a different story."[41]

The political verdict heralded the end of MacArthur's independence in Japan. Late in April, Draper and his business committee recommended that nearly all of SCAP's economic agenda ought to be amended. Reparations, deconcentration, and support for unions should be replaced by a program of redeveloping industry and boosting exports. In May, Draper dispatched the Deconcentration Review Board to Tokyo to insure Mac-Arthur did not "interfere with the economic recovery of Japan." The board overturned or softened most of the 325 orders to restructure monopolies. Draper also steered the EROA recovery bill through Congress during May and June. It appropriated several hundred million dollars in grants and commodity credits for the rehabilitation of Japanese industry.[42]

George Kennan lost no time in shepherding a revised strategic agenda through the national security bureaucracy. On June 2, he submitted a comprehensive policy plan (NSC 13) to the National Security Council. Proclaiming recovery as a "prime objective," it favored economic growth as the surest way to achieve political and military security. Kennan's plan would curb reforms and transfer authority to the Japanese government.

In October, the National Security Council and president approved a slightly amended NSC 13. The State and Defense departments agreed to terminate reparations, eliminate most restrictions on Japanese industry, and promote export-oriented production. The departments drafted a directive ordering SCAP and the Japanese government to allocate raw materials in ways that promoted export production over domestic consumption, limited social welfare spending, and compelled workers to accept a reduced living standard. Other provisions called for balancing the internal budget, limiting government credits to nonessential industry, imposing wage controls, and making Japan the economic hub of a revived Asian trading network.

Following his election (in which the Democrats also regained control of Congress), Truman committed his prestige and authority to the new program. On December 10, 1948, he issued a nine-point economic directive to SCAP and appointed a special emissary, banker Joseph Dodge, to implement the program. Like Kennan and Draper, Dodge previously played an instrumental role in devising the German "reverse course" and now enjoyed the administration's full support in Japan.[43]

Although MacArthur denounced these changes and disputed the right of the American government to impose policies, his actions belied his tone. From mid-1948 through early 1950 he often criticized orders or delayed their implementation, but ultimately submitted. The populist rhetoric of the months before the primaries gave way to vintage MacArthur. Hardly missing a beat, SCAP now promoted a far more conservative economic and social agenda. The occupation administration turned its powers against left wing labor and political groups and made a public alliance with Japanese conservatives.

MacArthur told Eichelberger he had finally resigned himself to an "indefinite" tenure in Japan. He would deny his enemies the pleasure of his resignation and planned to remain on the job and on the army's "active list forever—until I die." When the Eighth Army commander announced his own intention of retiring, MacArthur told him (falsely) that George Marshall and Dwight Eisenhower had blocked his promotion. Eichelberger, who knew this was not true, wrote in his diary that MacArthur remained a "strange character who probably wonders why he has so few friends and eternally blames the other fellow."

Hosted to a round of social events in Washington, the retiring general quickly discovered how low MacArthur's star had sunk. Marshall, Eisenhower, and General Omar Bradley urged him to write an exposé of the occupation, assuring Eichelberger that it would make him the most sought-after job candidate in Washington. Eisenhower described MacArthur as a "man of no character" who had "spent a life of hate and envy" directed at those closest to him. Ike felt that MacArthur "hated him all the nine years he served with him" since he "envied any quality a man could have which he didn't possess."[44]

After April, even low-ranking SCAP veterans found themselves tainted when they returned to Washington. Technical experts seeking jobs with Marshall Plan agencies and other assistance programs found the door closed once they revealed past ties with SCAP. One frustrated veteran informed his friends remaining in Tokyo that both Democratic and Republican officials labeled him one of the group of "long haired boys . . . who have helped General MacArthur put over his socialistic schemes." Republican candidate Dewey, as much as Truman, was "plenty sore at the Big Chief" and intended to take revenge after November.[45]

Ironically, MacArthur's sharp turn to the right in 1948 reduced the policy differences between himself and the Truman administration. Members of the British liaison mission in Japan reported the beginning of an "almost hysterical anti-communist witch hunt." The Japanese press was encouraged by SCAP to publish lurid accounts of underground Communist threats. Warnings of terror on the left replaced completely talk of militarism and fascism. Under pressure from MacArthur's intelligence chief, General Charles Willoughby, SCAP staff and Japanese officials were ordered with "teutonic rigidity" to "look left" for danger.

A British diplomat in Tokyo (with the unlikely name of Ivan Pink) noted how Willoughby's Counter Intelligence Corps cultivated right-wing informants who worked to enhance the power of Japanese conservatives by feeding the Americans an abundance of scare stories. The CIC interpreted even peaceful union activity as subversive "communist agitation."[46]

With MacArthur's approval, Willoughby slipped his chains completely after 1948. He began a search for conspiracies among Japanese leftists, suspected Comintern agents, and American liberals. The intelligence chief led an investigation into the remarkable Soviet spy Richard Sorge who, posing as a Nazi sympathizer in Tokyo, had furnished vital information to Moscow—much of which Stalin ignored. (Sorge was arrested in Japan in 1941 and executed in 1944.) Ignoring the fact that Sorge's target was the Axis enemy, Willoughby labored to prove an ongoing conspiracy which connected the Soviet spy to leftist American journalists like Edgar Snow and Agnes Smedley, who had criticized MacArthur and Chiang Kai-shek. Convinced that the old Sorge spy ring was responsible for revolutionary victories in Asia, Willoughby fed information to the House Committee on Un-American Activities during the late 1940s and early 1950s. In 1948, MacArthur permitted Willoughby to purge from SCAP ranks a number of New Deal liberals whom they suspected of Communist sympathies or of being Jewish, often synonymous in their opinion.[47]

As the occupation drifted to the right, unions composed of government workers in such public enterprises as the railroads, communications, and education played a leading role in strikes and demonstrations protesting the new economic policies. MacArthur responded during 1948 by stripping all government employees of their right to strike or bargain collectively. He condemned unions for impeding the "efficiency of government" and trying to "subvert the public interest" on behalf of "anarchy, insurrection and destruction." Privately, he described labor groups as a "licentious minority" in league with the Soviets. Crushing the government employee unions, he boasted, had "preserved the integrity of representative government," effectively split the Socialist party, and cleared the path for private employers to curb the labor movement.[48]

On occasion, MacArthur still sparred with the State and Defense departments for trespassing on his turf or insulting his status. He threatened to ignore all messages not sent by the Joint Chiefs, denying the right of civilian service secretaries to communicate with him. He still raised the possibility that direct orders from the JCS or the president might not be binding since he served as the "Supreme Allied Commander . . . subject solely" to the Far Eastern Commission and Allied Council for Japan! By switching between his American and international hats, MacArthur claimed the right to follow orders selectively. Yet, so long as he did not actually defy commands, the administration ignored his startling assertions.[49]

Under the guidance of economic czar Joseph Dodge, new budget and industrial policies breathed life back into the *zaibatsu*, curbed inflation, drove down workers' living standards, and set about restoring Japan as a major industrial exporter. Dodge envisioned Japan as a regional power, which exchanged its manufactured products for Southeast Asian raw materials. As Southeast Asian colonies achieved independence, they would enter Tokyo's economic orbit and be linked, through Japan, to a global capitalist economy. The Japanese noticed how the plan resembled the ill-fated "Greater East Asia Co-Prosperity Sphere," with America now providing the military muscle. Ultimately, it required the economic stimulation of the Korean War to revive Japanese exports. American military procurements, not (initially) civilian exports, primed the pump of recovery.[50]

During 1949 and 1950, MacArthur hung on in Tokyo as something of a figurehead, although he and the administration sustained the fiction of his authority. So long as he followed instructions, it seemed easiest to leave him in place. SCAP staff shrank in size while the Japanese government assumed greater control of routine administration.

With increasingly less to do in Tokyo, MacArthur turned his attention to other parts of Asia. He expressed growing fear of the Chinese revolution and increasing interest in protecting Taiwan. The bitter divisions within the Truman administration over how to respond to the Communist surge provided an opening wedge. Finally, the escalating cold war in Asia rescued the aging commander from the shadows and thrust him for one last time into the limelight.

11
The Boundaries of Chaos:
The Cold War in Asia

In July 1950, *Time* magazine lamented the "dizzyingly swift descent from the triumph of V-J Day to the desperation" of American troops dying in South Korea. As "ruthless" Communist tyrants throughout Asia enslaved their peoples behind a bamboo curtain, "lazy minded" American diplomats "lulled" the nation into a "false sense of security." Some of Truman's advisers even suggested that "Asian communists were not really communists!"

Only Douglas MacArthur, "whose job it was to police the boundaries of chaos in Asia, was not fooled." Unlike the naive "cookie pushers" in the State Department, "never for one minute did he believe the U.S. secure in the face of the Red Advance." He tried to alert "scores of American visitors to Tokyo" of impending disaster, but Washington dismissed his warnings as "political." Having "largely ignored" its most qualified Asia expert, the nation squandered victory in the Pacific.[1]

MacArthur's strategic vision was neither so incisive as his supporters claimed nor as obtuse as his critics feared. The general had a fluid concept of security in Asia and often reversed the relative importance of China, Japan, Korea, Formosa (Taiwan), and the Philippines to American interests. Nor did he specify whether Communist military strength, the economic consequences of the loss of Asia, or the political repercussions of revolution posed a greater challenge. Frequently, MacArthur combined public bellicosity with private moderation. Yet, an aide recalled, anytime higher authority questioned his claims, he would say "you don't understand the Oriental mind." Unless you do as I say, "blood will run in the streets."

MacArthur sincerely favored a demilitarized Japan, speaking of the nation as the "Switzerland of Asia." (Switzerland, of course, was well armed. He had used the same phrase during the 1930s to boast of the growing strength of a poorly armed Philippines.) Before the Korean War, he opposed Japanese rearmament as well as plans to retain extensive American bases following a peace settlement.

The general had not, of course, converted to pacifism. The Japanese people overwhelmingly opposed rearmament while MacArthur considered it unnecessary. Since 1945 Washington had annexed as "strategic trusteeships" an extensive array of Pacific islands. This offshore perimeter stretched from the Aleutians, through the Marshalls and Marianas, to the Ryukyus, Okinawa, and the Philippines. Air and naval power on these island bases could deter or defeat any amphibious thrust from China or the Soviet Union. Since the Soviets were not likely to attack Japan directly, it seemed unwise to make the crowded nation a target.

For all his rhetoric about Asia's key role in future events, MacArthur seldom commented on the turmoil in Indochina, India, and the East Indies. However, he had plenty of advice and criticism for American actions in China, especially when Truman placed George C. Marshall in charge of mediation efforts there.

The Chinese Trap

China's smoldering civil war reignited in August 1945. The Truman administration assisted Chiang's initial moves to suppress the Communists by having American ships and planes redeploy Nationalist armies and by sending 50,000 marines to protect north China. These actions failed to stem the Communist advance but threatened to involve American forces in clashes with the Chinese Communists or Soviet occupation troops in Manchuria. In December, Truman asked George C. Marshall to seek a political settlement of the civil war. General Marshall arranged the mutual withdrawal of American and Soviet forces from China but found it impossible to forge a political coalition between the warring Chinese factions, each of whom expected to win in a showdown.

In January 1947, after a year's work, Marshall proclaimed a plague on both camps and abandoned his mission. Truman brought him back to Washington to replace James Byrnes as Secretary of State. Marshall joined those in the administration who considered China a lost cause and a tar baby. Japan now loomed as the most vital American asset in Asia.[2]

MacArthur voiced no sympathy for Marshall's efforts in China. In July 1946, he told Navy Secretary James Forrestal that while Chiang Kai-shek and his government "might not be the best in the world," they "were on our side and should be supported." He accused State Department China specialists of "disloyalty" for exposing Kuomintang corruption.

In a discussion a few months later, MacArthur claimed that Marshall went to China only to win a place "with Truman in the next [presidential] campaign." After listening to these diatribes from his boss, Robert Eichelberger concluded that MacArthur would "be happy to see the ship of state sink beneath the waves providing the other guy [Marshall] went down with the ship."[3]

During his quest for the 1948 Republican presidential nomination,

MacArthur cooperated with congressional Republicans who demanded aid for China as their price for supporting the European Recovery Program. In March, at a crucial juncture in congressional debate on the Marshall Plan, and on the eve of the Wisconsin primary, the occupation commander sent a statement to Congressman Charles Eaton, of the Committee on Foreign Affairs, endorsing military aid to China. He described that embattled land as the "fundamental keystone to the Pacific Arch." The Truman administration, he complained, ignored the "global picture" and thought only about Western Europe. An "integrated solution" to the cold war required, as a first step, defending China.

Some Republican politicians queried the general about becoming a special "military adviser to Chiang Kai-shek" if the GOP retook the White House in 1948. He told British diplomats in Tokyo "he would not be adverse to adding this responsibility to his present one in Japan." As for the idea of leading American troops into Manchuria to drive out the Chinese Communists, MacArthur boasted he "could accomplish this in six months by air action with a force of 1000 suitable aircraft." He might then occupy Manchuria as a "trustee" for the Nationalists and, thereby, safeguard all Asia from communism.[4]

Whether opportunistic or not, MacArthur's focus on China put him at odds with the priorities of the Truman administration. While he predicted grave consequences from the "loss of China," Marshall, Acheson, and Kennan judged China a marginal prize in the cold war. George Kennan, who typified this outlook, described the nation as a vast poorhouse of little value either to Moscow or Washington. "If I thought for a moment," he wrote in 1948, "that the precedent of [aid to] Greece and Turkey obliged us to try to do the same thing in China, I would throw up my hands" in despair. Wasting money on China would starve the effort to keep "Russian hands" of Europe and Japan, the only centers of "military-industrial power on the surface of the earth" whose loss to the Soviets could alter the strategic power balance.[5]

Beneath his often lurid rhetoric, MacArthur avoided specifics. China (and later Taiwan) should be defended against communism, but he did not exactly suggest that American troops do the job. Instead, he called for boosting aid levels, and, on occasion, using naval or air power. He hedged his bets by cautioning against involvement in "hopeless situations." After his political setback in the spring of 1948, MacArthur dropped the China issue for a year. Later, when the Republican party and some military officials raised the cry to defend Taiwan, the general climbed back on the bandwagon with vigor.

The Korean Vortex

After June 1950, MacArthur's name became almost synonymous with Korea. His ambivalence toward that nation comprises yet another remarkable aspect of his mythology. Until shortly before Tokyo's surrender, the United States had not planned to send occupation forces into the Japanese colony of Korea. Washington had engaged in inconclusive discussions with the Soviets about a postwar trusteeship but voiced no objection to having Russian forces liberate the peninsula. By August 1945, fears of Soviet designs in northeast Asia prompted Truman to soften peace terms offered Japan and to demand an occupation zone in Korea south of the 38th parallel. Either because Stalin cared little about Korea or hoped that flexibility might induce Truman to grant Russia similar rights in Japan, he accepted Washington's demand.

Southern Korea fell within MacArthur's zone of operations. In September, he designated General John R. Hodge to land his XXIV Corps in the south, largely because his unit was the nearest available. Hodge possessed few political skills, knew almost nothing about Korean conditions, and operated under vague guidelines. Concerned primarily with preserving "order," he dissolved a left-leaning provisional government (the Korean Peoples' Republic) and utilized Japanese colonial authorities and Korean collaborators as local administrators. MacArthur's contribution to the political turmoil came a short time later when, despite State Department objections, he arranged the return to Seoul of an elderly nationalist exile living in America, Syngman Rhee. This provided Rhee with a mantle of legitimacy even though diplomats condemned him as a reactionary, petty tyrant.

Gradually, the Soviets and the Americans sponsored client regimes reflecting emerging East–West tensions. Despite some halfhearted attempts during 1946 to create a cross-zonal trusteeship, Moscow and Washington each supported antagonistic Korean clients and compromise failed. Hodge permitted rightists to manipulate elections for an interim legislature in October 1946. In the north, former guerrilla leader Kim Il Sung organized a Communist government that same year. By 1947, the temporary division looked increasingly permanent.[6]

Although they hoped to prevent Communist domination of the peninsula, the Joint Chiefs of Staff and MacArthur considered the Korean occupation a tiresome sideshow draining scarce resources. Maintaining occupation troops amidst postwar demobilization strained army logistics. In September 1947, the service chiefs reported that "from the standpoint of military security, the United States had little strategic interest in maintaining the present troops and bases in Korea." In case of general war, American forces would be cut off and overwhelmed.[7]

The State Department considered it politically important to sustain a non-Communist Korean regime regardless of its military utility. Diplomats argued that the two Koreas represented a test case between Moscow and Washington. Success in developing a stable and prosperous south would demonstrate America's determination to help allies resist subversion. In 1947, the department nearly included Korea in a major assistance package linked to the Greek–Turkish aid program; in 1948, Congress began providing money for Korean economic development. Despite later charges of neglect, before June 1950 the State Department waged an uphill battle against the Defense Department and congressional Republicans to support South Korea.

In April 1948, Truman approved a National Security Council report that defined American policy in Korea until mid-1950. The United States would not abandon the southern regime since the extension of Soviet control would "enhance the political and strategic position of the Soviet Union with respect to both China and Japan." Denying Seoul enough military and economic assistance to defend itself "could be interpreted as a betrayal by the U.S. of its friends and allies in the Far East," causing non-Communist governments to realign themselves with Moscow. Although Washington could not guarantee Korea's security, it would try to extend enough aid to deter or defeat any Communist assault.[8]

In 1948 Washington convinced an accommodating United Nations to supervise Korean elections. (The Soviets possessed a veto in the Security Council, but the United States routinely dominated the General Assembly.) In May, a handful of observers witnessed the voting in the south (none were permitted to enter the north), which selected an assembly. The conservative legislature soon drafted a constitution and chose Rhee as president.

As Washington urged, the UN General Assembly recognized the legality of the new Republic of Korea without specifying whether its mandate applied to all or only the southern half of the divided land. When the Soviets blocked UN membership for the Seoul regime, the United States returned the favor by vetoing admission of the communist Democratic People's Republic of Korea.

In August 1948, American occupation forces transferred internal political authority to the Republic of Korea. Visiting Seoul, MacArthur proclaimed that the 38th parallel "barrier must and will be torn down. Nothing shall prevent the ultimate unity of your people as free men of a free nation." While journalists listened, the general told President Rhee that if the Communists attacked the south, he would "defend it as I would California."

Endemic social and political violence swept the new nation, and the Rhee regime routinely jailed and mistreated opponents. Still, by late 1948, the situation appeared stable enough to allow the phased withdrawal of American troops. By the summer of 1949 only a small advisory mission remained. Moscow had already recognized Kim's government in Pyongyang and removed its occupation troops. Thereafter, two rival regimes led by committed nationalists each claimed the right to rule the entire peninsula. Kim and Rhee both pledged to unify the country by any means, including war.[9]

MacArthur had always resented sharing authority with the State Department in Korea. He complained about the "coffee drinking diplomats" who caused wars and forced him to "clean the mess up." The internal violence that racked South Korea from 1946–48 amused MacArthur, his military secretary recalled, as it reflected badly on the State Department.

Almost everyone familiar with the situation in Seoul agreed that General Hodge performed dismally. A member of MacArthur's staff described him as "one of the sillier men I've met in my life." Hodge sent desperate messages to Tokyo pleading with MacArthur for advice, for more resources, and for guidance on how to handle the political infighting among Koreans. MacArthur "repeatedly would wire back saying he was too busy, too preoccupied in Japan" and "hadn't the time" to visit Korea or attend to its problems. Some suspected that MacArthur allowed the situation to deteriorate because he "wanted the State Department to stew in its own juice."[10]

In March 1949, as part of the military establishment's pressure on the administration to hasten the withdrawal of American troops from Korea, the general granted an interview with British journalist G. Ward Price. He traced carefully what he called the American "line of defense" in the Pacific. The perimeter ran from the Philippines, through the Ryukyus and Okinawa, then on through Japan, the Aleutian islands, and Alaska. Pointedly, he excluded the Republic of Korea, Taiwan, and the Chinese mainland. (In January 1950, Secretary of State Acheson traced a similar line in a speech in Washington. Even before the Korean War began, both MacArthur and Republican leaders accused Acheson of encouraging a Communist assault by omitting South Korea.)

Although MacArthur may have secretly urged Rhee to unify Korea, publicly he limited his support of the south. In September 1949, he told a congressional delegation that "South Korea is in no danger of being overrun by North Korea." He predicted that any invasion of the south would occur only as part of a general Communist offensive against the West, not as part of a civil or regional conflict. MacArthur also agreed with military and civilian analysts who opposed giving heavy offensive weapons to Rhee's army. Such aid would spur a dangerous arms race, increasing the likelihood of war on the peninsula. Before June 1950, MacArthur gave little public indication that he considered Korea especially vital to the United States.[11]

Formal responsibility of SCAP for Korea lapsed with the end of the occupation in 1949. Nevertheless, MacArthur and Willoughby maintained an influential presence. Willoughby organized a Korean Liaison Office with the job of gathering information on North Korea, China, and the Soviet Union. (MacArthur prevented much of a CIA presence in his realm until after the Korean War began.) Rhee and American officials in Tokyo also utilized a network of private mercenaries, soldiers of fortune, and business men as a liaison between SCAP and the South Korean government.[12]

Japan and the Defense of Asia

By early 1949, containment in Europe and Japan rested on a surprisingly solid foundation. The German and West European economies responded well to the Marshall Plan. American assistance played an equally important part in revitalizing Japanese industry. Despite urging from the Pentagon, the Truman administration avoided large-scale military involvement in China and kept some distance from Chiang Kai-shek's rump regime on Taiwan. Secretary of State Dean Acheson (who replaced Marshall in 1949) emphasized aid for the emerging states on China's border, stretching in an arc from Japan to India. Acheson told Congress that bolstering this "great crescent" would contain Chinese communism without direct American military involvement. Eventually, he thought, the Southeast Asian states could replace China as Japan's major Asian trading partner.

In May, President Truman accepted a recommendation of the National Security Council proposing an end to the "regime of control" in Japan. Yet his contentious civilian and military advisers developed wildly contradictory approaches to the problem. Although Truman followed Acheson's advice on most Asian questions, he had a nagging habit of agreeing with the last adviser he spoke with.

Japan's apparent stabilization convinced Acheson and his advisers to press ahead with a peace treaty. In fact, the Secretary of State insisted, only a restoration of sovereignty could ease the "growing restiveness of the Japs." Formation of a central police force and provision for a leased American naval base would assure the nation's security. Extending the occupation, Acheson warned, would drain American resources and drive Tokyo toward an accommodation with the "Commie system in Asia." As an interim step, he urged SCAP to transfer more authority to American civilians and the Japanese government.[13]

The military establishment, including the Joint Chiefs of Staff, Defense Secretary Louis Johnson (who replaced James Forrestal early in 1949 when mental illness forced his resignation), and Tracy Voorhees (Draper's successor as army under secretary), preferred to continue the occupation. The JCS argued that due to the "debacle" in China and the "developing chaos on the Asian mainland together with its communist trend," America could not afford to lose firm control of Japan. Bases there provided "us with staging areas from which to project our military power to the Asiatic mainland and to the USSR." These considerations outweighed any concern with Japanese opinion. Defense planners even urged Truman to press Tokyo to rearm and deploy its own army, despite the antimilitary clause of the new constitution.[14]

The division over Japan reflected an even deeper split over China policy. Acheson and his Asia policy advisers thought that patience and moderation might wean the Chinese Communists away from links with the Kremlin. The military dismissed this argument as naive and proposed a more confrontational approach, including the economic isolation of China, the defense of Taiwan, military aid to French Indochina, and the retention of forces in Japan. Less concerned than civilian planners about Japan's regional economic role, military bureaucrats viewed containment in Asia largely as a military problem.

Douglas MacArthur found fault with both positions. When he urged a Japanese peace settlement as early as March 1947, the entire administration rejected his advice. Two years later, he still favored a treaty but opposed retaining American bases, fostering Japanese rearmament, or turning SCAP authority over to civilians in preparation for a treaty. "By no stretch of the imagination," MacArthur informed Acheson, could Washington make him transfer power to American diplomats or to Japanese politicians. He again claimed "international status," which could only be altered by unanimous agreement among the wartime allies, including the Soviet Union. The Truman administration risked "catastrophic consequences" by dismantling prematurely his "strong spiritual front" against the "communist advance."

While neither man was comfortable admitting it, MacArthur and Acheson shared a common outlook regarding Japan. Now that administration policy had put Japan on the road to economic recovery, the secretary of state joined the occupation commander in calling for a nonpunitive treaty and the departure of most American forces. However, the general still feared that outsiders would steal the credit for everything accomplished since 1945. The Japanese people, he immodestly reported to Acheson, held him in "respect bordering on veneration." No other individual or agency could arrange a settlement. Given a free hand to arrange a treaty, he even promised to extract from the Russians a binding pledge to respect Japan's neutrality. Like most Japanese (but unlike most Americans) MacArthur deemed it vital to include both China and the Soviets in a treaty.[15]

The occupation commander searched for common ground between the antagonistic State and Defense positions. As it became clear that no one in Washington favored involving the Soviets in drafting a treaty, he dropped the idea of consulting Moscow. Earlier, MacArthur accused the Joint Chiefs of being willing to "scuttle the Pacific," as evidenced by their insensitive demand that the Japanese rearm. Now, to appease the Defense Department, he called for maintaining a small American "military force in Japan for the indefinite future." He offered himself as the "neutral chairman" of a peace conference to be held in Tokyo.[16]

MacArthur broke with the State Department (and supported the Pentagon) on one major issue. He dispatched Colonel Stanton Babcock to tell the Joint Strategic Survey Committee of the JCS that he considered it "of the greatest importance that Formosa not fall under communist control." (At the time, most Americans referred to the island of Taiwan as Formosa. Both names are used below.) "By hook or by crook, we must keep it out of Communist hands." MacArthur proposed several schemes, including putting the island under American trusteeship, under the control of some other "safe nation," or even returning it to the Japanese (e.g. SCAP).[17]

Taiwan had other defenders in high places. Although most China specialists in the State Department accepted the fact of a Communist China, many still saw advantage in supporting an independent, non-Communist government on Formosa. For one thing, such a regime could compete with Peking for the loyalties of the millions of Chinese throughout Southeast Asia. However, these diplomats rejected Chiang Kai-shek's desire to reconquer the mainland and even favored replacing his clique on Taiwan with a more liberal leadership.

During 1949–50, the Truman administration considered several proposals to support a coup against Chiang with the idea of bringing to power a group reconciled to an independent Taiwan. At one point even George Kennan urged a "Teddy Roosevelt like" move to seize the island and put it under the administration of MacArthur or the Japanese. The general's ideas, then, were not particulary outlandish.[18]

Taiwan aside, the strongest opposition to a Japanese settlement came from the Defense Department. Louis Johnson, Tracy Voorhees, and the service chiefs demanded not only American bases and Japanese rearmament, but, incredibly, Soviet and Chinese acceptance of these provisions as a precondition for a treaty. "Continuation of our dominant [military] position in Japan" outweighed all other considerations, the Joint Chiefs insisted.

The Pentagon did not really fear Soviet designs on Japan. Rather, it worried that a settlement would limit access to forward air and naval bases in case of war with the Soviet Union and China. Also Defense planners wanted a Japanese army to supplement limited American ground forces for combat in Asia. Beyond these considerations, the Pentagon viewed a Japanese peace treaty as a bargaining chip to compel the president and the State Department to alter their relatively restrained China policy. Military representatives informed Acheson they would support a settlement with Japan only when the State Department agreed to defend Taiwan, spurn recognition of Communist China, and maintain substantial American forces in Japan. The defense establishment made their support for an end of the occupation conditional upon the willingness of Truman and Acheson to harden policy toward China and Taiwan.[19]

Given President Truman's uncertainty, the administration remained deeply divided over how to contain communism in East Asia. Moderates argued with hard-liners about whether a benign policy could lure Peking away from Moscow and whether Japan should forge economic links to China. Would such trade speed Japan's recovery or make it a pawn of the Communist powers?

Before the Korean War, Truman endorsed contradictory approaches, reflecting his own divided instincts as well as the split among his aides. For example, the president approved private trade with China, but tolerated the Nationalist blockade of mainland Chinese ports. He suspended military aid to Taiwan, but continued diplomatic and economic ties to Chiang's regime. For a time he permitted American diplomats to remain in Peking, but declined to recognize the Communist government.

Secretary of State Dean Acheson, Truman's closest foreign policy adviser, despised the Pentagon's muscular anti-communism as well as the veneration of Chiang voiced by the mostly Republican "China bloc" in Congress. Although a vigorous ideological opponent of communism himself, Acheson ridiculed talk of reversing China's revolution. The secretary approved limited measures to sustain a non-Communist regime on Taiwan but opposed direct American intervention, which might alienate Asian nationalism or push China even closer to the Soviet Union.

During 1949 and 1950, General MacArthur and Defense Secretary Louis Johnson worked hard to loosen Acheson's grip over Asian policy. A successful insurance executive in private life, Johnson earned Truman's favor by raising desperately needed campaign funds in 1948. He made it known that he considered the defense post a stepping stone for higher national service. Curiously, Johnson combined bombastic rhetoric (especially toward China) with a campaign to reduce military spending, then hovering around $12 to 13 billion annually.

Even though Johnson opposed a Japanese treaty and wanted to retain military bases there, MacArthur saw tactical advantage in supporting him against Acheson. The defense secretary, like the occupation commander, favored defending Taiwan as a symbolic bastion against communism. If the Chinese Communists took the island, MacArthur warned, "our whole defensive position in the Far East was definitely lost." America's "defensive line" would be forced back to "the west coast of the continental United States." Chinese or Russian planes operating from Taiwan would dominate East and Southeast Asia. Of course, given America's numerous Pacific bases, this seems a great exaggeration.

The general endorsed a proposal made by Claire Chennault, the World War II flying ace and military adviser to Chiang, that Washington create a five-hundred plane air force, piloted by American volunteers, and a small navy to "blockade and destroy China's coastal cities." Late in 1949 the Central Intelligence Agency discovered that MacArthur had permitted some Japanese aviators to travel to Taiwan to assist the Kuomintang air force.

MacArthur may have hoped that his vigorous support for Taiwan would swing Johnson behind a Japanese settlement. Out of gratitude, perhaps, the Defense chief might also push Truman to give MacArthur responsibility for Taiwan or even China. In spite of his private activities, officially the general recommended only increase levels of military assistance to Taiwan, not an American defense commitment.[20]

MacArthur and Johnson scored few successes before early 1950. Acheson won the president's support during 1949 for a series of National Security Council (NSC) recommendations (drafted by the State Department) designed to limit conflicts with China. The administration decided against supporting anti-Communist forces on the mainland, offering Taiwan military protection, or prohibiting Sino-American trade (except for strategic items).[21]

MacArthur waffled on the issue of China trade. Publicly, he endorsed blockades and boycotts while privately admitting the critical importance of the "China market" to Japan's recovery. Until the Korean War, the occupation commander tried to have it both ways, demanding authority to ban Sino-Japanese trade while actually encouraging the mutually profitable commerce.[22]

Johnson tried to outflank the State Department by demanding that the National Security Council develop a "comprehensive plan" and "new courses of action—political, economic and military—which might be undertaken" to push back communism in Asia. Moderation, he complained, had led to the loss of China and growing threats to the rest of the region.

In July 1949, the administration proposed to Congress a Mutual Defense Assistance Program (MDAP) which would fund both weapons and defense production in Europe and Asia. Although drafted primarily to support NATO, by the time the bill passed in October, it contained a provision (Section 303) giving the president $75 million to conduct covert operations to contain communism "in the general area of China." Not surprisingly, Johnson, like Acheson, wanted his department to get control of this money and program. According to Drew Pearson, even MacArthur joined the queue, telling Truman the money could delay indefinitely Chinese Communist expansion. The general proposed funding "Chinese Warlords who still have big armies in the field." (Acheson and Truman toyed with the idea, but the warlords' armies dissolved by year's end.)

The congressional China bloc intended the aid to help the Nationalists defend Taiwan or to retake China. Several even demanded that General MacArthur be given extraordinary control over all special defense expenditures in East Asia. But Secretary of State Acheson explained the administration would use the secret fund to "prevent the spread of this communist menace throughout Southeast Asia," and to promote cooperation among the nations stretching in a great arc from Japan to Pakistan.[23]

State Department officials considered MDAP an ideal vehicle to promote trade and political links between Japan and Southeast Asia. A stable perimeter of front line states could deter Chinese expansion without direct American involvement. As in Europe, these Asia specialists saw containment as an ongoing political and economic program, not merely a mechanism for arming allies. When Truman signed the final bill in October, the fight between Acheson and Johnson to control MDAP funds began in earnest.

The secretary of state worked on two levels to build support for a moderate program. He appointed a special blue-ribbon panel, the "Far Eastern Consultants" (chaired by Ambassador-at-Large Philip C. Jessup), to devise "programs of action" for "halting the spread of totalitarian communism in Asia." The consultants met with Truman and members of Congress, urging civilian control of MDAP programs in Asia. Acheson also ordered the publication in August of the thousand-page "China White Paper" (officially called *United States Relations with China, With Special Reference for the period 1944–1949*), a detailed chronicle of the Nationalist debacle. Although the volume's introduction was larded with anti-Communist phrases, the report savagely condemned Chiang's record and argued against future American efforts on his behalf.[24]

Meanwhile Defense consultants in the National Security Council followed Johnson's lead by drafting a remarkably bellicose plan to roll back Communist power in Asia. By October, they drafted a forty-page report warning of a Soviet effort to build new Asian "Co-Prosperity Sphere" incorporating China, Japan, and Southeast Asia. By combining the industrial base of Japan and the resources of Southeast Asia, the report warned, Russia would gain a tremendous advantage over the West. To prevent this, the study called for defending Taiwan, arming anti-Communist forces in Southeast Asia, and committing American power to the "rollback of Soviet control" everywhere in Asia. Success depended on appointing a "high ranking individual" to direct "all United States efforts against Communism in China" and to wage "cold war in Asia." MacArthur seemed the leading candidate for the job.[25]

State Department officials immediately set about to frustrate these proposals. They released the White Paper, lobbied members of Congress; met with leading members of the business, press, and academic communities; and held the president's hand whenever he seemed tempted by Louis Johnson. The advocates of restraint spoke of the need to encourage China's deviation from the Kremlin. Shielding Taiwan would only drive the mainland into Russia's arms. America should try to "capture" Asian nationalism by ending the occupation of Japan and supporting non-Communist independence movements in Southeast Asia.[26]

Truman's visceral anti-communism often clouded his understanding of Asian politics. Several times in 1949 and 1950 the president rebuffed tentative feelers extended by Peking and even toyed with suggestions from Johnson and MacArthur that he support anti-Communist Muslim guerrillas in western China or assign the American navy to blockade North China. Some of Truman's anger came in response to the arrest in Manchuria of Consul General Angus Ward, charged by the Communists with espionage. (He was released after several months.) On these occasions Acheson and his aides had to walk the president back from the brink of intervention. Truman explained that while he agreed in a "broad sense" with the moderate agenda, he still worried about its effectiveness in stopping communism.[27]

During the final months of 1949, Acheson's staff redrafted the controversial NSC document, now called NSC 48. In a tactical move, State Department personnel retained its flamboyant anti-Communist rhetoric while gutting plans for attacking China, defending Taiwan, or preventing trade with China.[28]

MacArthur joined the debate on two levels. Through various intermediaries, he supplied information to pro-Taiwan, anti-administration Republicans. Drew Pearson learned from several representatives that the General "had more to do with stirring up congressional fervor over Formosa than even the State Department realizes." MacArthur also endorsed Defense Department proposals to utilize MDAP funds for confronting China and protecting Taiwan. If the island fell, he warned, the Communists would breech the entire Pacific defense perimeter. In September, he warned a group of congressman visiting Tokyo that Moscow planned to subdue Europe by first conquering Asia. As an antidote, he endorsed Chennault's plan for American volunteers to shield Taiwan and attack China. MacArthur and Willoughby permitted a number of Japanese veteran pilots to go to Taiwan and also encouraged the efforts of American entrepreneurs such as Chennault, Preston Goodfellow, William Donovan, and William Pawley to sell weapons to the Nationalists. These activities prompted rumors on the island that SCAP had developed a secret plan to "thwart a Communist takeover," whatever formal American policy might be.

When Senator H. Alexander Smith (R–New Jersey) visited Tokyo at the end of September, MacArthur and Willoughby plied him with details of alleged State Department conspiracies to spread "smear stories" about Chiang Kai-shek. The general blamed the Communist victory in China on the fact that Truman, Marshall, and the navy denied him postwar authority outside Japan. Dismissing Korea as an unimportant outpost, MacArthur argued that "Formosa must never be allowed to get into the hands of hostile groups." He appealed for the senator's help in assisting Chiang.

Smith and MacArthur discussed another idea popular among China bloc Republicans—naming the general special supreme military commander in the Far East. The two men actually devised a list of subordinate commanders (General Albert Wedemeyer, Admiral Oscar Badger, General Claire Chennault) who might serve under MacArthur. Not surprisingly, all were strong supporters of the Chinese Nationalists.[29]

Although he maintained strong public support for Chiang Kai-shek, MacArthur acknowledged privately that the generalissimo's unpopularity, at home and abroad, hindered efforts to defend Taiwan. Accordingly, late in 1949, the general initiated an effort to arrange Chiang's removal. Ambassador Wellington Koo reported that officials in Taiwan heard that MacArthur wanted them to press Ching to "take a trip abroad" and to pass power to K. C. Wu. Shortly afterwards, General Willoughby visited the island to discuss the possibility of an "officer from General MacArthur's staff" serving as a special adviser. Koo, Hollington Tong, and other Kuomintang officials guessed that MacArthur "must have got, if not instructions, at least suggestions from Washington to try to persuade the G'mo to leave Formosa."[30]

In December, Johnson sent Army Under Secretary Tracy Voorhees on a mission to Tokyo. Defense officials wanted MacArthur to stop pressing for a Japanese settlement treaty and to concentrate his energies on changing the administration's China policy. The general (who despised Voorhees for his past interference with SCAP's economic program) refused to relent on Japan. He agreed, however, that a Communist takeover of Taiwan would prove a fatal blow to American security in the Pacific. Even "Japan in enemy hands" would not be so dangerous. MacArthur recommended that Truman simply declare the island neutral territory, forbidding attacks upon it by the Communists or from it by the Nationalists. This, he felt, would secure the area without committing American forces. Later, the United States could sponsor an "independent" Chinese regime on Taiwan. As MacArthur undoubtedly knew, State Department officials had pondered similar ideas.[31]

MacArthur, Louis Johnson, and the Joint Chiefs all agreed that preserving a non-Communist Taiwan would delay international recognition of Peking and deflect Chinese military pressure away from Southeast Asia. On December 23, the defense secretary and service chiefs appealed to the president to approve "limited military measures," including the dispatch of naval units, to defend Taiwan. They also asked permission to "make an immediate survey" of the island's future military needs—presumably by MacArthur.[32]

Although Acheson also harbored some hope of saving Taiwan (preferably without the use of force and without Chiang), he sensed that the military establishment and MacArthur raised the issue mostly as a smoke screen behind which to attack the administration's entire Asia/China policy. The secretary of state guessed that Johnson and the JCS hoped to convince Truman to support the hard-line version of NSC 48 when it came up for review late in December. In an angry exchange with the service chiefs, Acheson argued that if they really wanted to halt communism in Asia, they ought to support his efforts to end the Japanese occupation and strengthen the "neighbors of China" in Southeast Asia. Diverting aid to or risking military involvement in Taiwan or China, he insisted, would weaken the crescent of Southeast Asia and leave Japan a "pensioner of the American taxpayer."

JCS Chairman General Omar Bradley and his colleagues (Generals Hoyt Vandenberg and J. Lawton Collins, and Admiral Forrest Sherman) quickly distanced themselves from MacArthur's assertion that Asia's future hinged on Taiwan. The island had some, but not overriding, military importance. The chiefs admitted that their talk of military intervention reflected a desire to do something dramatic with MDAP money—and thus lay claim to more of it.

Acheson's tongue-lashing proved quite effective. When the NSC met later that day to debate NSC 48, Johnson had sensed defeat and cut the meeting for an out-of-town lecture. The council and president quickly endorsed the defanged version of NSC 48. The report reverberated with calls to halt and beat back communism in Asia, but carefully avoided commitments to any such program. The administration endorsed limited "diplomatic and economic" assistance to Taiwan, not military protection. No substantial aid would go to anti-Communist forces in China proper. Diplomatic ties with Taiwan might continue, but, when it was "clearly in the U.S. interest to do so," Washington might recognize the Peking regime. In the interim, the United States would not oppose its allies recognizing or trading with the Communist government.[33]

Truman and Acheson hastened to implement the moderate Asian agenda. On January 5, 1950, the president announced an end to "military aid or advice to Chinese forces" on Taiwan. (The Nationalists could still use their own funds to purchase arms.) Five days later, during executive testimony before the Senate Foreign Relations Committee, the secretary of state spoke up for Asian anti-colonialism and emphasized the importance of assisting non-Communist nationalists. He urged the senators to pay more attention to the "real center" of Asia, "these other countries" (other than China) along the "crescent or semicircle which goes around Japan, at one end, and India, at the other." Acheson recommended spending most MDAP funds and the $100 million that remained from the 1948 China Aid Act to develop a regional economy.

In a public speech to the National Press Club on January 12, the secretary developed this theme. Instead of refighting China's civil war, the United States should cultivate moderate Asian nationalists by providing economic assistance. Acheson described a Pacific defense perimeter (a line the United States would defend at once with its own forces) that included the Aleutians, Japan the Ryukyus, and the Philippines. In addition, he pledged economic and political assistance to Japan, Korea, and the emerging states of Southeast Asia. Despite later criticism, Acheson did not write off South Korea. He promised it financial aid and, if attacked, military support through the United Nations. The time had also come, the secretary declared, to implement either a peace treaty or some "other mechanism" to restore Japan's sovereignty.[34]

These initiatives silenced, temporarily, the hard-liners. Testifying before a Senate committee a few weeks later, Defense Secretary Johnson and General Omar Bradley minimized Taiwan's importance. Even MacArthur relented a bit. In January, he told Ambassador-at-Large Philip C. Jessup (then on an Asian tour) that Louis Johnson and the Joint Chiefs were fools for exaggerating the Communist military threat in Asia and blocking a Japanese peace settlement. The general applauded Acheson's proposal to assist Japan and Southeast Asia, noting its resemblance to the "basically sound concept behind the Japanese Co-Prosperity Sphere." Economic aid, not military intervention or bases in Japan, the general agreed, would contain Asian communism.

The fundamental problem in Asia, MacArthur told Jessup, stemmed more from the Defense Department than from the Soviets or Chinese. The ignorant Louis Johnson had bullied the bungling Joint Chiefs into opposing a settlement with Japan and demanding exorbitant base rights. In fact, Tokyo's security could be assured by a small garrison and an agreement with the Soviets.

Why, MacArthur asked Jessup, did Truman not simply impose his will on the recalcitrant Defense establishment? If the administration backed a Japanese settlement, the general pledged his full support against Johnson and the Joint Chiefs. Pounding the table, he declared that "when decisions were made by the Secretary of State," they were "decisions of the United States and he would loyally carry them out." If Acheson came through with a treaty, he implied, he would soften criticism of Truman's China policy. In effect, MacArthur tried, simultaneously, to win the State Department's support for a Japanese treaty by taking a softer line on China even as he worked to secure Defense Department support for a treaty by taking a tougher line.[35]

During February, Acheson and his advisers outlined to Truman a vision of a "Pacific Collective Security arrangement" linking Southeast Asia to Japan and the nations of the Pacific basin. Defense officials still spoke of an indefinite Japanese occupation and of expanding American military influence on the islands. After a visit, in February 1950 to Tokyo, the Joint Chiefs informed Truman they "were still strongly of the opinion that it is premature to make a treaty at this time." One of the service chiefs lamented that the homeland of the "former enemy appeared to be not only the strongest bastion but about the only tangible thing left of the fruits of victory in the Pacific."[36]

Infuriated by this impasse, MacArthur denounced the Joint Chiefs, Louis Johnson, and Tracy Voorhees for wrecking American policy in the Pacific. In discussions with diplomatic officials, he impugned the intelligence, honor, and even West Point grades of his fellow officers. The Defense Department, he charged, cared nothing about Japan except as a staging area for "forward air operations against Russia in time of war."[37]

Chances for a treaty diminished still further as relations with China deteriorated. Conflicts between Washington and Peking eroded Acheson's influence with Truman, raised questions about the direction of Chinese communism, and opened a breach easily exploited by administration critics, including MacArthur.

In January 1950, Chinese authorities threatened to seize part of the American consulate in Peking, alleging the property had been acquired through an "unequal treaty" extorted from the Ch'ing Dynasty. Possibly, the Communist government thought this tactic would scare Washington into extending diplomatic recognition. (Threats against the British consulate were dropped when London recognized the People's Republic a few days later.) Truman and Acheson, however, considered the action criminal. When consular staff were placed under restrictions and denied the right to confidential communications, Acheson decided to avoid further provocations by removing all diplomatic personnel.[38]

These tensions coincided with a barrage of attacks on the administration fired off by Republican politicians. The little-known junior senator from Wisconsin, Joseph McCarthy, launched the most sensational charges during a speech on February 9, in Wheeling, West Virginia. He pilloried a subversive State Department clique, led by Ambassador Philip C. Jessup and Foreign Service officer John S. Service, for "losing" China to communism. The senator coordinated his efforts with Chinese Nationalist representatives in Washington. Initially, the administration dismissed McCarthy as merely boorish. However, his unrelenting accusations soon took on an dependent life and polarized foreign policy along increasingly partisan lines.

In mid-February, while newspapers trumpeted McCarthy's lies, the Soviet Union and China signed a thirty-year friendship pact, pledging Soviet aid and the return of Chinese territory taken in 1945. The signatories promised to resist "aggressive action on the part of Japan or any other state which should unite with Japan, directly or indirectly, in acts of aggression." Moscow and Peking also demanded to participate in any Japanese settlement.[39]

China's harassment of American diplomats, Republican charges of disloyalty, and the Sino-Soviet pact made it difficult for Acheson to stay the moderate course. Peking's defense arrangement with Moscow could be interpreted as a veiled threat justifying the bases in Japan demanded by the military. Cooperation between the two Communist giants also undermined predictions of an imminent Sino-Soviet split. If China and Russia were allies, critics asked, why curry Chinese favor by abandoning Taiwan?

Influenced by these concerns and pressed by Louis Johnson, Truman backpedaled and permitted the Defense Department to resume limited arms shipments to Taiwan. At the same time, a network of private aid groups, centered around Claire Chennault, Preston Goodfellow, and former OSS head William Donovan, funneled additional weapons to the Nationalists. They sued to take possession of a fleet of aircraft in Hong Kong that the Nationalists claimed to have sold to private investors but which Peking insisted remained Chinese government property. (The Chinese pilots wanted to return with their planes to China.)

Donovan traveled to London, Hong Kong, and Tokyo early in 1950 seeking support in his legal bid to acquire the planes, which would form the core of a clandestine air force run by Chennault. MacArthur offered encouraging words, while Charles Willoughby promised to cooperate in establishing spy networks in China and to assist an anti-Communist guerrilla campaign.[40]

Preston Goodfellow, an American close to South Korea's Syngman Rhee as well as to MacArthur, shuttled between Seoul, Tokyo, and Taipei during early 1950. He spoke of organizing a "foreign legion" of Chinese Nationalist and Korean troops to undertake operations against North Korea and China. In January and again in March, Goodfellow informed Ambassador Koo that MacArthur's "reaction to carrying out an underground activities program on the mainland was entirely sympathetic." Willoughby and others on the SCAP staff suggested that channeling funds for these activities through Tokyo would get around the administration's "deep rooted objection" to providing direct assistance to a Taiwan still controlled by Chiang.[41]

Acheson struggled to maintain his grip on policy by tacking right. In mid-March he made a speech condemning the Chinese Communists as Soviet puppets threatening "adventures beyond their borders." He promised military assistance should China attack its non-Communist neighbors. The State and Defense departments dispatched an American advisory mission to Saigon to oversee the distribution of money and weapons to French and Vietnamese anti-Communist forces fighting the Vietminh.[42]

In a simultaneous effort to appease Republican critics, Acheson replaced his assistant secretary for Far Eastern affairs, W. W. Butterworth, with Dean Rusk, a diplomat more tolerant of the Chinese Nationalists and one on good terms with Congress and the Pentagon. Also, in April, he invited a leading Republican foreign policy spokesman, John Foster Dulles, to serve as a top adviser on Asian policy and the Japanese peace treaty.[43]

Rusk and Dulles envisioned a grand compromise involving Taiwan and Japan. They suspected the military establishment and congressional Republicans might accept a Japanese settlement if the administration agreed to shield Taiwan. Also, Rusk told Acheson, as long as the island held out, it would attract the "major attention and efforts of the Communists' military forces," sparing Southeast Asia.

Although Acheson disliked Dulles personally (when serving briefly as an appointed senator from New York, he had attacked the administration), the secretary was delighted to find that his new top adviser favored a quick settlement with Japan. Dulles considered a treaty (not rearmament) the best way to bind Tokyo to the West. He also favored linking Japan closely with Southeast Asia, Australia, and New Zealand, all of whom opposed the Defense Department's plan to rebuild Tokyo's military power.

Yet Louis Johnson and the Joint Chiefs spurned every proposal made by Acheson during the spring. The Defense Secretary insisted that the Japanese people were content and that "the only propaganda for a peace treaty . . . came out of the Department of State." General Omar Bradley and Admiral Forrest Sherman warned that the departure of Americans from Japan could upset the military balance in Asia and make Tokyo a Communist target. Air force and army chiefs, Generals Hoyt Vandenberg and J. Lawton Collins, claimed that, despite his public assertions, even MacArthur opposed ending the occupation. He craved the publicity of a peace conference more than a settlement.[44]

Actually, while MacArthur had bent over backward to accommodate Defense Department concerns for postoccupation security, he still favored a settlement. The Japanese public and government strongly favored an end to the occupation. Even though Prime Minister Yoshida preferred not to accept bases and wanted the Soviets and Chinese included in a peace treaty, he gave private assurances he would accept any "practical arrangement the United States might consider necessary" for insuring Japanese security and returning sovereignty.

In May, Yoshida dispatched two close advisers, Shirasu Jiro and Finance Minister Ikeda Hayato to Washington. They warned American officials that their countrymen had grown restive and susceptible to Soviet blandishments. The Japanese government would accept reluctantly a settlement excluding the Soviet Union and China. If Washington insisted, they would even tolerate postoccupation bases. At the same time, Minister Ikeda added, the Japanese wanted to see real evidence of American strength in East Asia. Talk about abandoning Taiwan and an uncertain commitment to Indochina and South Korea frightened conservatives in Tokyo. Although eager to trade with Communist China, they feared the extension of revolution. The current drift in Asia policy, Ikeda declared, left the Japanese "desperately looking for firm ground." They were "skeptical on just what and when and where the United States would stand firm, and in particular with respect to Japan."[45]

Dulles seized on Ikeda's offer (of bases) and threat (of a separate deal with Moscow) to link a solution of the Japan and China–Taiwan problem. All Asians, Dulles asserted, saw Mao's victory in China as heralding a "shift in the balance of power." They awaited a sign revealing America's decision to stand firm or to "fall back and allow doubtful areas to fall under Soviet Communist control." Moscow and Peking, he suggested, played a psychological game, relying on a few symbolic victories in marginal areas to spread panic among American allies. However, if "at some doubtful point we quickly take a dramatic and strong stand that shows our confidence and resolution," he wrote, the United States could prevent this "series of disasters."

Taiwan was precisely the sort of "doubtful" point where "a stand must be taken." By "neutralizing" the island, America would impress the "eyes of the world." Failure to act would push the fence sitters in Asia, the Middle East, and Europe into the neutral or Communist camp. "National prestige" demanded that the administration play an "indispensible part in sustaining the free world" by protecting the island.[46]

Dean Rusk, Acheson's liaison with the Defense Department, took Dulles's memorandum as a cue to press harder for a commitment to Formosa. In meetings with General John H. Burns, an assistant to Johnson, Rusk endorsed plans to expand military sales to the Nationalists and to initiate "covert action in support of resistance on Formosa." The time had come, Rusk told his own colleagues, to "draw the line in Asia" by defending Taiwan, increasing aid to Southeast Asia and Korea and pushing through a Japanese settlement.

To prevent a Sino-Soviet breakthrough, Rusk submitted an action plan to Acheson and Truman. Washington should arrange for Chiang Kai-shek's "retirement." His successors on Taiwan would request UN protection and, pending resolution of the island's status, the Seventh Fleet would "prevent action by or against Formosa which would disturb the peace." MacArthur, of course, had advocated similar moves since late 1949.[47]

Ambassador Wellington Koo later recollected that during the spring MacArthur also explored the possibility of military cooperation between Taiwan and South Korea. Besides his personal knowledge of the general's activities, in 1970 Koo learned from General Ho Shi-chi, formerly a Chinese member of the Allied Council in Tokyo, that in May and June 1950, Ho, MacArthur, and retired Admiral Charles Cooke (then a military and political adviser to Chiang and a conduit of information to the congressional China bloc) requested that Chiang deploy several thousand Nationalist soldiers in South Korea in anticipation of a northern invasion. Chiang demanded so much payment that the "discussions were still going on when the war in Korea actually broke out." If true, this suggests that MacArthur had embarked on a private foreign policy in Asia before June 1950.[48]

The general made no secret of his support for and involvement in other activities pursued by Cooke, Chennault, Goodfellow, and Donovan. Besides intelligence gathering and lobbying with Congress, these included plans for a volunteer American air force, for transferring weapons from Japan to Taiwan, and for arming anti-Communist guerrillas on the mainland. Cooke and MacArthur were especially disturbed by the possibility of the Chinese Communists or Soviets capturing Taiwan and utilizing it as a jet fighter base to control Southeast Asia.

When the Joint Chiefs still opposed using American forces to defend Taiwan, MacArthur made his strongest pitch yet. On May 20, he likened the island to an "unsinkable aircraft carrier and submarine tender" that anchored America's Pacific line of defense. Should Taiwan pass into Communist hands, the United States would have to fall back to Hawaii and the West Coast. General Bradley found this warning so stark he brought it "to the personal attention of the President."[49]

Asia's fate, MacArthur told several officials in Tokyo, hinged on keeping Communist forces off Taiwan. It made no sense to defend anyplace else if Washington wrote off the island. The crisis was so severe that he no longer saw an "urgent need" to conclude a treaty with Japan. Voicing new fears of a Communist breakthrough, the general spoke of compelling Tokyo to make substantial postoccupation bases available to the United States and ordered a tough crackdown on the Japan Communist Party.

Did MacArthur actually consider Taiwan the pivot of the Asian balance of power? Had events really convinced him that the United States should retain bases in Japan and postpone a treaty? The general, it seems fair to say, did not judge these cases on strictly military merits. Most likely, he hoped to strike a deal linking a Japanese settlement with a command on Taiwan.

Whatever the terms of an eventual settlement with Tokyo, Japan had become a professional dead end. The Washington-controlled recovery program constrained nearly all his action. Even if he presided over a peace conference, the moment of great political symbolism had already passed. Nothing short of a heroic stand at some Asian outpost could breathe new life into the fraying MacArthur legend—or free him from hated civilian control. Taiwan's importance to American security was questionable; as a tonic for the aging general, it had no competition. Almost desperately, he coveted responsibility to defend the Nationalist redoubt.

During May, MacArthur again tried to pressure the Defense Department into a quick withdrawal from Japan. He told journalists that the United States could end the occupation easily without compromising Japan's security. As the "Switzerland of Asia," the world would respect its neutral status. This "Japanese Switzerland Neutrality stuff," commented Robert Eichelberger, outraged Pentagon planners. His former boss "must be nuts to get home," or out of Tokyo, "to put out that stuff." It was one more example of "what a selfish 'to hell with the other fellow' he is." Eichelberger persuaded General Bradley and Defense Secretary Johnson that MacArthur's views now, as earlier, reflected his political aspirations.[50]

As word spread of MacArthur's efforts to link the Japan treaty to protection of Taiwan, Ambassador Koo jumped on the bandwagon. He urged Defense officials and Dulles to encourage MacArthur to send additional military supplies, planes, and pilots from Japan to Taiwan. The general, Koo hoped, would have a free hand in "making such equipment and supplies available to Formosa without necessarily obtaining the formal approval of Washington."[51]

As Koo spoke, Dulles and the Defense officials planned separate visits to Tokyo. Dulles hoped to coordinate a Japanese settlement with the containment of China, while the military delegation would press MacArthur to postpone a treaty and make Japan a more "active ally" in their plan to end "Russian domination of Manchuria and China."

State Department planners now envisioned a treaty linking Japan economically and politically to the states on China's periphery. A prosperous, pro-American Japan, Dulles argued, would be more useful to Washington than a heavily armed but dissatisfied ally. If United States "stood fast" in Korea and Taiwan, he believed, the Soviets would hold back and make it possible to protect Japan with a "defensive guarantee" and a small American military presence. However, if the Communists took Korea or Taiwan, the United States would be compelled to offer greater protection to Japan. Defense officials did not dispute the minimal Soviet threat. They opposed neutralization and justified a longer occupation, retention of bases, and rearmament on other grounds: Japan's utility as a forward basing area in case of war with the Soviet Union of China.[52]

The Dulles party arrived in Tokyo on June 17. Almost immediately, the diplomat departed on a four-day visit to South Korea where he toured the 38th parallel and made a rousing anti-Communist speech to the legislature in Seoul. Louis Johnson and the Joint Chiefs landed in Tokyo the same day as Dulles, but the two delegations avoided much contact. Almost immediately, Johnson ignited a public row by delivering a "fifteen minute harangue" in which he denounced Dulles as an "impractical man who approached the world's problems with a religious, moral and pacifistic attitude."[53]

One member of Dulles's party, Arizona newspaper editor William R. Mathews, passed on some frightening information. In Korea, President Rhee "as much as said" to the journalist that he intended within the year to "take the offensive and take over North Korea." An anonymous American in Seoul with links to Taiwan (perhaps Cooke or Goodfellow) predicted that presidents Rhee and Chiang would coordinate their activity against North Korea and the Chinese mainland in order to draw the United States into their struggle. Within a short time, Mathews told Dulles, "we are going to have a couple of hot potatoes on our hands."[54]

MacArthur urged the two delegations to seize the offensive in Asia by offering Japan a settlement (providing for a limited American military presence) and shielding Taiwan. The policies were integral, he argued, since Soviet air and naval forces operating from Taiwan could isolate and subdue Japan and Southeast Asia. The island's symbolic importance surpassed even its military value. Asians forced to "make a choice between Communism and the West" saw it as the place to draw a "line beyond which Communist expansion will be stopped." Unless Washington was prepared to "abandon" Asia, he should be sent to survey the "military, economic and political requirements" for holding Taiwan.[55]

As MacArthur spoke, the administration had begun to move in his direction. The Defense Department developed elaborate contingency plans to protect Taiwan and use it as a base of operations against the mainland. Even the State Department seemed ready to meet MacArthur, Johnson, and the Joint Chiefs partway. Early in June Acheson told British Ambassador Sir Oliver Franks he would find some method—short of outright military intervention—to shield Taiwan. Assistant Secretary Dean Rusk explored a variety of options on Acheson's behalf. Rusk and representatives of the military, CIA, and Nationalist army pushed forward with plans to depose Chiang and, possibly, place the island under MacArthur's command.

Acheson hoped that a non Communist Taiwan would help deflect Chinese expansion away from Southeast Asia. Once Chiang was out of the equation, he thought it might be possible for the mainland and island regimes to reach at least an informal understanding. However, State Department officials still feared giving MacArthur license. The "moment General MacArthur debarks on Formosa," Acheson's advisers warned on June 23, "we will be committed to the island and its leaders." This made it imperative to remove Chiang as soon as possible.[56]

Louis Johnson and Omar Bradley still opposed a Japanese settlement, despite the shift on Taiwan. Any treaty, they feared, would limit the ability of American armed forces to utilize Japan. MacArthur and Prime Minister Yoshida repeatedly told the visiting Pentagon officials that Japan's contribution to containment should come from reactivating industries vital to the "reconstruction of American armaments." Building up an army and handing over extensive base rights would prove counterproductive. But this argument fell on deaf ears.

On June 23, MacArthur conceded defeat and abandoned several of his most cherished policies. If the Pentagon insisted, he would quit talking about a neutral Japan. The "entire area" of the country, he now agreed, "should be regarded as a potential base" for substantial American forces. In consideration for this, MacArthur recommended that Washington provide Tokyo an additional $300 million per year to close the gap in foreign trade balances. The general also agreed to press the Japanese to expand military production and possibly rebuild their armed forces. Although pleased by MacArthur's concessions, Bradley and Johnson left Tokyo without promising to support a treaty.[57]

The impasse left MacArthur, Japan, and China policy in a bureaucratic gridlock. American security in the Pacific seemed hostage to events on Taiwan and the military's desire for unlimited base rights in Japan. MacArthur and his critics could scarcely guess that within days fate would deal the general an imposing military hand as well as the most complex political dilemma of his remarkable career.

12
The Korean War Begins

On June 25, 1950, "Mars' last gift to an old warrior," as MacArthur later called the Korean conflict, thrust the seventy-year-old general back into the limelight of politics and war. During the following eight months he achieved notoriety surpassing that of the previous decade. At first, his tactical boldness routed the enemy's forces and intimidated domestic critics. Eventually, his strategic recklessness provoked Chinese intervention and a clash with his civilian and military superiors. Convinced that his enemies lay along the Potomac as well as the Yalu River, the general sought to exorcise both sets of demons. Ultimately, he provoked his own recall, wounded the Truman administration, and cast a generation-long shadow over American policy in Asia.

Origins of the War

The immediate origins of the Korean War remain shrouded in mystery. A chronic pattern of cross-border incursions and internal violence had prevailed for a year before the large-scale North Korean attack of June 1950. Kim Il Sung may have believed that a quick thrust would incite a rebellion in the south and topple the Rhee regime before America responded. In his memoirs, Nikita Khrushchev recalled Kim proposing this to Stalin in 1949. The Soviet and Chinese leadership apparently gave their tepid support to this adventure, hoping it would prove a limited but useful blow against American influence in northeast Asia.[1]

American policy toward Korea probably confused friend and foe alike, as Congress, the State, and Defense departments all sent mixed signals about future commitments. In January 1950, Republicans in Congress temporarily killed a Korean aid program, piqued at the administration's reluctance to support Taiwan. The Joint Chiefs and MacArthur made no secret of their determination to withdraw American forces from Korea and,

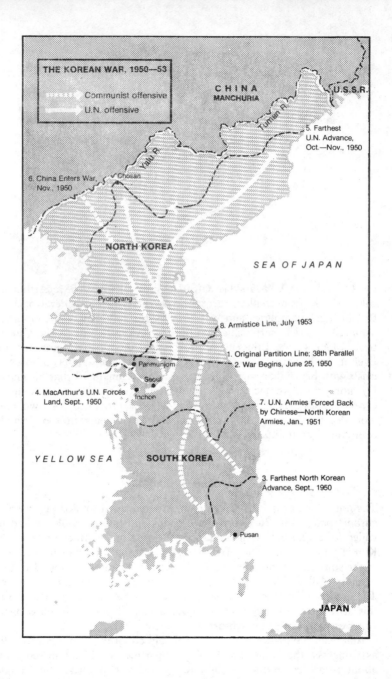

THE KOREAN WAR, 1950—53

Communist offensive

U.N. offensive

CHINA
MANCHURIA

U.S.S.R.

Tumen R.

5. Farthest
U.N. Advance,
Oct.—Nov., 1950

6. China Enters War,
Nov., 1950

Yalu R.

Chosan

NORTH KOREA

SEA OF JAPAN

Pyongyang

8. Armistice Line, July 1953

1. Original Partition Line; 38th Parallel

2. War Begins, June 25, 1950

Panmunjom

Seoul

4. MacArthur's U.N. Forces
Land, Sept., 1950

Inchon

7. U.N. Armies Forced Back
by Chinese—North Korean
Armies, Jan., 1951

YELLOW SEA

SOUTH KOREA

3. Farthest North Korean
Advance, Sept., 1950

Pusan

JAPAN

in case of general war, to write off the entire peninsula. Despite the flap over the January 1950 "Defense Perimeter" speech of Acheson (who, like MacArthur, relegated Korea to a secondary security status), the State Department pushed a modest Korean assistance package through Congress. In March and April, Acheson warned Communist powers against crossing the 38th parallel. American military and diplomatic advisers in Seoul urged limiting the types of weapons provided lest an overconfidant President Rhee invade the north.

American intelligence sources reported a steady North Korean military buildup in the months before the war. Yet analysts in Tokyo and Washington predicted that the Communist regime could not subdue the south without direct Soviet or Chinese help—which few saw as likely. Many experts felt the greatest threat to the Rhee government came from the fierce internal political opposition in Seoul. The domestic crisis deepened in May after Rhee's opponents scored gains in parliamentary election.

As soon as reports of the North Korean attack reached the outside world (June 25, local time; June 24, Washington time), Douglas MacArthur and members of the Truman administration saw a range of threats and opportunities. Most analysts in Washington considered the invasion a Soviet-directed probe of vulnerable free world boundaries, in effect, a test of American credibility toward its allies. If Washington failed to repel this challenge, the Communists would move to intimidate Japan, the Middle East, Europe, and Southeast Asia.

In many ways, President Truman relished the opportunity to prove his resolve to the Russians, the American public, and his Republican critics. As veteran journalist Joseph C. Harsch remarked at the time, "never before" in twenty years in Washington had he "felt such a sense of relief and unity pass through the city" as when Truman moved to defend South Korea.[2]

MacArthur's reactions to the war fluctuated between giddy predictions of victory ("all he had to do was send a few Americans over there and the North Koreans would run") and desperate warnings that the Chinese were on the verge of driving American influence back "to the West Coast." Publicly, he lavished obsequious praise on the president, while he privately condemned Truman and his advisers as appeasers.

Somewhat like Truman, MacArthur saw the conflict as a marvelous opportunity to bloody the enemy (be it North Korea, China, or the Soviet Union), liberate a peripheral area of the Soviet empire, and reverse the pattern of retreat he saw in recent American policy. While the administration (with occasional lapses) saw the war as one of limited means and ends, the general perceived a chance to confront boldly and reverse the Red tide in Asia. On a personal level, he guessed that either victory or martyrdom might finally catapult him into the presidency.

Attack and Intervention

Word of the North Korean attack officially reached Washington on Saturday night, June 24. President Truman, then visiting family in Missouri, received the news in a telephone call from Secretary of State Acheson, who had already requested a meeting of the UN Security Council. Truman arranged to return to the capital the next evening.

Since the withdrawal of occupation forces in 1949, the United States had only vague contingency plans for Korea. Several hundred military advisers assisted the Republic of Korea Army (whose troops were known as ROKs), but they were not a combat force. NSC and Defense and State department doctrine called for the protection and evacuation of American nationals in case of war, but left unresolved whether or to what degree Washington would respond. The depleted condition of American nuclear weaponry and conventional forces in 1950 made it difficult to envision a major commitment of troops to the Korean peninsula. MacArthur's Far Eastern Command possessed four infantry divisions and a small air force equipped mainly with World War II vintage planes. The Seventh Fleet provided more substantial naval support.

Despite the lack of direction and modest forces at his disposal, the general did not look to Washington for guidance. He had, of course, advocated the early withdrawal of occupation troops from Korea and assessed the country as having little strategic relevance. However, he and Willoughby continued to operate intelligence networks there and as noted earlier, may even have had a foreboding of the invasion. Although John Foster Dulles (then in Tokyo) found him confused and uncertain for several hours over how to respond to the attack, William Sebald described the general as "surprisingly ebullient and very confident that all he had to do was send a few Americans over there and the north Koreans would run." He chortled that if "only I could send the 1st Cavalry Division, why heavens, you'd see those fellows scuddle up to the Manchurian border so quick you would see no more of them." MacArthur told John Allison, an assistant to Dulles, that the north's move across the 38th parallel might be nothing more than a "reconnaissance in force." If "Washington will not hobble me," he boasted, "I can handle them with one hand tied behind my back." He underestimated the abilities of North Korean troops nearly as much as he had the Japanese.[3]

MacArthur waited several hours before notifying Washington of the fighting. (Ambassador John Muccio communicated the news the night before.) Possibly, he delayed conferring to prevent Washington from questioning his dispatch of a shipload of munitions—under air cover—to South Korea. By the time the Pentagon sent tentative approval for such a shipment (pending the president's OK), MacArthur had already acted. Thus his pattern of anticipating directives and forcing the administration to follow began at the war's outset.[4]

The UN Security Council convened on Sunday, June 25, to discuss the

Korean question. For several months, the Soviet Union had boycotted council meetings, protesting its refusal to seat the People's Republic of China. Whether intended or not, this boycott hurt Peking's case and led some observers to suspect that Moscow preferred an isolated China. In any case, no Soviet delegate was present to veto an American-sponsored resolution condemning North Korea for a "breach of the peace," calling for a cease-fire and the withdrawal of forces, and requesting member states to assist the UN "in the execution of this resolution."[5]

Later that day, the president returned to Washington to consult with senior diplomatic and military advisers. En route to the meeting, he told Assistant Secretary of State James Webb, "By God, I'm going to let them have it." Unconcerned with details, Truman framed the question as one of how, not whether, to intervene on behalf of South Korea. Still, he probably hoped that indirect assistance would allow the Koreans to repulse the invaders. The president also wanted to make certain the Soviets were not using the attack as a smoke screen for moves elsewhere. He hoped to forge a consensus within the administration and with Congress before taking decisive steps.[6]

At a conference at the Blair House (the president's residence during the White House renovation), Acheson urged Truman to authorize MacArthur to provide supplies and air cover to South Korea. (The general, of course, had begun to do so.) The secretary of state "suggested that the President should order the Seventh Fleet to . . . prevent an attack on Formosa from the mainland." (Omar Bradley had just read to the group MacArthur's proposal on Taiwan, which contained this idea.) Acheson still distinguished between keeping the island out of Communist hands (while its status might eventually "be determined by the UN,") and allying with Chiang Kai-shek. He hoped the presence of the fleet would buy time and possibly defuse tensions by preventing attacks in either direction across the Taiwan Strait. The secretary also urged that "aid to Indochina should be stepped up."

During this forceful presentation, Johnson and the Joint Chiefs seemed almost reticent. They had traditionally placed a higher strategic value on Taiwan than Korea, and General Bradley and Admiral Sherman continued to talk along these lines. Sherman suggested extending MacArthur's authority over the island. At the least, Bradley asserted, MacArthur should be sent to survey Taiwan's defense needs. Acheson protested that although America had a real interest in keeping the island out of Soviet and Chinese hands, the United States should avoid any commitment to Chiang's clique. Oddly, he found an ally in Defense Secretary Louis Johnson, who spoke out against giving MacArthur "too much discretion" or of delegating "presidential authority" to a notoriously stubborn field commander.

Both the president, who ordered contingency plans to "wipe out all Soviet air bases in the Far East," and Acheson seem to have startled the Joint Chiefs, who were not accustomed to hear such strident talk from civilian leaders. General Bradley actually worried that the focus on Korea might obscure more pressing concerns in Japan, Okinawa, Taiwan, the Philippines, and Southeast Asia. He feared the attack in Korea might prove a diversion staged to distract the United States from more important targets. Alternatively, he and his colleagues worried that the Soviets had put on hold their "designs against Europe and the Middle East for an all out push in the Far East conducted by [Stalin's] satellites." Still, as Bradley later wrote Truman, "everyone present seemed to be of the opinion that the failure to take action to protect South Korea would be appeasement and History proved that one appeasement leads to another and this inevitably leads to war." Only near the end of his life did Bradley reexamine this premise. Back "in those days," he admitted, "we held the rather simplistic belief that *all* communist moves worldwide were dictated from Moscow and by Stalin personally."

Although aware of these concerns, Acheson and Truman worried more about the challenge to American credibility. They also knew, even if they did not say so, that partisan politics had to be factored into any decision. For months, Republican critics, including Senators McCarthy and Taft, had accused them of cowering before Asian communism and of abandoning China. From Tokyo, Dulles (the leading Republican within the administration) warned that "to sit by while Korea is overrun by unprovoked armed attack would start a disastrous chain of events leading most probably to world war." Truman and Acheson may have hoped that by drawing the line in Korea they could retain control of events and deflect pressure from both Congress and the Defense establishment for direct action against China. As one student of the issue argues, both men were reluctant to admit political motivation by themselves or by others. As a result, they portrayed the Korean crisis as "part of a global challenge directed from Moscow at American collective security systems."[7]

The president spent a day mulling over his options before resuming discussion at Blair House on June 26. Again taking charge, Acheson pressed for a military commitment. With the invaders advancing against light resistance, he urged that an "all out order" should "be issued to the Navy and Air Force to waive all restrictions on their operations in Korea." Truman even agreed to consider eventual operations against North Korea.

These actions, Acheson stressed, were linked closely to intervention in the Taiwan Strait, increased military assistance to the Philippines, and a decision to send a "strong military mission" to Indochina. To Acheson's horror, Truman proposed that Taiwan be declared "part of Japan" and placed under "MacArthur's command." The president revealed that Chiang had promised to "step out" of power if the Americans would defend the island. Perhaps he would do so now "if MacArthur were put in."

The secretary of state warned of dire international complications if Washington got "mixed up in the question of Chinese administration of the island." Restoring Taiwan as a virtual colony of Japan, of course, would, outrage most Asians. Also, Acheson shuddered at the thought of placing Chinese territory under MacArthur at a time when covert American efforts to depose Chiang continued. Truman and Louis Johnson chimed in to ridicule all Nationalist leaders as swindlers who had pilfered funds into foreign banks and real estate. To Acheson's relief, the president dropped his proposal.

Late that evening, the president confided his inner thoughts to assistant George M. Elsey. Standing in front of a globe, Truman said he was "more worried about other parts of the world" than Korea. Providing military assistance there and evacuating Americans "was easy and clear." But "what he worried about . . . was the Middle East." He "put his finger on Iran and said: 'Here is where they will start trouble if we aren't careful.' " Korea had become the "Greece of the Far East." By standing "up to them like we did in Greece three years ago, they won't take any next steps" in Iran or the Middle East. If "we don't put up a fight now," Truman cautioned, there was "no telling what they'll do." This convinced Elsey that the president "sincerely determined to go very much further than the initial orders that he had approved for General MacArthur."[8]

The next morning Truman and his aides briefed congressional leaders (although they declined to seek formal legislative approval) on these decisions. Washington had to act, Acheson explained, because the South Koreans were losing and the Europeans "appeared to be in a state of near panic." The North Korean invasion, Truman insisted, was "very obviously inspired by the Soviet Union." Should it succeed, the Russians would "keep right on going and swallow up one piece of Asia after another," continuing into the Near East and Europe. Besides supporting the Koreans, it was "equally necessary for us to draw the line at Indo-China, the Philippines and Formosa."

The president informed the group that American air and sea forces (though not, as yet, ground troops) would assist South Korea. The attack across the 39th parallel proved that "Communism has passed beyond the use of subversion to conquer independent nations and will now use armed invasion and war." By blaming communism, rather than the Soviets, Acheson explained, he sought to permit Stalin a graceful exit. Since Taiwan, in enemy hands, would pose a "direct threat to the security of the Pacific area and to the United States forces performing their lawful and necessary functions in that area," the Seventh Fleet would prevent any armed movement in either direction across the Taiwan Strait. Truman also announced an acceleration of aid to the Philippines and to Indochina and the dispatch of an American military mission to provide "close working relations" with anti-Communist elements in Indochina.

Later that day the UN Security Council (with the Russians absent) approved a second American-sponsored resolution calling on member states to "furnish such assistance to the Republic of Korea as may be necessary to repel the armed attack and to restore international peace and security in the area." This provided justification for American intervention as well as for MacArthur's command of a new theater of war.[9]

Most officials supported these actions. Former Chief of Staff Dwight

Eisenhower told top Army officials the administration might have erred only by making too "partial . . . appeal to force." Unless Washington did "everything possible," the country would soon face "a dozen Koreas." Ike urged Omar Bradley and his staff to move in a "dozen directions" and be "prepared for whatever may happen, even if it finally came to the use of the A-Bomb (which God forbid.)"[10]

MacArthur Takes Command

Following the second Blair House meeting, Army Secretary Frank Pace, Jr., authorized MacArthur to dispatch a survey team to Korea along with ammunition and equipment for ROK forces. Pace also instructed him to use air and navel forces to evacuate American civilians and protect Seoul. To assist this mission, MacArthur received control of the Seventh Fleet. The general boasted to William Sebald, he had already "sen[t] munitions . . . under fighter cover" to Korea. But, he added, "we need have no fear he would start a war."[11]

By June 27 (local time), the situation in Korea deteriorated. Even though Far East Air Forces (FEAF) fighters downed three North Korean planes, the ROK army collapsed as the Rhee government fled the capital. Both Sebald and John Allison were struck by the change in MacArthur's mood. He now appeared a "dejected, completely despondent man," who "spoke in terms of writing off Korea."[12]

That same day the general dispatched a survey team to the battlefront. When it recommended committing American troops, MacArthur decided to make his own inspections. On June 29, he and a few members of his staff flew to a small airfield south of Seoul where they witnessed the collapse of ROK resistance. At that moment, MacArthur later reported, he decided to request combat troops and mount an amphibious assault behind enemy lines.[13]

While the general drafted his appeal to Washington, the president met with his cabinet on June 29 to discuss a directive to MacArthur. Should air and sea operations be mounted against North Korea? What if Soviet forces appeared? Truman insisted that MacArthur's instructions contain no "implication . . . that we are going to war with Russia at this time." He favored "any steps we have to [take] to push the North Koreans behind the line [i.e., 38th parallel]," but opposed "getting us over-committed to a whole lot of other things that could mean war."

Defense advisers felt that imposing "any limitations" on the field commander might undermine his ability to hold a bridgehead. Truman countered that while it might be permissible to destroy air bases, supply dumps, and so forth, north of the 38th parallel, the American goal was to "restore order" to the boundary. Although MacArthur should "keep the North Koreans from killing the people we are trying to save," the president opposed granting authority "to go north of the 38th parallel." Acheson concurred, adding that American planes operating in the north must not stray over China or the Soviet Union. If they did, "no one would be able to foresee the consequences."

Unless the United States dispatched ground troops, Acheson argued, the south would be overrun, which would cause a political "disaster." He considered it "essential to give the commander on the spot whatever he need[ed]." If the Soviets entered the fighting, American units had every right to defend themselves. Given the relatively muted Russian response to events thus far, Acheson and Truman considered intervention unlikely. Since Moscow had complained mostly about American violations of Chinese territory, Truman predicted that "the Russians are going to let the Chinese do the fighting for them."

Both Acheson and Truman worried about the accuracy of MacArthur's reports. The president instructed Louis Johnson to order the general to submit "complete daily reports." It "was just as hard to get information out of" MacArthur now, he observed, "as it had been during the war."[14]

About 5 A.M. (Washington time) the next morning, MacArthur informed officials at the Pentagon that unless two divisions of combat troops were dispatched immediately, he would be unable to hold even a bridgehead. Army Secretary Pace telephoned Truman who authorized sending a regiment at once and promised a quick decision on the two divisions. He also expressed renewed interest in "accepting two divisions offered by the Chinese Nationalist Government," even though it would outrage Peking. Still, he reminded himself, we "must be careful not to cause a general Asiatic war," which might bring a Russian "attack in the Black Sea and the Persian Gulf," both "prizes Moscow has wanted since Ivan the Terrible who is now their hero with Stalin and Lenin."[15]

On June 30, the Joint Chiefs formally authorized use of American ground troops to hold the port of Pusan and permitted MacArthur to extend air operations over the north. But, once again, he had jumped the gun. Several hours earlier the general ordered air commander Stratemeyer to begin attacking (with "No Publicity") air fields north of the 38th parallel.[16]

Truman discussed these developments with members of Congress at a meeting on June 30. The president and General Bradley spoke against using any Chinese Nationalist troops. Bradley argued that most foreign forces would prove more trouble than they were worth. Truman asserted (incorrectly) that American units had no combat mission but were sent to keep open vital supply lines in Korea. Responding to Senator Kenneth Wherry's complaint (soon taken up by many Republicans) that he should have sought congressional approval, Truman cited his authority as commander in chief to act in an emergency. He promised to "tell Congress" if "any large scale actions were to take place."

The president stressed the importance of national unity and accepting the fiction that MacArthur acted solely "for the United Nations." Although the general "*was* obeying his [Truman's] order's" exclusively, the American government must "not let ourselves be put in that light to the rest of the world." Despite this charade, the Joint Chiefs insisted that MacArthur submit all reports to them, never to the UN. Their official history states that the JCS were determined to "prevent the UN from involvement in strategy or tactics." By the close of business on June 30, as Dean Acheson put it, "we were . . . fully committed in Korea."[17]

Truman and Acheson did not yet know that Louis Johnson had resumed his behind the scenes manipulations. On June 30, Johnson and his deputy, Paul Griffith, met separately with Wellington Koo. The Chinese ambassador praised the Defense chief as Taiwan's savior. Johnson declared that "it had been a hard struggle in the face of determined opposition from the State Department," but "he had won his fight." He boasted of turning the Korean crisis into a "change of policy vis-à-vis Formosa." Cryptically, he asked Koo to "tell Madame Chiang and Dr. Kung he had kept his promise."[18]

A few days later, Averell Harriman experienced at first hand Johnson's duplicity. During a meeting he attended in Johnson's office, the defense secretary made a call to Senator Robert Taft and "congratulated [him] on a speech he made a few days ago, criticizing the President for not consulting congress before acting in the Korea crisis." The secretary then turned to Harriman and promised that if "they could get Acheson out, he (Johnson) would see that Harriman was made secretary of state." Harriman, who considered Johnson a vulgar parvenu, replied "that he could not be bought that easily," and stormed out to report the incident to Truman. The president's staff considered this evidence of the defense secretary's mental instability and lust for power.[19]

A week later, the UN Security Council requested that the American government establish a unified military command. All participating forces would serve under a commander designated by the president. On July 8, Truman approved the one name forwarded by the Joint Chiefs: Douglas MacArthur. In an effusive response, MacArthur pledged his "complete personal loyalty to you [Truman] as well as an absolute devotion to your monumental struggle for peace and good will throughout the world."[20]

Truman probably held his nose when making the appointment. On July 1, aboard the presidential yacht *Williamsburg*, aides Eben Ayers and Charlie Ross heard an earful from their boss "about the Korean situation and about General MacArthur for whom the President has little regard or respect." Truman described MacArthur as a "supreme egoist who regarded himself as something of a god." Recalling the collapse of the Philippines in 1942, the president berated the general for allegedly deserting his command and escaping to become "a hero and dictator in Japan."

About the same time, John Foster Dulles returned from Tokyo and briefed the president about MacArthur's response to the war. He reported that the general's subordinates all cowered in fear of telling him of the attack. The diplomat finally brought the message himself. Concerned by MacArthur's refusal to face reality, Dulles recommended "hauling" MacArthur "back to the United States." The president probably agreed, but remarked that "the General is involved politically in this country," had been considered as a Republican presidential candidate, and that "he could not recall MacArthur without causing a tremendous reaction in this country where he has been built up to heroic stature."[21]

Whatever his personal feelings, the president should have hesitated before naming MacArthur commander. The general was now past seventy and certainly not in peak form. He already had two full time jobs, as occupation commander in Japan and as overall commander of American forces in the Far East. Piling on additional responsibilities would strain even the most capable officer and raise serious questions about how he could adequately perform the several roles. It appears that little thought was given to retaining MacArthur as SCAP while selecting a younger general to take charge in Korea.

Perhaps Truman really did fear the political repercussions of passing the general over. Given the reaction to his later firing in April 1951, this was no idle concern. The president may have hoped that placing MacArthur in charge (and giving him responsibility) in Korea, would defuse Republican attacks on the administration's Asia policy. With many members of Congress edgy about lack of consultation, neither Truman nor his advisers were eager to provoke additional Republican wrath by slighting their favorite officer.

The War Abroad and at Home, July–September

During the first few weeks of July, North Korean forces rolled steadily south against crumbling ROK resistance. Although many civilians fled before the advance, a sizeable portion of the populace assisted, or accepted, the invading army. Politically and militarily, the Republic of Korea collapsed like a house of cards. Although no mass uprising occurred, the Seoul regime proved incapable of rallying internal support.

Forced to commit piecemeal his limited forces from Japan, MacArthur counted on holding a line around the port of Pusan. Despite the desperate situation on the ground, the Far East Air Forces quickly established air supremacy, which was never seriously challenged.

American forces were ill prepared for the battles they faced. The Eighth Army had become flabby, poorly trained, and understrength during the Japanese occupation. Although the general could not be faulted for Washington's failure to augment men and weapons before mid-1950, he shared responsibility for the lax standards that had set in over the previous few years. Also, in spite of his alleged expertise, MacArthur clearly had no idea about the fighting qualities of his enemy.

The Far East Air Force suffered similar problems. Since 1945 the nation's air power doctrine focused upon strategic nuclear bombing of the Soviet Union, relying upon aging B-29s based in Britain. The best tactical fighters and bombers did not go to Japan. North Korea's limited air power, and China's decision not to commit major air strength, allowed the FEAF to control the air throughout the war. Even when FEAF resources increased after 1951, air power never proved decisive. North Korean cities and industry were attacked more heavily than enemy forces.[22]

MacArthur continually escalated his demands for men, equipment, and support. On July 8, he estimated a need for over four divisions; the next day he doubled the request. Such a commitment, he knew, would not only strip Japan of American troops, but would involve virtually the entire combat force then fielded by the United States. Worried Pentagon planners sent Generals J. Lawton Collins and Hoyt Vandenberg to Tokyo in mid-July to assess the situation. Collins recalled that MacArthur's plea for more men "gave me the impression of addressing not just his immediate listeners, but a larger audience, unseen." At the same time, the UN commander revealed plans for an amphibious assault behind enemy lines.

Although they could not meet all his demands, Truman and the Joint Chiefs worked hard to augment MacArthur's forces. National Guard and Army Reserve units were mobilized quickly. Congress appropriated an extra $10.5 billion in military funds and, by year's end, raised total defense spending from $13 to about $50 billion. By the end of July, some 45,000 troops reached Korea and the total rose to nearly 150,000 a month later. By late September, MacArthur commanded nearly a quarter million troops, virtually all of the nation's ground combat units.[23]

American B-29 bombers obliterated many urban and industrial targets north of the 38th parallel during August. Even though fierce ground combat continued along the Pusan perimeter throughout that month, American and ROK forces held the line. By early September, the momentum of battle changed. The defenders already outnumbered the attacking army and enjoyed a rapidly growing advantage in tanks, artillery, and aircraft. American supply ships and planes moved from Japan without Soviet or Chinese challenges in the air and sea around South Korea. MacArthur promoted twin strategies he deemed essential for ultimate victory. The first involved a daring amphibious landing behind enemy lines to crush Communist forces in a pincers. He also prodded the administration to alter its policy toward Taiwan and China. The Nationalists, he insisted, could play a vital role even if their troops were not used in Korea. The more Communist forces deployed along the Taiwan Strait, the fewer Peking would have available to shift north or consolidate its grip on the country. MacArthur had no use for the distinction between shielding Taiwan and supporting Chiang on or off the island.

For some time after June 25, Acheson and his allies in the administration continued to work for the generalissimo's removal. Presidential aide Matthew Connelly, aid administrator Harlan Cleveland, and Ambassador Phillip Jessup told Wellington Koo and other Chinese officials that only when Chiang left the scene (voluntarily or otherwise) would "the U.S. go all out to help [Nationalist] China without hesitation."[24]

In a message to Congress on July 19, the president declared that the

United States had no territorial ambitions toward Taiwan, did not intend to utilize it as a base for attacking China, and hoped that the UN could play a part in determining its future status. With his usual hyperbole, MacArthur sent a note to Truman calling this a "great state paper, in ultimate effect perhaps the most significant of modern times, for it means that the United States has determined that the Pacific areas shall be free." The general saw in it a "turning point of this era's struggle for civilization" and exclaimed his immense pride in "serving under your leadership."[25]

MacArthur seemed primed to read more into the president's declaration than Truman intended. In mid-July, Generals Collins and Vandenberg were perplexed by the fact that MacArthur seemed fixated on events in Korea and "less concerned with the world situation . . . than were the Joint Chiefs." He blithely told them that a battlefield victory on the peninsula would "check Communist expansion everywhere and thus obviate the necessity of our being fully prepared to meet aggression elsewhere," especially in Europe. He wanted to "grab every ship in the Pacific and pour the support into the Far East." "Delays or half-way measures," he warned, would prove disastrous.

Once he had sufficient forces to halt the enemy offensive, the UN commander intended to begin a counterattack with an amphibious landing behind enemy lines. After destroying Communist troops on *both* sides of the 38th parallel, he would "compose and unite" Korea—even though he had no authority yet to do so. Vandenberg worried this might provoke Chinese intervention and require "an advance into Manchuria." MacArthur doubted it. But since Manchurian and Siberian links with Korea depended on tunnels and bridges, "he saw a unique opportunity for the use of the atomic bomb to deal a crippling blow to these supply routes."

(In an independent effort to deter expanded Chinese and Soviet assistance to the North Koreans, during late July and early August the president approved the temporary deployment of nuclear-capable B-29 bombers to bases in the United Kingdom and Guam. The planes carried only partial bombs, as the fissionable cores remained in the United States under Atomic Energy Commission control. Washington expected Soviet intelligence to learn of the aircraft movement and draw the appropriate warning. As the Korean battlefield situation improved in September, these specially configured B-29s were withdrawn from the Pacific.)

Although MacArthur's reference to atomic warfare (his first of several) did not become public knowledge, his other ideas leaked out. On July 9, Drew Pearson learned from several sources that the general "proposed . . . bombing of the South Manchurian railway carrying Russian supplies to the North Koreans." These reports must have reached Soviet and Chinese ears nearly as quickly as Pearson's.

MacArthur told General Collins in July that he intended to visit Taiwan. Like the administration, he opposed accepting Chiang's offer of Nationalist troops for Korea. They would only "be an albatross around our necks for months." Instead, he wanted more American troops and a free hand to control events. MacArthur concluded his meeting with the visiting army chief with an inspirational declaration. "We win here or we lose everything; if we win here, we improve the chances of winning everywhere."[26]

Two weeks later, on July 27, the NSC accepted a proposal made by Louis Johnson and the JCS to begin reconnaissance flights along the China coast to observe any buildup threatening Taiwan. Before Acheson intervened, the council almost approved a companion plan authorizing the Nationalists to launch preemptive attacks against the mainland in spite of the "neutralization" policy. MacArthur, who supported both proposals, decided to visit Taiwan himself and notified the JCS of his plan.

Even "if he [Chiang] has horns and a tail," MacArthur told William Sebald, so long as the generalissimo was "anti-communist we should help him." The time had come for the "State Department to forget its vendetta . . . and assist him." The Joint Chiefs suggested postponing the visit but left it up to MacArthur.[27]

On July 31, the general and a dozen officers arrived in Taiwan and closeted themselves for two days with Chiang's entourage. Before going into closed session, however, MacArthur took Chiang's arm and in earshot of reporter Burton Crane asked, "Well, what do you people want?" At the end of the talks, MacArthur announced that he had come to assess the island's defense needs. Praising the generalissimo's "indomitable determination to resist Communist domination," he announced "arrangements have been completed for effective coordination between American forces under my command and those of the Chinese Government" to meet any attack.

In lieu of an official record, the best evidence of what transpired comes from the recollections of MacArthur's staff and Ambassador Koo. They assert that no secret deals were made. MacArthur urged the Nationalists to accept American guidance in the distribution of military supplies, brushed aside the offer of troops, and declined to discuss "political" questions.[28]

Even if he had not strayed technically from official policy, MacArthur's cavalier style enraged many in Washington. For example, the general told diplomat William Sebald that he had "no intention of providing details" to the State Department since the talks were his "sole responsibility." Perhaps more important, MacArthur's highly visible embrace of Chiang undermined ongoing efforts to remove Chiang. The generalissimo used the visit to convince supporters and opponents that his leadership did not endanger, but was vital for, American aid.

State Department suspicions of a MacArthur–Chiang deal increased when, following his return to Tokyo, the UN commander recommended sending jet fighters to Taiwan. Worried that the planes would be used to attack China, Acheson convinced Truman (on August 4) to instruct defense secretary and MacArthur that the Taiwan Strait remained neutralized. "No one other than the President as Commander in Chief," Truman declared, possessed "authority to order or authorize preventive action against concentrations on the Chinese mainland."

MacArthur responded contritely to the president's complaints about the aircraft transfer and the Taiwan visit. He assured Truman that he understood "thoroughly the limitations on my authority as theater commander and you need have no anxiety that I will in any way exceed them." He blamed any misunderstanding on false reports by various sources. For the moment, this settled the issue. A few days later, the White House did not object when MacArthur dispatched a survey team to the island. It soon proposed a $271 million military aid package and joint defense measures.[29]

Nevertheless, the president and the Joint Chiefs had lingering doubts about their field commander. Rumors arose concerning the general's emotional stability and health. About this time, T. J. Davis, MacArthur's aide during the 1930s, reportedly told members of the JCS staff about the general's stormy sexual life and suicide threats. Such allegations may have prompted Truman, in August, to delegate an old friend, General Frank Lowe, to report on "General MacArthur's physical condition and ability to withstand stresses incident to his duty." Lowe remained in the theater until April 1951, got on well with MacArthur, and reported very little about any "stress."[30]

At the president's direction, White House adviser Averell Harriman and General Matthew Ridgway accompanied Lowe. Harriman had the task of convincing the UN commander to mute his criticism of administration policy. Although MacArthur greeted him cordially and pledged his loyalty, "for reasons which are rather difficult to explain," Harriman wrote, they failed to come "to a full agreement on the way he believed things should be handled on Formosa and with the generalissimo." MacArthur praised Truman's Korean policy as a "magnificent," decision that would save the world from "communist domination." He outlined for both Harriman and Ridgway plans for an amphibious invasion that would destroy North Korean forces before winter.

MacArthur, Harriman reported, still ridiculed the administration's China policy and boasted of his own credentials as the greatest American expert on Oriental psychology. Westerners, he explained, "hate to die." With "Orientals, life begins with death. They die quietly, folding their arms as a dove" and with that "MacArthur folded his arms and sighed." Washington should delegate the task of handling Chiang to an expert like himself.

MacArthur attributed the State Department's distorted view to reports filed by the consul in Taipei, Robert Strong. A trenchant critic of the Nationalist regime, Strong had enraged MacArthur and Chiang by his probing investigations of their activities before and after the war began, the mysterious financial manipulations of several Chinese leaders and their American friends, and the links connecting MacArthur to anti-Communist soldiers of fortune such as Claire Chennault and Admiral Cooke. (Chennault's Civil Air Transport received contracts directly from MacArthur's headquarters to carry out covert missions in Korea and surrounding areas.) MacArthur's pressure contributed to Strong's replacement by Karl Rankin, a pliant pro-Nationalist diplomat.

The general promised he would "obey any orders he received from the President" and would not permit Chiang to drag the United States into a war with China. But, Harriman noted, MacArthur made the promise "without full conviction." At one point the UN commander even suggested "it might be a good idea" to permit Chiang to attack China. Either he would win or you could "get rid of him that way."

MacArthur argued that increased military assistance to the Nationalists would delay UN membership for China and thereby fracture Mao's support and destabilize the regime. In turn, this would "encourage a split with the Soviets." Instead of "kicking Chiang around," the administration should take advantage of the "great potentialities" for covert operations in China. While he did not tell Harriman, MacArthur had probably approved joint American–Nationalist guerrilla operations against the mainland.[31]

General Ridgway also found MacArthur obsessed by the China question. In the midst of their consultations on the proposed Inchon landing, Ridgway recalled, the UN commander kept changing the subject. Although MacArthur no longer thought the Communists would invade Taiwan, he told Ridgway that he got "down on my knees" and "prayed nightly that they will." He would then go to the front and "deliver such a crushing defeat it would be one of the decisive battles of the world—a disaster so great it would rock Asia and perhaps turn back communism."[32]

Truman ignored these red flags when he told journalists on August 10 that "General MacArthur and I are in perfect agreement." MacArthur issued a press release that same day complaining that his trip to Taiwan had been "maliciously misrepresented to the public by those who invariably in the past have propagandized a policy of defeatism and appeasement in the Pacific." Unnamed enemies, "10,000 miles away," schemed "to promote disunity and destroy faith and confidence in American purposes and institutions and American representatives at this time of great world peril." Harriman's visit, he reported later, "left me with a feeling of concern and uneasiness that the situation in the Far East was little understood and downgraded in high circles in Washington."[33]

The Taiwan issue exploded again in mid-August when an official of the Veterans of Foreign Wars asked the general to send a message to the group's Chicago convention, scheduled for the end of the month. On August 20, he responded with a bitter critique of administration policy designed to dispel "misconceptions currently being voiced concerning the relationship of Formosa to our strategic potential in the Pacific." Drawn largely from earlier, still classified reports, he described Taiwan as an "unsinkable aircraft carrier and submarine tender" and Chiang as a vital ally.

MacArthur attacked the "threadbare argument" made "by those who advocated appeasement and defeatism in the Pacific that if we defend Formosa we alienate continental Asia." His intimate knowledge of oriental psychology proved that Asians "respect and follow aggressive, resolute and dynamic leadership." They had contempt for "timidity and vacillation." Truman had shown courage in sending the fleet to protect Taiwan, thus "light[ing] into flame a lamp of hope throughout Asia." But, he implied, it would take a greater leader to carry that lamp into China.[34]

The White House learned of the statement on August 25, when the September 1 issue of *U.S. News and World Report* went to press. The general's action outraged Acheson and Harriman, and they convinced Truman to convene a meeting the next day of top officials. MacArthur's security breech in publishing classified materials also infuriated the Joint Chiefs. The president "decisively repudiated the statements MacArthur had made" as they were in "direct contradiction" of administration policy. Truman, Acheson, and Harriman interpreted the message as an effort to force the administration into making an "unlimited commitment toward Formosa" despite public assurances that Washington had no designs on the island. The secretary of state brought the president a report predicting that the Soviets and Chinese would use MacArthur's statement as an anti-imperialist rallying cry to mobilize Asian leaders against the United States. America's friends would conclude that "we have an uncontrollable military commander" able to "deliver U.S.—China policy over to the Kuomintang." Chiang would carry out some "provocation" against the mainland while the Chinese Communists would conclude that Washington intended to use Taiwan as a "springboard for an American or American-backed attack against" them.

Despite Truman's insistence that he do so, Louis Johnson asked if "we dare" send MacArthur "a message that the President directs him to withdraw his statement." Johnson feared issuing anything stronger than a "request" that MacArthur "cancel" it. During the following hours, Johnson begged Harriman and Acheson to ask Truman not to compel him to send a formal order to Tokyo. In fact, the White House staff learned from Deputy Defense Secretary Steve Early that Johnson even opposed sending an oral reprimand. Presidential aide George Elsey recommended sending the withdrawal order along with an official statement on China policy. Unless they "nailed down MacArthur" he would try to "wiggle out" of compliance by claiming ignorance of "policy on Formosa." A frustrated Truman finally dictated a letter to Johnson and ordered him to send it to the general.[35]

Even though MacArthur complied with the order, the contents of the VFW letter were published widely. He also continued to defend the substance of the remarks as "personal opinion" to which he was entitled. Truman cushioned the rebuke by sending the general a personal note elaborating on American objectives toward China and Taiwan as expressed in the president's statement of July 20 and a letter sent to the UN on August 25.

The incident left MacArthur increasingly resentful of Truman and those close to him. Speaking to William Sebald, he "burst into a long tirade about a State Department clique attempting to undermine his position." MacArthur "pilloried the State Department and said that one of these days he intended to blast them wide open." Truman felt equally bruised, revealing later that at the time he contemplated removing the general from his Korean command.

The president displaced at least some of his resentment on Louis Johnson. The defense secretary's bizarre behavior when faced with confronting MacArthur, coupled with his undermining Acheson and encouraging Taft, proved intolerable. Early in September, Truman fired him and asked the venerable George C. Marshall to come out of retirement and assume the Defense post.

The appointment angered many Senate Republicans who still blamed Marshall for the "loss of China." Senators William Jenner, Joe McCarthy, and their ilk attack the famed general's record in war and peace, describing him as a "living lie." Although MacArthur said nothing about the appointment, Marshall's return could not have pleased him. (Adding insult to injury, Truman soon appointed Eisenhower NATO commander.) Marshall made a tremendous effort not to allow his personal feelings to influence his handling of MacArthur. But aside from that, age had rendered Marshall's judgment less acute than during the previous war. As a result, he exercised less restraint over MacArthur than Truman anticipated.[36]

Inchon and Beyond

As he began planning a counterattack in July, MacArthur appointed his chief of staff, Lieutenant General Edward M. Almond, commander of the newly designated X Corps. Almond expressed some concern about adding this operational command to his already substantial responsibilities in Tokyo. His boss told him not to worry since "we'll all be home by Christmas."[37]

On September 15, a huge naval flotilla and marine units assaulted the port of Inchon, on Korea's west coast, almost 200 miles north of the Pusan perimeter and only 20 miles from Seoul. At the same time, the Eighth Army launched an offensive from Pusan and in a few days the enemy was in full retreat. Communist forces fled toward the 38th parallel, and by September 28 American units recaptured Seoul. The sudden reversal of military fortunes had a dramatic impact upon both MacArthur and the Truman administration.

Ever since the summer when MacArthur conceived this operation, nearly all the military and civilian officials briefed on it had raised objections. Inchon's extreme tides, mud flats, and high sea wall made amphibious operations exceptionally risky. MacArthur retorted that these factors provided the advantage of surprise. Also, the port's northern location would speed the recapture of Seoul. The Joint Chiefs urged MacArthur to land further south, nearer Pusan. With practically all American ground combat units already committed to Korea, a failure would leave the United States extremely vulnerable. It seemed incredible to bet everything on one operation and the sweep of tides.

Undaunted, MacArthur compared his plan to that devised by British general Wolfe during the assault on Quebec in 1759. The "prestige of the Western world," he declared, hung in the balance. "The Communist conspirators" had elected to "make their play for global conquest" in Asia and there they must be fought. Dismissing events in Europe as a war of "words," MacArthur told his colleagues that the "fate of Europe" would be decided in Asia. Even though he refused to inform them of vital operational details, the general's drive and optimism persuaded the Joint Chiefs to approve his plan.[38]

Inchon proved a deceptively easy victory. The port was only lightly defended and the landing went largely uncontested. The apparent collapse of the Korean People's Army (in fact, no senior officers were captured, many troops escaped, and large reserves remained in the north) changed the military balance on the peninsula and the political dynamic in Washington. MacArthur's will alone seemed enough to vanish enemy troops and bureaucratic critics. As General Collins noted, the "success of Inchon was so great" and the "subsequent prestige of General MacArthur was so overpowering" that the Joint Chiefs "hesitated thereafter to question later plans and decisions of the general which should have been challenged." In retrospect, Inchon marked the final important victory in the streak going back to New Guinea. Following Inchon, Collins observed, MacArthur marched "like a Greek hero of old to an unkind and inexorable fate."[39]

Even before the battle turned, the Soviet Union and China had indicated a desire to limit the war. In August, the Soviet delegate returned to the UN Security Council. Moscow condemned American policy in general terms but issued no specific threats. Meanwhile, Soviet spokesmen hinted about a compromise in discussions with British and American diplomats. The Chinese criticized bitterly Washington's decision to shield Taiwan but also decided to stay out of the Korean conflict so long as North Korea was not threatened. British Foreign Minister Bevin and Indian diplomats suggested several formulas for restoring peace at the 38th parallel. These involved giving Peking a UN seat and, possibly, arranging the transfer of Taiwan to China.

The Truman administration spurned all talk of a diplomatic solution before and after Inchon. Truman and Acheson feared Republican criticism if they "compromised" with communism just as MacArthur prepared to trounce the enemy. The administration understood that the rapidly growing military budget would permit the strengthening of NATO and, perhaps, the rearmament of Germany. In effect, the war played midwife to the NSC-68 rearmament program. In October, George Marshall actually voiced regret that the war might end before the American people gained a "full understanding of the problems we face ahead of us." The massive rearmament now under way might be curtailed if peace returned too quickly in Korea.

Even though most Republicans supported the War (while often criticizing the lack of congressional consultation), they continued to attack the administration's China policy. When Marshall replaced Johnson at the Pentagon, Senator Taft and several other Republicans charged the shift represented a plan to sacrifice the Chinese Nationalists. Either out of belief or calculation, Secretary of State Acheson (whom *Life* magazine in September labeled "the symbol of appeasement of communism everywhere in Asia") rejected any idea of linking peace to Chinese control of Taiwan. This would only "whet communist appetites and bring on other aggressions elsewhere." As a result, the State Department discouraged any initial peace talks.[40]

Truman and his advisers were reluctant to explore negotiations for another reason: they were evolving plans to occupy the entire peninsula and extend the authority of the Republic of Korea to the Chinese and Soviet borders. As early as mid-July, MacArthur indicated his desire to unify Korea. Like the general, most Pentagon planners saw little danger of Soviet or Chinese retaliation. A handful of State Department officials—especially former PPS head George Kennan—urged caution and suggested a deal restoring the status quo in Korea, neutralizing Japan, and arranging a UN plebescite for Taiwan. Kennan saw that a policy of rolling back communism had replaced the more measured containment policy and also worried about the extreme latitude given MacArthur. Intelligence analysts in the CIA warned that China and the Soviet Union were unlikely to tolerate destruction of the North Korean regime.

A final decision to cross the 38th parallel awaited the American counteroffensive in mid-September. By then, however, the United States was on a roll and the idea of halting at the prewar boundary evaporated. Informal statements by Syngman Rhee and American officials during the summer suggested that the 38th parallel no longer had significance and that the restoration of peace required Korean unification. Early in September the NSC approved the concept of conducting ground warfare in the north, with the question of political unification deferred. Since the NSC defined America's goal as insuring the "complete independence and unity" of Korea, this all but insured a large scale move north of the parallel.[41]

As American forces cleared South Korea during late September, both MacArthur and the Truman administration saw a unique opportunity to roll back communism and inflict a major symbolic defeat on the Soviet bloc. Truman and Acheson still viewed this victory in limited terms. They would punish aggression, humble the Communist bloc, strengthen pro-American states on China's periphery, and restore domestic and international faith in American commitments. MacArthur, as we will see, thought in far more expansive terms. This tide of optimism obscured the general's military judgment and clouded the administration's perception of Chinese and Soviet reactions.

On September 27, the Joint Chiefs authorized MacArthur to cross the 38th parallel with the object of destroying all Communist forces. He could continue operations so long as no signs existed of actual or impending Soviet or Chinese intervention and with the understanding that only ROK troops should be used in the provinces bordering Manchuria and Siberia. The next day he informed the chiefs that the Eighth Army would advance overland to Pyongyang while X Corps traveled by sea to mount an amphibious assault on the east coast port of Wonsan, 100 miles above the old border.

In what can only be called an unfortunate attempt to reassure his old rival, Defense Secretary George C. Marshall added a personal message. "We want you to feel unhampered strategically and tactically," Marshall wrote, "to proceed north of the 38th parallel." MacArthur interpreted the secretary's encouragement as a license. He responded by declaring "all of Korea open for our military operations unless and until the enemy capitulates." Troops of the ROK crossed the border on October 1, followed by American forces a few days later.[42]

MacArthur expressed slight gratitude for Marshall's confidence. Instead, he complained privately about Washington's excessive caution and "unwarranted interference in the conduct of the war." British diplomat Alvary Gascoigne found him ready "to take on the Chinese or even the Russians" if necessary. He "spat blood at Dean Acheson." Gascoigne's graphic dispatches to the Foreign Office described him as "pretty dangerous."[43]

The China Factor

The decision to unify Korea confirmed China's fears about America's new enthusiasm for sweeping away Communist regimes in divided countries. To this point, public declarations be Premier Chou En-lai (Zhou Enlai) and other officials stressed Peking's fury at America's resumed protection of Taiwan and its campaign to keep the People's Republic out of the UN. Now, besides their fraternal support for Kim Il Sung, Chinese leaders became concerned with the growing threat to Manchuria.

From Mao Tse-tung on down, Chinese leaders agreed they must respond to America's thrust across the 38th parallel. Combined with United States activities in the Taiwan Strait and Indochina and the incipient rearmament of Japan, American strategy seemed bent on encircling China. Peking reacted cautiously to the initial dispatch of U.S. forces to Korea. During July and August about 90,000 additional troops were transferred to northeast China augmenting some 120,000 already there. China also dispatched some antiaircraft units south of the Yalu River to protect bridges connecting Korea with Manchuria. MacArthur's highly publicized visit to Taiwan on July 31, Chinese historians report, convinced Peking that at least the general, if not Truman, planned to escalate the war against China.[44]

Even before the Inchon landing, the Chinese press asserted the indivisibility between Chinese and Korean security. Following the successful counterattack, Mao and his colleagues debated seriously how they should respond. When American forces crossed the 38th parallel, one source quoted the party leader as declaring "now we cannot afford not to engage the

Americans." By meeting the enemy on Korean soil, Mao hoped to shield Manchuria and keep the war limited. Other Chinese accounts from the period confirm Peking's fear that certainly MacArthur—and possibly the Truman administration—wanted to test China's mettle and see if it could be defeated or intimidated. Mao "believed it would be better to engage American forces where the risks and consequences were more manageable." By engaging in a forward defense of Manchuria, China would protect its heartland and send a strong warning to Washington. All Chinese sources stress that the decision to intervene massively came in response to the American move to destroy North Korea and the growing uncertainty over whether MacArthur would observe any limits.

China had little time to prepare for its move south. It lacked a well-coordinated logistic system and modern weapons, faced a technologically powerful enemy, and had only vague promises of future Soviet military support. The commander of the Chinese People's Volunteers (terminology providing Peking a possible "escape clause"), Peng Dehuai, bemoaned these hazards in a speech given in October 1950. But, like Mao, concluded "we have to fight now."[45]

Initially, Mao and Peng devised a plan to hold a line north of Pyongyang, rather than trying to push the Americans completely out of North Korea. This, they hoped, would compel Washington to rethink its strategy of advancing to the Yalu. Then negotiations on restoring the boundary could begin. However, the speed of the UN offensive frustrated their scheme. In talks to their colleagues during October, both Chou and Mao rejected the idea of a passive defense along the Yalu. Such a plan would require a huge investment of resources and troops for an indefinite period. All northeast China, Chou explained, would "be in jeopardy." But by halting the enemy in the "Korean quagmire," America would be unable to "extend its aggression to China."

Clearly, the Peking leadership hoped Washington would react rationally to a measured intervention. Once the Americans understood the high cost and limited rewards of approaching Manchuria, they would abandon the policy and, possibly, reassess their entire East Asian strategy. Still, the Chinese remained unsure exactly who among the Americans made policy. Would the president, Acheson, Marshall, the Joint Chiefs, or MacArthur play the key role? This uncertainty probably convinced the Chinese to make as dramatic a commitment as possible.[46]

On October 3, Chou informed Indian Ambassador K. M. Panikkar that if American troops (as distinct from the ROK army) crossed the 38th parallel, they would "encounter Chinese resistance." Throughout the summer China moved troops northward, and by early October, American intelligence analysts estimated as many as a half million regular soldiers might be concentrated in Manchuria. Propaganda blasts against the United States also changed at the end of September. Peking now threatened armed (as distinct from political) resistance to United States forces north of the 38th parallel. In spite of this, American authorities downplayed the "China factor."[47]

President Truman and Dean Acheson dismissed Chou's warning to Panikkar as a "bald attempt to blackmail the United Nations" against approving America's efforts to unify Korea. MacArthur also discounted Chou's warning as "pure bluff." The general told Alvary Gascoigne that if Peking really wanted to deter Washington, it would not have used the left-leaning Panikkar as a channel.

> In any case MacArthur claimed he had plenty of troops [to deal] adequately with the Chinese and even with the Russians if they should prove so foolish as to enter the arena at this stage. The Chinese, he said, had neither troops nor equipment nor air power to take him on—if the Chinese had come in three weeks ago it might have been another matter. . . .

> If the Chinese came in, MacArthur would immediately unleash his air force against towns in Manchuria and North China, including Peking—he knew Chou must know that, and must realize his (MacArthur's) vastly greater potential in the air, on the ground, and on the sea. Chou's statement to Panikkar was just blackmail.[48]

The general's contempt for Chinese sensibility and power surpassed that of most American officials. Since Washington had no designs on Manchuria, why would Peking risk a thrashing on behalf of North Korea? MacArthur judged Kim Il Sung a Soviet, not Chinese, puppet. If China simply wanted to help Kim conquer the south, he wondered why the PRC or Soviet Union had not joined the battle before Inchon, when they might have driven American troops off the peninsula?

Policymakers satisfied themselves that the Chinese were bluffing and that Soviet peace feelers only meant that Moscow had decided to abandon its Korean client. MacArthur, like Acheson, revealed an ambivalent attitude toward Chinese conduct. Despite their contempt for the PRC as a Soviet puppet, both considered it capable of acting of its own volition and in its own interest. MacArthur thought Mao might actually be more expansionist than Stalin, making the Russian reluctant to back up his junior partner if America retaliated. Acheson, who worried more than did MacArthur about the Sino-Soviet defense pact, thought the Chinese blamed the Soviets for starting the Korean War and would not intervene so long as their own territory was not threatened. Neither Acheson nor MacArthur took seriously the ties of ideology, geography, and culture that bound together China and North Korea. Beside these factors, America's increasing military support for Japan, Taiwan, and French Indochina provided Peking many reasons to assist its neighbor and engage in a forward defense of Manchuria.

As the war progressed, MacArthur revealed some important differences with his own government. The JCS had urged a halt in the drive north if Chinese or Soviet intervention either ocurred or appeared imminent. If MacArthur's forces encountered new enemy troops, he was to take defensive action and await direction from Washington. Yet, the general claimed authority to "immediately unlease his air force" against China if that nation's troops entered the war. (NSC 81/1, drafted before Inchon, noted the possibility of attacking China if Chinese forces entered *South*, not *North*, Korea. But it stressed that the United States should not "permit itself to become engaged in a general war with Communist China.") In contrast, MacArthur appeared eager to flex his muscles in a war of "liberation."[49]

On October 7, the United Nations General Assembly (where the Soviets had no veto) passed an American-sponsored resolution calling for the unification, by force if necessary, of Korea. Once again, the American-dominated UN provided a belated fig leaf for actions already undertaken by the United States. As American troops crossed the 38th parallel, Mao ordered the Chinese People's Volunteers to "march speedily to Korea and join the Korean comrades in fighting the aggressors."

Although intelligence failed to detect the bulk of these troops movements, China's public declarations stirred some concern in Washington. The Joint Chiefs drafted a message, approved by Marshall and Truman, outlining how MacArthur should respond to Chinese intervention anywhere in Korea. He was to continue operations "so long as, in your judgment, action by forces now under your control offers a reasonable chance of success." Of course, leaving things to his judgment meant imposing no real restrictions. However, he should "obtain authorization from Washington prior to taking any military action against objectives in Chinese territory."[50]

MacArthur's confidence bordered on elation. Flouting military doctrine, he divided his forces for the northward trek. The Eighth Army would seize Pyongyang and move north. Over the opposition of his own subordinates, he decided that X Corps should launch an amphibious assault on Wonsan, on Korea's east coast. This complicated operation quickly bogged down and tied up the X Corps until late October. It also proved worthless since ROK ground units captured the city while the marines waited off shore for mine fields to be cleared. As the Bob Hope USO show performed for bored troops, substantial North Korean forces and their political leadership escaped from the south and joined a large, partially mobilized reserve along the Yalu.

Wake Island Conference

As MacArthur's forces approached the 38th parallel, the White House staff (especially press aide Charles Ross, administrative assistant George Elsey, and Averell Harriman) fretted over the domestic political situation. Republican pressure and McCarthyite slurs had already prodded the administration to move into North Korea. Truman's staff searched for some way to generate positive political capital for the Democratic Party, which faced a tough congressional election in November. The "thought was," Charles

Ross later explained, that "it was good election year stuff" to send Truman to confer with his victorious general in Korea. To his credit, the president opposed the plan as "too political, too much showmanship." He saw no "real need . . . for him to confer with MacArthur." But early in October, the White House staff convinced their boss to play along. Ross found it persuasive to remind Truman about Roosevelt's meeting with MacArthur in Honolulu before the 1944 election. The conference was set for October 15 on Wake Island.[51]

In his memoirs, the president listed several important reasons for conferring with the general. They had never met and he hoped that a face-to-face session would alleviate earlier strains. Truman also remarked that since MacArthur had lived in the Orient for almost fourteen years, he had "had lost some of his contacts with the country and its people." Presumably, this reunion would restore the general's roots.

Despite this rationale, Acheson, Marshall, and all the Joint Chiefs save General Bradley declined invitations to attend. Truman brought along Dean Rusk and Philip Jessup from the State Department, Army Secretary Frank Pace, and White House adviser Harriman, in addition to two plane loads of staff and reporters. En route to Wake the president wrote family members that he was off to "talk to God's right hand man."[52]

MacArthur showed equally little enthusiasm about the meeting. Told only that he should appear at the appointed time and place, he first tried to bring along his own bevy of reporters. When the White House objected, he decided to travel with the smallest staff possible. MacArthur told William Sebald that he would take only two or three aides, a "simple soldier" in contrast to the president who needed a large supporting cast. The "whole business" of the conference was a "political junket anyway," and MacArthur noted the similarity to the 1944 Honolulu meeting with FDR. Before leaving Tokyo, the general's aides listened to him denounce "the State Department, Truman and Communists." He would remain "on guard against the evil people still in control."[53]

Despite many apocryphal stories about the Wake conference, the private discussions between the general and president and the formal sessions went quite smoothly. Substantive business lasted less than two hours and most of the participants merely repeated known facts. No one devoted more than a couple of minutes to discussing any topic and no important new subjects were raised. MacArthur answered familiar questions by repeating statements he had make in official reports.

The notes taken by several participants and collated by General Bradley (MacArthur later claimed they were transcribed secretly and then "doctored"—but had not objected to the transcript given him shortly after the meeting) reveal the superficial tenor of the talks. The UN commander set the pace by predicting a quick end to the war, with formal fighting over by Thanksgiving. Most of the Eighth Army would return to Japan by Christmas with several of its divisions available for service in Europe. Had the Soviets or Chinese intervened earlier, MarArthur observed, it might have proved "decisive." He now saw "very little" chance of major inter-vention. The president must have been pleased to here this, as well as to have MacArthur's predictions on the record.

General Bradley seemed relieved that the group had gotten on so well and that MacArthur confirmed that "the Korean War would soon be over." Now, the JCS chairman hoped, the administration "could proceed" with the more vital "task of building NATO." This concern dominated the thoughts of the Washington contingent.[54]

Truman reported that in a private talk the general affirmed much of what he told the rest of the group. He also apologized for the flap over the VFW message. The president told the other conferees that he and the UN commander were now in complete accord on the China issue and no need existed to discuss it further. MacArthur assured Truman "he was not in politics in any way—that he had allowed politicians to make a 'chump' (his word) out of him in 1948 and that it would not happen again."

The general later described the meeting as a pleasant, if irrelevant, jun-ket. He recalled asking Truman whether he intended to seek reelection in 1952. The president responded by asking if he would run. "If you have a general running against you, his name will be Eisenhower, not Mac-Arthur," came the answer. Ike, Truman chortled, knew nothing of politics. In words that sound more like those of MacArthur than of Truman, the president allegedly declared that an Eisenhower administration "would make Grant's look like a model of perfection."[55]

Five hours after arriving on Wake, Truman handed MacArthur a box of candy (for his family), pinned a medal on his chest, and both parties flew home. Speaking in San Francisco two days later, the president de-scribed MacArthur as a "very great soldier." MacArthur sent a message of appreciation, praised the results of the Wake conference, and voiced a hope the two men had built a "strong defense against future efforts of those who seek for one reason or another (none of them worthy) to breach the understanding between us."[56]

China Intervenes

Despite the kind words, Truman's military and diplomatic advisers contin-ued to worry about MacArthur's plans and behavior. General Bradley knew that the UN commander had contempt for nearly all officials in Washing-ton, civilian or military. He also felt (perhaps with a bit of hindsight) that MacArthur's decision to divide his forces and send an amphibious force to Wonsan appeared the triumph of "vanity" over "good sense." Had a "ma-jor at the Command and General Staff School" proposed to divide his forces, tie up his logistics, and delay vital operations for three weeks to assure a dramatic (and pointless) amphibious assault, Bradley later wrote, "he would have been laughed out of the classroom." But MacArthur hoped the special "aura of glamour" of amphibious operations and one more "bold and decisive strike in the enemy's rear" would yield headlines and editorials expounding on his "military genius."[57]

Despite the foul-up at Wonsan, by the third week of October some ROK units reached the Yalu River. On the twentieth MacArthur flew from Tokyo to the front where he declared the "war is definitely coming to an end." Plans were made to send excess units home or to Europe. Friendly UN members were asked to reduce or cancel promised assistance. MacArthur predicted that a single American division would suffice to occupy the north.

Anxious to mop up enemy remnants, the general disregarded JCS orders to utilize only ROK forces along the Chinese border. On October 24 he announced his intention of utilizing "any and all ground forces" available to "secure all of North Korea." Meekly, the Joint Chiefs requested an explanation of why he had violated their prohibition against sending American units to the frontier. "Undoubtedly," the chiefs wrote, he had "sound reasons for issuing these instructions." But as "your action is a matter of some concern here," they wished to learn what they were.

MacArthur explained that he interpreted earlier instructions merely as a statement of broad policy, not a "final" order. South Korean troops were inadequate and, he continued, Marshall's assurance that he should feel "unhampered" in proceeding north of the 38th parallel provided all the latitude he needed. Without much basis, he claimed "the entire subject was covered in my conference at Wake Island." Rather than argue, the JCS dropped the subject.[58]

On October 29, an editorial in The New York Times predicted that, aside from the possibility of some "unexpected developments on the frontier," Americans could "now be easy in our minds as to the military outcome" in Korea. As the editorial went to press, Chinese troops shredded several ROK units and then turned their wrath upon American forces. Chinese prisoners taken in combat revealed many details about Peking's expanding intervention. However, as the official JCS history of the war notes, both MacArthur and the Joint Chiefs believed that since "Communist China would have little to gain by intervening in Korea at this time," it was illogical to conclude that Peking would commit more than token forces. Reacting with "no particular sense of urgency," intelligence officials in Tokyo and Washington estimated that only 15,000 to 20,000 Chinese troops crossed the Yalu River. In fact, nearly 200,000 had already done so.[59]

The sudden appearance of Chinese forces worried the Joint Chiefs enough to request an appraisal from the UN commander. On November 4, MacArthur made an unusual admission that the pattern of intervention confused him. He did not know if Peking intended to fight a major war with the United States or whether it would make only a token, face-saving effort on behalf of a Communist ally. Since it was "impossible at this time to authoritatively appraise" Peking's motives, he saw no reason to alter his mission of destroying enemy forces and unifying the country.[60]

Chinese commander Peng Dehuai later described this limited intervention as a ruse "to show ourselves to be weak" and to feed MacArthur's arrogance to "lure him deep into our areas." MacArthur must have been more anxious than he admitted since the next day he ordered his air commander to prepare a massive bombing campaign to "destroy every means of communication and every installation, factory, city and village" between his front lines and the Yalu River border. This included several bridges linking north Korea and China, but spared some hydroelectric facilities and dams servicing Manchuria. The attack, scheduled to begin on November 7, would require, as MacArthur said, that "combat crews . . . be flown to exhaustion if necessary."

The general neglected to inform Washington of his plan, even though it involved operations along the Chinese border and, quite probably, intrusions into Manchurian air space. When the air commander, General Stratemeyer, told Air Force Chief Hoyt Vandenberg of the impending attack, alarm bells rang all over Washington. Vandenberg and his fellow chiefs relayed word to the State Department where Acheson exploded. The United States had promised the British to consult with them before undertaking any action affecting Manchuria. Also, the administration was trying to persuade the UN to pass a resolution calling upon China to remove its troops from Korea. Air attacks along the Yalu, Acheson feared, would further antagonize Peking and violate his pledge against carrying the war to China without allied approval. The president agreed that the risks of widening the war and offending London outweighed possible gains. Truman ordered no bombing within five miles of the border, although the rest of the air campaign could take place as planned.[61]

MacArthur threatened to resign before allowing an aide to "talk" him out of it. He then sent a scathing retort to the JCS. The day before, MacArthur urged Washington not to panic. Now he screamed that "men and material . . . are pouring across the Yalu bridges" threatening the "ultimate destruction of forces under my command." Only destroying the bridges could halt the attack. "Every hour of delay" would be paid for in "American blood." He insisted he had no intention of getting into a war with China and demanded that the Joint Chiefs bring his protest to Truman. If overruled, MacArthur asserted, he foresaw a "calamity of major proportion for which I cannot accept the responsibility." The general had, of course, passed the buck to Truman one day before the congressional elections.

Even as he pledged to respect China, the UN commander released a public statement blasting the People's Republic. MacArthur charged that by sending troops to Korea, Peking had "committed one of the most offensive acts of international lawlessness in historic record." Even more enemy troops were gathering in what he called the "privileged sanctuary" of Manchuria. The "privilege," he implied, might soon be lifted.[62]

The Joint Chiefs met in special session to respond to this bombshell. General Bradley and his fellow chiefs were outraged by MacArthur's hec-

toring tone and appeal, over their heads, to the president. "Right then—
that night—the JCS should have taken the firmest control of the Korean
War and dealt with MacArthur bluntly," Bradley later wrote. But they
feared that if they ordered MacArthur to move south to a more defensible
line across the "narrow waist" of north Korea, it was "certain to evoke
another burst of outrage, perhaps a tumultuous resignation and angry public
charges of appeasement"—all on election eve.

MacArthur's tantrum worked, as Bradley and Truman approved the
bombing offensive on grounds of military necessity. Contritely, the president and Joint Chiefs requested that the general observe the border and
not attack power plants on the Yalu considered especially important by
China. MacArthur once again urged the Joint Chiefs not to panic and
argued that the best way to discover China's capabilities and intentions
was by launching a final "offensive effort" toward the Yalu.[63]

The bombing campaign flattened nearly everything in the border zone—
except the bridges (and the off-limits hydroelectric facilities). When the air
offensive ceased on December 5, planes had knocked down only four of
the dozen spans across the Yalu. MacArthur blamed this on harassment
by Chinese fighter planes operating out of Manchurian "sanctuaries." But
many technical problems hindered the operation. Also, American fliers frequently violated Manchurian air space, despite the policy against doing so.

As for the elections (in which leading Republicans accused Truman of
"communist coddling"), the Democrats fared badly, losing twenty-eight
House and five Senate seats. Congressional Democrats complained that
MacArthur had staged the crisis of early November as a pro-Republican
ploy that "cost a million votes" to the party in power. Equally menacing,
Senator Joe McCarthy claimed credit for defeating several key Senate
Democrats.[64]

Still anxious to assuage MacArthur, Marshall sent the general another
personal message on November 7. The defense secretary stressed that he
and the entire administration empathized with the theater commander's
difficult position and sought to help him in all possible ways. Marshall
suggested that the Chinese genuinely feared for the safety of dams and
hydroelectric projects along the Yalu and might consent to limit their
intervention if America backed off a bit.

MacArthur dismissed this concern as irrelevant, attributing Peking's intervention to a combination of Chinese "character and culture" and Communist "lust" for expansion. He blamed Peking, not Moscow, for escalating the Korean War and for planning aggression throughout Asia. China,
he suggested, had replaced the Soviet Union as America's main enemy.
Fortunately, the United States could hit this target with relatively impunity.[65]

On November 7, in the midst of these exchanges, Chinese troops suddenly broke off contact and faded away. They may, as Peng claimed, have
merely baited the trap or needed time to reinforce units in preparation for
a winter offensive. Possibly, they wished to offer Washington an opportu-

nity to reassess policy and seek negotiations. The battlefield lull persisted for almost three weeks. Dean Acheson later realized that this hiatus represented the "last chance to halt the march to disaster in Korea." All Truman's advisers knew something was terribly wrong, but they "muffed" doing anything about it.[66]

Sensing jitters from Washington, MacArthur informed the Joint Chiefs on November 9 that any change in his mission of occupying all Korea would "completely destroy the morale" of his forces. South Korean troops might even turn against their American allies. Any thought of negotiating with China or accepting a British proposal to establish a demilitarized buffer along the Yalu represented the "Munich attitude" prevalent in London. Abandoning "any portion of Korea to the aggression of the Chinese Communists would be the greatest defeat of the free world in recent times." The British, he argued, had lost the "respect" of "all the rest of Asia" by granting Peking diplomatic recognition. The United States faced a choice whether to mimic the defeatist British or press on to "complete victory."[67]

Over the ensuing two weeks, military and diplomatic advisers weighed MacArthur's claims against contrasting evidence. China, they knew, might feel genuinely threatened by the destruction of North Korea and loss of its border power stations. Alternatively, Peking and Moscow might be laying a trap to engage the United States in a wider war in Korea, enabling the Soviets to strike in Europe or the Middle East. In either case, it seemed reckless to commit American forces to a major land campaign against China.

On the tactical level, administration officials sympathized with MacArthur's desire to strike against aircraft operating out of Manchuria. (The Chinese air force—many of whose planes were initially flown by Russians—confined its operations to the border region, avoiding interference with American supply lines, airfields, or storage facilities in South Korea or Japan. From Peking's perspective, Japan and South Korea were "privileged sanctuary" for UN forces.) During November, Washington discussed with its allies permitting "hot pursuit" into Manchuria, only to find the British and French terrified that an expanded war in Asia would strangle NATO in its cradle.[68]

In an attempt to defuse tensions, Truman and Acheson issued public statements pledging to respect Chinese territory. Yet MacArthur's operations and rhetoric, as well as growing American support for Taiwan, undermined these gestures. Nevertheless, in mid-November, China accepted a UN invitation to send a delegation to discuss the Taiwan situation. Although this signaled a possible breakthrough, bad weather across the Soviet Union Delayed the travel of the Chinese representatives who arrived in New York just as MacArthur resumed his drive to the Yalu River.

At this juncture, many civilian and military officials considered it advisable to stop MacArthur's offensive short of the Yalu. A measure of restraint, after all, might demonstrate America's good faith toward China and serve as the prelude for negotiations at the United Nations. Soviet-bloc diplomats appeared to encourage this possibility through informal remarks. One, for example, suggested a deal in which American forces would stay out of a buffer zone south of the Yalu, thus assuring the security of Manchuria. America would have a free hand in the remainder of Korea. At the same time, Washington would cease support for Taiwan, standing aside as the Chinese moved to reunite the island with the mainland.

But, as the official JCS history notes, "at no time" during these meetings "did anyone suggest that General MacArthur should be ordered to cancel or delay his proposed offensive." Omar Bradley admitted the president's advisers lacked the courage to act on their own instincts and "let ourselves be misled by MacArthur's wildly erroneous estimates of the situation and his eloquent rhetoric, as well as by too much wishful thinking of our own."[69]

Acheson later laid most of the blame for inaction on the Pentagon. He could do nothing unless the Joint Chiefs and Marshall spoke forcefully to Truman about the danger of a wider war. "But it would have meant a fight with MacArthur . . . and his relief under arguable circumstances. So they hesitated, wavered, and the chance was lost."[70]

MacArthur expressed confidence in his ability to handle any Chinese threat. In mid-November he told William Sebald that by destroying everything between UN lines and the Yalu, as well as the bridges across the river, he would halt further infiltration and prevent Chinese troops from living off the land. When UN forces reached the border, and China realized it had lost the opportunity for imperialist expansion, Peking would abandon the field.

When Chinese and North Korean troops were pushed across the Yalu, the general told Sebald, the "Korean campaign would be at an end." Although he intended to respect the border, "should the planned operation fail and Communist forces continue to stream into North Korea from Manchuria," he saw "no alternative from a military point of view to bombing key points in Manchuria." Then, the "fat would be in the fire" because it would probably "bring about a countermove by Soviet Russia." This could "only lead to a spreading of the war and he therefore hoped that it would not be necessary to resort to such drastic action."[71]

Concerned about the dangers of a wider war, John Foster Dulles added his voice to those urging caution. The leading Republican within the administration wrote MacArthur on November 15 that he disagreed vigorously with the many Republicans who "beat the drums for reckless action . . . on the mainland of Asia." Dulles applauded the "miracle" in Korea, but hoped the general would not "make the Asian mainland the area for testing the relative strength of the Free and Communist World."[72]

MacArthur's almost casual acceptance of these risks seems remarkable. Even if American forces did not approach the Yalu, China felt it imperative to intervene once forces crossed the prewar boundary. MacArthur's provocations, Washington's repeated willingness to follow his lead, and growing cooperation with Taiwan confirmed fears in China of encirclement and a strong tendency to ignore Peking's fundamental security interests. Military intervention, Chinese leaders believed, might be the only way to bring Washington back to its senses. Yet, neither the president nor his advisers possessed the imagination to grasp the Chinese perspective or the will to stop the general.

As danger signs appeared, MacArthur, became, if anything, more belligerent. He boasted to Ambassador Muccio that his air offensive had turned part of Korea into a "dessert." No more than 30,000 Chinese had entered the country, he claimed, since they could not elude his air attacks or intelligence analysts. Even as he spoke, nearly 300,000 Chinese and about 65,000 North Koreans stood poised to attack.[73]

While MacArthur planned his "end the war" campaign, the Joint Chiefs huddled nervously. What if Chinese troops continued to fight after UN forces reached the Yalu or if the offensive faltered? Should Peking be told to "quit or we would have to hit them in Manchuria?" Nearly as afraid of MacArthur as of the Chinese, the Joint Chiefs asked him to consider withdrawing to a line south of the Yalu *after* reaching and clearing the border.

As his units prepared to move, the general dismissed the idea as a "betrayal" of the Korean people and a sign of weakness in the face of Communist aggression. Peking knew he had no hostile designs, so why give in to a bluff? On November 24, he launched a "massive compression envelopment campaign of North Korea" and predicted "we can get the boys home by Christmas."[74]

MacArthur's successor, Matthew Ridgway, thought the general envisioned himself as "the swordsman who would slay the communist dragon" and save Asia. Perhaps this motivated his "reckless drive to the borders of Manchuria" and added "luster to his dream of victory." Unfortunately, on November 25, the Chinese People's Volunteers swarmed out of the Korean mountains changing the course of the war, American policy, and MacArthur's life.[75]

13
An Entirely New War

Late on November 25, 1950, nearly 300,000 Chinese and perhaps 65,000 North Korean troops began a massive counterattack, quickly halting the "end the war offensive." By November 27–28 the magnitude of the battle became clear to officials in Washington and by December 1, UN forces had taken 11,000 casualties, raising the spectre that MacArthur's command might be destroyed or pushed off the peninsula. American goals shifted almost immediately, as a quest for an armistice near the 38th parallel replaced talk of unifying Korea. Fearing the worst, the Joint Chiefs thought American troops might have to be evacuated in a Dunkirk-like operation.

Actually, many of the Chinese Volunteers consisted of recently surrendered and recycled Nationalist troops or peasants press ganged into service. Their initial success in routing the far better equipped American forces surprised many of the Chinese and their commanders. Although American reports described the attackers as "hordes" coming in "human waves," the numerical balance of forces in December was not wildly out of line. The problem lay more in faulty leadership, training, morale, and tactics in the Eighth Army and X Corps.

Also, the pattern of the Pacific War repeated itself in Korea. When MacArthur could bring to bare overwhelming power and technological superiority (as at Inchon), he achieved dramatic results. But when faced with a determined enemy in roughly equal numbers, or confronted by surprise, he lost his edge and fared rather badly. At such moments, as in the Philippines in 1942, he blamed enemies at home for his difficulties.

American policy suffered a series of jolts punctuated by panic from late November until mid-January, when the Chinese offensive bogged down. Truman, Acheson, Marshall, and the Joint Chiefs decided quickly to abandon their inflated hope of unifying Korea. More concerned with avoiding a general war, saving lives, and holding Europe and Japan, they disowned liberation for the more familiar comforts of containment. At this level, the Chinese calculation proved successful.

As the crisis unfolded, the Truman administration faced at least five dilemmas: (1) the Chinese military sweep southward; (2) hysterical pronouncements from MacArthur prophesying defeat unless Washington expanded the war; (3) political attacks from congressional Republicans who championed MacArthur and called for firing Acheson; (4) protests by the British and French who feared a larger war in Korea would cripple NATO and open Western Europe to Soviet attack; (5) a congressional debate over rearmament, sending troops to NATO, and the strategy of limited war.

Immediately, the protagonists began pointing fingers of blame. The Truman administration publicly charged the Chinese with calculated aggression and privately accused MacArthur of endangering his forces while courting global war; the European allies criticized Washington for provoking Peking, failing to curb MacArthur, and undermining NATO; Republican leaders labeled the Europeans appeasers, defended MacArthur, and demanded Acheson's scalp; MacArthur complained that Washington's refusal to expand the war against China condemned his forces to defeat and rewarded Communist aggression. The debate continued along these lines up to and beyond April 1951.

MacArthur presented the administration—and the American public—with a simple interpretation of developments. He charged the Chinese with unprovoked aggression that altered completely the Korean War. Determined to unify Korea, he demanded authority to attack the "source of aggression"—China—through air and naval bombardment and by utilizing Chinese Nationalist troops in Korea and against the mainland. He also considered the use of atomic weapons. In classified reports and public interviews, the general warned that only these actions could prevent the Chinese from destroying American forces and conquering all Asia.

In contrast, President Truman, Secretaries Acheson and Marshall, and the Joint Chiefs feared getting "sewed up" in the "wrong war" against the "second team." After November 28 they sought primarily to preserve American troops, arrange an armistice, avoid a wider war with China and the Soviet Union, and speed the buildup of NATO. Committing additional men, weapons, and prestige to Korea undermined these priorities. Civilian and military planners contemplated operations against China, but only as a last resort to cover an evacuation or to use in retaliation if the Communist powers launched air attacks against South Korea or Japan. They rejected the idea of attacking China or employing atomic weapons and Nationalist troops as part of a "win the war" strategy.

President Truman and Secretary of State Acheson reset their diplomatic priorities with impressive alacrity. Disputing Peking's assertion that its forces acted in defense of their homeland, Acheson and Truman denounced China as a Soviet puppet state whose leaders could not be weaned from Moscow. Although this harsh and simplistic attitude poisoned the chances for any improvement in Sino-American relations, it had a curiously restraining effect upon military strategy. The president and his chief advisers worried that extending the fighting into China (by means of air and naval attacks or use of Nationalist troops) would probably bring the Soviets into the struggle. Besides the obvious danger of global war, an expanded conflict in Korea would tie down American power in a strategic backwater, alienate many UN members, and leave Europe and Japan dangerously exposed. These considerations led the administration to favor a negotiated settlement in Korea more or less along the lines of the prewar boundary, so long as it did not require unacceptable concessions to Peking.

In retrospect, it seems clear that officials in both the State and Defense departments overrated the unity of interests between Peking and Moscow. While this misperception caused them to cease attempts at improving relations with China (aside from seeking an armistice), it increased their determination to limit American involvement in Korea. In a sense, the advocates of restraint and limited war had the good fortune to be "right for the wrong reasons." In other words, fear of Soviet intervention in Korea and elsewhere convinced Truman, Marshall, Acheson, and others to reject the temptation of complete victory.

While American and European leaders came to see the Korean war as a dangerous diversion, the UN commander saw the issue rather differently. In his mind only one critical position existed in this or any war: the one where he exercised command. As during the Pacific War, he feared that enemies at home preferred to see "MacArthur lose, rather than America win." Aware of the divergent interests between the Soviet Union and China, he dismissed the likelihood or consequences of a wider conflict. Russia, the general insisted, would not risk world war on behalf of its client. Nor would an expanded war in Asia jeopardize Europe. In contrast to his rivals in Washington, MacArthur's vision proved "wrong for the right reasons."

His suggestion of a breech between China and the Soviet Union found expression in the subsequent Sino-Soviet split. However, this insight caused the general to advocate a reckless and aggressive, rather than a flexible and conciliatory, approach. If, regardless of American escalation, Moscow would not enter the Korean war or alter its global policies, than the United States could punish Peking with impunity. Instead of settling for a stalemate, American power could inflict a defeat on China, which would insure Korean unification. MacArthur only implied his ancillary belief that America air and naval strength should assist a Chinese Nationalist assault on China.

In his first substantive report (November 28) on the new situation, MacArthur informed the Joint Chiefs that Chinese forces had entered the war in "great and ever increasing strength," dashing any hope of limiting the conflict to a fight against Korean troops. Because of China's "subterfuge," he faced "an entirely new war." Absolving himself of all error, he passed responsibility to Peking and Washington. The general insisted that China's limited intervention (a month earlier) sought to halt the UN advance short of the Yalu so that Communist forces could prepare for a much larger offensive the following spring. Luckily, his recent attack (which he soon began calling a "reconnaissance in force," rather than an "end the war offensive") had forced the enemy to strike prematurely. Still, as MacArthur admitted, the Chinese probably intended the "complete destruction of all United Nations forces in Korea." (A few days later he declared this their definite aim.)

He had done "everything possible within [his] capabilities" but his command "now faced . . . conditions beyond its control and its strength." Events required a response that "broaden[ed] the potentialities to world embracing considerations." In a follow-up message, the general requested permission to "negotiate directly" with the government on Taiwan for the dispatch of Nationalist troops to Korea. He needed authority to broaden the war, or the Truman administration would be responsible for a disaster.[1]

To their credit, both the civilian and military leadership in Washington resisted MacArthur's effort to stampede them through his real or calculated panic. Although equally concerned with the fate of American units surrounded by the Chinese in North Korea, they began to recover their perspective once the blush of easy victory had faded from MacArthur's communiques. Korea was hardly the only, or even most important, theater of the cold war. Western Europe and Japan outweighted the peninsula's strategic importance. If nothing else, China's boldness compelled Washington to reconsider the inflated costs of liberation and rollback as compared to containment.

When first apprised of the developing disaster in Korea, Truman considered "declaring a complete emergency" and going before Congress. (He dropped the idea the next day.) China's actions, he observed bitterly, resulted in large part from "the vile vicious campaign of lies" carried out by "Senator McCarthy and others . . . who had been attacking Secretary of State Acheson and others." Truman noted that *Pravda* carried a story about deep divisions in Washington.[2]

At a special meeting of the National Security Council held on November 28, the president along with his top military and civilian advisers assayed the costs of the "entirely new war." Secretaries Acheson and Marshall, as well as the Joint Chiefs, discussed the need to salvage American forces and prestige in Korea while avoiding a general war with China and the Soviet Union. A consensus emerged around several points. All agreed that China's intervention increased the danger of a world war. If so, it made little sense to commit more forces to Korea. (After December, the Joint Chiefs refused to send any substantial new units to Korea. MacArthur received only some National Guard divisions for duty in Japan.) Instead, the United States ought to push ahead with the military buildup called for in NSC-68, enhance NATO, and avoid getting "sewed up" (as Marshall put it) in Korea. In retrospect, General Bradley felt that the decision to implement the NSC-68 arms buildup constituted the single most important result of the meeting. "From this day onward," he wrote, the momentum for general rearmament gathered steam, and the JCS, "while keeping a sharp eye on Korea," focused on the drive for "ever greater force levels" elsewhere.

Truman agreed with the idea of holding a defensible line somewhere on the peninsula and then negotiating an armistice or orderly withdrawal. Acheson spoke of a need to "make life harder" for the Chinese by condemning them in the UN and imposing trade restrictions, but he cautioned against military action affecting Manchuria. The secretary of state, like his military counterparts, feared that attacks upon Chinese territory might well expose South Korea and Japan to air and sea assault by China and the Soviet Union. Operations against Manchuria should only be undertaken to cover an evacuation. The conferees saw no military benefits—but many political costs—in utilizing Chinese Nationalist forces. Similarly, they rejected MacArthur's plea for substantial reinforcements. (In fact, almost no combat units were currently available.) The United States, they felt, would be lucky to get an armistice that restored the 38th parallel.[3]

Over the next few days, as the Chinese pushed forward and American forces retreated, the administration and MacArthur jockeyed for position. Acheson and the Joint Chiefs wanted MacArthur to withdraw his forces in an orderly fashion and make a stand at some defensible point such as the "narrow waist" above the 38th parallel. The general retorted (in a tone that Bradley felt "treated the JCS as if we were children") that his forces were not adequate and faced destruction. He could only stop the Chinese if provided with additional American and Nationalist Chinese troops and if allowed to hit Manchuria.

Officials in Washington rejected this strategy on several grounds. American allies in Europe and Asia, Acheson stressed, "were in a virtual state of panic." They felt "United States leadership had failed" and considered the "present difficulties" the "fault of General MacArthur's actions." The government must reassert authority and prove its ability to limit the war by holding a defensive line south of the Yalu while seeking a negotiated armistice. MacArthur's plea for Chinese Nationalist troops, Bradley felt, suggested he "was monumentally stupid, had gone mad, or had rejected JCS and administration policy to keep the war localized and was willing to risk an all out war with China, regardless of the consequences."

Disputing MacArthur's prediction of doom, the Joint Chiefs thought the bulk of American forces could be moved south in good order. (The situation of the X Corps, in the northeast, remained tenuous until evacuation in mid-December.) Fortunately, the Chinese had not committed the bulk of their air power based in Manchuria. As General Ridgway noted later on, the Chinese (and Soviets) appeared to have an "unwritten agreement" that left "our airfields in Korea and Japan immune from attack." American air and naval strikes against Chinese territory, or the utilization of Nationalist troops from Taiwan, the Joint Chiefs feared, would invalidate this agreement. If so, America's vital "privileged sanctuaries"—the undefended airfields and ports of South Korea and Japan—would become easy targets.

CIA Director, General Walter Bedell Smith, expressed the fears of his colleagues when he declared that the Soviets hoped to "bleed us to death in Asia and thereby defeat the armament effort in Europe." At the same time, Truman, Acheson, and others agreed that abandoning Korea would endanger Japan and all non-Communist Asia. The United States needed to preserve a foothold in Korea without stumbling into a general war or sacrificing Japan and Western Europe. In effect, this meant reaching an agreement with Peking that restored something like the status quo ante.

Yet MacArthur's inflexibility put the administration in a painful bind. In message after message during early December, he raised the possibility of destruction unless Washington widened the war. In a December 1 interview with *U.S. News & World Report* and in telegrams to other news publications, he congratulated himself for unmasking China's hostile intentions. He also condemned as an "enormous handicap, without precedent in military history" the "limitations which prevent unlimited pursuit" of Chinese forces. MacArthur blamed this policy, in part, on the "selfish" and "short sighted" ideas of Western European leaders. Returning to an old theme, he declared that if the battle against communism were not won in Asia, "it will be fought, and possibly lost, on the battlefield of Europe." In a message to Washington, the general described his forces as so "mentally fatigued and physically battered" that without reinforcements or a change in strategy he foresaw their "final destruction."

JCS chairman Bradley suspected that the shattered moral of American forces stemmed more from bad leadership than from Chinese power. Nevertheless, he and his colleagues drew back from challenging the theater commander. On December 3, as the Joint Chiefs wrung their hands and pondered how to mollify MacArthur, General Matthew Ridgway (then the army's deputy chief of staff) asked General Hoyt Vandenberg why "don't the Joint Chiefs send orders to MacArthur and *tell* him what to do." Vandenberg only shook his head. "What good would that do? He wouldn't obey the orders. What *can* we do?" ("Relieve him," Ridgway grumbled in reply.) MacArthur's defeatism, fear over the fate of the hard-pressed X Corps, and concern that an evacuation might require air strikes against Manchuria brought the Joint Chiefs, in Bradley's opinion, to a "low point" in the Korean war.[4]

Of course, the president, as well as the JCS, had the ultimate power to sack a recalcitrant or incompetent officer. Truman later wrote that he "should have relieved general MacArthur then and there" but did not want to give the impression he was being fired because "the offensive failed." In a note to himself, the president fumed that "General MacArthur as usual has been shooting off his mouth. . . . I must defend him and save his face even if he has tried on various and numerous occasions to cut mine off. But I must stand by my subordinates." Truman temporized by issuing a pair of directives, on December 6, instructing all civil and military officials to clear policy pronouncements with the State and Defense departments and avoid communicating their opinions directly to the press. Little noted at the time, violations of these directives later contributed to MacArthur's recall.[5]

Truman's anxiety probably contributed to several misstatements at a November 30 news conference in which he raised the spectre of atomic war against China. Responding to a series of questions about the possibility of extending military operations into Manchuria, the president remarked that use of atomic weapons had "always been" under consideration and that the "military commander in the field will have charge of the use of the weapons, as he always has." Whether Truman simply erred or sought indirectly to warn Moscow and Peking about the risks of escalation, reporters had a headline and the administration a new problem.

In addition to the fact that the Atomic Energy Act gave the president sole authority over the use of nuclear weapons and that no operational bombs (as distinct from pre-positioned bomb components) were currently deployed near Korea, the Joint Chiefs had already decided against this approach. MacArthur had raised the issue in July and, in August and September, nuclear-capable B-29s had been sent temporarily to Guam. Following China's initial intervention, early in November, both State Department and JCS planners studied possible use of the atomic bomb in Korea. Nuclear strikes, they decided, might be useful in covering an evacuation, but the terrain in Korea and lack of concentrated targets there (not to mention the limited American atomic arsenal) made use of the bomb impractical. Tactical atomic artillery shells were not developed until 1953. In any event, Truman's remarks were not connected to the movement of atomic weapons to Pacific bases.[6]

Although the White House quickly issued a correction of the president's remarks, the damage had been done. Amidst the concern for American soldiers and the threat of a wider war, Senator Joe McCarthy attacked Truman for allowing Acheson and a "crimson clique in the State Department" to squander American lives. Other Republicans and conservative Democrats endorsed the flamboyant red baiter's call for firing both Acheson and Marshall.

Opinion polls taken during December revealed that 49 percent of the public disapproved of Truman's leadership while only 36 percent approved. (The approval rating dropped below 30 percent in the spring.) By early January two out of three Americans felt the United States should leave Korea while about half stated intervention had been a mistake to begin with. Pressed domestically, the president announced plans for a large increase in defense spending and began drafting a declaration of national emergency, issued on December 16.[7]

MacArthur's behavior and Truman's apparent slip of the tongue, Acheson told his colleagues, had put America's foreign friends in a panic. Although hardly sympathetic toward China, the secretary of state and Joint Chiefs realized that Peking had not (despite MacArthur's claims) committed its full strength to the war effort. Both Bradley and Acheson thought it vital to attempt a cease-fire *before* initiating air and naval strikes against the PRC. The trick was arranging a deal with China without bargaining over Taiwan, French Inodchina, a UN seat, or the Japanese peace treaty—all issues Peking insisted were related to the Korean situation. Acheson complained that MacArthur either ignored or misrepresented the fact that China was the "wrong nation" to fight. It was only the "second team, whereas the real enemy is the Soviet Union." The administration sought to strike a balance midway between abandoning Korea and expanding the war, as either extreme threatened America's global interests. Yet, the "spirit of defeatism emanating" from MacArthur jeopardized efforts to contain the Chinese offensive and negotiate from a "position of strength."

Early in December, Dean Rusk informally broached the idea of placing Army Chief General J. Lawton Collins in "supreme command in Korea" while permitting "General MacArthur to concentrate on his duties in Japan." Lacking the courage of this conviction, top officials elected to send Collins only on a "fact-finding" mission. While awaiting this report, Truman and his subordinates had to comfort British Prime Minister Clement Attlee, who arrived in Washington on December 4 for emergency talks.[8]

During four days of intense discussions, Attlee tried to convince Truman, Acheson, Marshall, Bradley, and others that a wider war in Korea would destroy the Western alliance. Viewing himself as the representative of responsible anti-communism (in contrast to the messianic American strain) the prime minister pressed American leaders to woo China to the peace table by offering it a UN seat and, possibly, some sort of deal on Taiwan. Besides paving the way for a cease-fire, Attlee thought these actions might wean Peking from the Soviet bloc.

Acheson, formerly an architect of moderation toward China, retorted that he had "probably been more bloodied" than anyone in seeking to encourage Titoism in China. (Ironically, Mao later reported that not until China proved its mettle by intervening in Korea did Stalin "come to believe that we were not Yugoslavia, not Tito.") Washington had "tried to be friends with the Chinese" for fifty years, but the Communists had only responded with "unprovoked aggression." The secretary would "not pay anything" for a cease-fire. The other designer of containment, George Kennan, agreed; the United States owed Peking "nothing but a lesson." Similarly, Truman refused to offer "rewards for aggression." Chiang had "so many converts here," the president observed, that coupling concessions with negotiations was "political dynamite in the United States." Despite their contempt for Chiang, Marshall and Bradley explained that from the "military viewpoint it [would be] very dangerous to give up Formosa." Truman added that he "had to consider the political situation here" before arranging any deal with China.

American officials informed Attlee that, while they preferred to avoid a wider war, if Chinese forces pushed south of the 38th parallel Washington might undertake efforts to "stimulate anti-communist resistance within China itself, including the exploitation of Nationalist capabilities." These threats seemed intended more to demonstrate American military resolve than to achieve any particular result in China.

The British leader reported to his colleagues that when he first arrived the United States was considering a "limited war against China" by means of an "economic blockade and stimulating internal trouble in China." In such a case, he threatened to abandon support of the American policy in the UN and elsewhere. Attlee believed, with some basis, that he raised important "doubts about the value of a limited war." Although Truman refused to budge on the issues of Taiwan or UN membership, the president promised that America would settle for the preservation of South Korea rather than the unification of the peninsula. He did not rule out action against China or use of the atomic bomb, but promised to consult with London before expanding the war.[9]

To a large degree, General Collins's report to Truman and Attlee on December 8 alleviated the tensions between the two allies. The army chief surveyed the situation in Korea and met twice with MacArthur during the first week of December. He quickly concluded that the X Corps could be evacuated from the north and merged with the Eighth Army. The combined force could hold a perimeter around Pusan indefinitely—even at current troop levels—despite MacArthur's dire assertions.

Yet when Collins queried the UN commander on December 6, MacArthur stuck to his line. Unless he received authority to strike at Manchuria, blockade the China coast, and utilize fifty-thousand or more Chinese Nationalist troops, he would have to evacuate all Korea. If allowed to attack Manchuria, blockade the China coast, use Nationalist troops, *and* employ tactical atomic bombs ("if technically appropriate"), he could hold a portion of Korea. Should the Chinese voluntarily halt at the 38th parallel, MacArthur recommended accepting a temporary armistice at that line. However, even then he wanted more American and Chinese Nationalist troops. Without them, the United Nations command "should pull out of Korea."

Collins now saw things differently. He told Truman and Attlee that the American and UN member forces were not in "a critical condition today" and could be pulled back to a defensible bridgehead. In fact, General Bradley commented, it seemed clear that a line could be held "with the present force supplemented by the normal flow of replacements." Officials heaved a collective sigh of relief at what Bradley called this "ray of sunshine." Now, "we had options to discuss other than catastrophe." The Joint Chiefs instructed MacArthur to evacuate X Corps by sea and to pull the Eighth Army back to near the 38th parallel. These moves were completed in mid-December with little interference by the Chinese, who were already having difficulty extending their offensive south.

The American military position stabilized a bit when, on December 25, General Matthew Ridgway took command of the Eighth Army, replacing General Walton Walker who had died in a traffic accident a few days earlier. Even though the enemy soon recaptured Seoul, U.S./UN forces slowed their advance and began to regroup successfully. MacArthur gave Ridgway great latitude, encouraged him to hold a line as far north as possible, and watched with evident surprise (as did the Chinese) as the new commander quickly rallied his supposedly shattered army.[10]

Battlefield success in November–December both thrilled Chinese commanders and, as Peng Dehuai admitted, worried them a bit. They had humbled the world's greatest power and humiliated their most determined American adversary, MacArthur. By seizing control of most of North Korea, they held a strong hand for any armistice negotiations. At the same time, Peng realized that America had lost territory and equipment, but not all that many soldiers during the retreat. Also, as Chinese lines moved south, tremendous logistic problems developed. American air superiority prevented large scale resupply efforts so that food and other supplies had to be moved in small amounts often on the backs of troops.

The Chinese military commander worried that either Washington or MacArthur had cynically ordered a "purposeful southern retreat" as part of a "luring deep strategy," much like the Chinese maneuver of November. Once the People's Volunteers overextended their lines, they would be pummeled by American air and artillery power. Peng urged his superiors to slow the offensive in mid-December, giving his forces time to "consolidate." Mao disagreed, however, arguing that a renewed push would force both the United States and United Nations to bargain on Chinese terms.[11]

As the fighting paused for a bit, attention focused on efforts by non-aligned UN states to halt the war. One resolution called for an unconditional cease fire, while a second favored a cease-fire followed by a conference in which China could address the questions of Taiwan, UN representation, and the withdrawal of foreign troops from Korea. Although they wanted an armistice, State and Defense officials feared that an unpoliced agreement, without a demilitarized zone and careful monitoring, would favor the Chinese. However, a majority in the UN General Assembly voted for the resolution on December 14. Reluctantly, the administration went along as this proposal seemed less threatening than the call for an Asian political conference. In any case, Acheson, Marshall, and the Joint Chiefs announced that the United States would only accept a cease-fire that included a closely monitored freeze of forces along the 38th parallel as well as a prisoner exchange.

China's special representative at the UN, General Wu Hsiu-chuan, refused to discuss a Korean armistice apart from other regional questions and left for home. On December 22, Chinese Premier Chou En-lai rejected any cease-fire proposal not mandating an end to American military involvement in Taiwan and Korea or not giving Peking a seat in the UN. Pulling a rhetorical phrase from MacArthur's quiver, Chou declared that the 38th parallel had been "obliterated forever" as a "demarcation line of political geography." China intended to unify the peninsula by force. Apparently, Chou and Mao still hoped this threat (or actual accomplishment) would force their adversaries to bargain. When the UN passed a second cease-fire resolution the next month (which proposed a political conference following, rather than preceding, an end to hostilities), Peking again refused, charging that Washington only wanted "breathing space" to rebuild momentum and would not address seriously China's diplomatic grievances. American officials greeted this rejection with relief for they feared domestic political complications if cease-fire talks were linked to diplomatic negotiations on the fate of Taiwan.

China's insistence that regional issues be addressed before a cease-fire agreement, coupled with American determination to avoid political haggling and to negotiate from strength, killed a real opportunity for a halt in the bloodletting. The prolonged war cost China dearly in lives (including Mao's son) and treasure. While the United States suffered far fewer casualties than its adversary, the stalemate had a profound impact on American politics.[12]

The administration, in mid-December, moved to rally faltering domestic support. In rapid succession, President Truman announced emergency economic controls, expansion of the army, a quadrupling of the defense budget to over $50 billion, Eisenhower's appointment as NATO Supreme Commander, and a freeze on most trade with China. On December 16, he declared a state of national emergency giving the federal government additional control over the economy. Without much success, Truman tried to assure skeptical Republicans that his national emergency declaration was not a prelude to global war. Senator Taft responded that the American people simply could not comprehend what the administration had in mind since most of the military buildup was destined for Europe, not Korea. Taft's criticism initiated a formal challenge (the "great debate" of early 1951) that he and other Republicans soon raised to administration policy.[13]

Lurching from Crisis to Crisis

The brief respite of late December quickly gave way to new battlefield pressures. For example, Seoul again fell to Communist troops early in January. Even though the bulk of American forces had moved far enough south to break contact with the enemy (critics contended that in many cases they fled without putting up a fight), the Chinese were massing for a renewed offensive with the goal of driving the UN command off the peninsula. Officials in Washington worried that the enemy might even succeed in trapping and annihilating American and ROK forces. Rather than risk the loss of prestige and such a large proportion of American combat forces, the Joint Chiefs gave serious thought to evacuating South Korea before being pushed out or destroyed.

Throughout most of December and early January, the Joint Chiefs edged toward recommending a complete withdrawal from Korea. As before the war, they still considered the peninsula a strategic backwater and thought it more important to preserve American ground forces for the defense of Japan and Western Europe. For example, Admiral Sherman and General Vandenberg discussed the idea of issuing an ultimatum that China halt its attack at the prewar boundary. If it refused, American ground forces would be pulled out while naval and air attacks pummeled China. These operations, they felt, would not deplete the arsenals available to defend Europe and Japan.

Truman and his diplomatic advisers rejected this approach. While likely to reduce American casualties, the action would expand the scope of the war, probably torpedo truce talks, and insure the physical loss of Korea to communism. Writing off Korea, even if followed by punishing blows to China, would still call American credibility and guarantees into question. A wider war against China, as Acheson told Attlee, might be in the cards but only if the United States were forced out of Korea.[14]

Although State Department officials felt the United States should retain a foothold in Kora for political-diplomatic reasons, military officials feared such a stand would endanger forces needed elsewhere. The diplomats pressed the military to agree that at least some portion of South Korea could be held at an acceptable cost. In response, Defense officials wanted the State Department to admit that only political considerations justified remaining in Korea.

The president, Acheson, Marshall, and the Joint Chiefs finessed their differences at the end of December. Abandoning any thought of unifying Korea by force, they instructed MacArthur to inflict punishment on Chinese ground forces while preserving the integrity of his own units. If faced with a Chinese onslaught below the 38th parallel, he was to pull back to a series of defensive positions and eventually withdraw his forces to Japan. Washington made clear that he should consider protection of that nation his overriding concern.[15]

MacArthur realized that the administration's priorities in Korea and East Asia were diverging rapidly from his own. Early in December a British intelligence operative, W. K. Rickett, began discussions with CIA director Walter Bedell Smith, Allen Dulles (John Foster's brother and one of Smith's deputies), Joseph Davies (FDR's controversial ambassador to Moscow and a sometime friend of Truman), Claire Chennault (whose private, CIA-affiliated airline transported supplies to Korea and Taiwan), and Chinese Nationalist Ambassador Koo regarding a carrot-and-stick plan to alter Peking's policies. As reported by Koo, the Americans and British contemplated "buying off" selected Chinese generals and speaking directly with Mao Tse-tung about China's willingness to break with Moscow. If Mao or other officials agreed, the Western powers would consider providing aid, bringing China into the UN, and making a settlement in Korea acceptable to Peking. To pressure China (or punish it in case of refusal), the Americans would also resume assistance to anti-Communist guerrillas on the mainland. Koo indicated that Chiang would play along, expecting Peking to balk at the idea and thereby set the stage for greater aid to him. On December 19, John Foster Dulles told the ambassador (who routinely leaked material to Tokyo and Republicans) that although Washington "would not recognize the Chinese Communist regime and would not turn over Formosa to them," it would consider "seating the Chinese Communists in the United Nations and withdrawing United Nations forces from Korea" if a "suitable face-saving formula for everybody" could be found.

The plan for a "buy off" or a guerrilla campaign ran into complications when Nationalist authorities raised questions about Rickett's ties to the British government and the degree of American support he enjoyed. The scheme assumed new dimensions at the end of December when one participant, Joseph Davies, explained the full price of the offer. Davies claimed to have spoken with Truman and others who suggested that Chiang Kai-shek "pack up all his private belongings and resources" since Washington now had "the idea of taking over Formosa and making it a [American-administered] trust territory . . . with someone other than the Generalissimo as the ruler." The British had already suggested this as a way of mollifying Peking's demand that a Taiwan settlement be linked to a Korean armistice.[16]

Even though this particular scheme faltered, Koo confirmed that the "idea of using Nationalist forces on the mainland" began late in 1950 and continued for some time. The "administration continued to experiment," Koo wrote, with a "Taiwan based guerrilla program." It had no "real expectation that anything would come of such an experiment, but they wished to satisfy public and congressional pressure."[17]

Unlike the administration, MacArthur thought in terms of escalation, not compromise. When queried by the Joint Chiefs regarding how he would respond to enhanced Soviet or Chinese intervention in Korea, he submitted, on December 24, a "list of retardation targets which he considered would require 26 [atomic] bombs." An additional four bombs were requested for use against invasion forces and four others "to be used on critical concentrations of enemy air power," presumably in Manchuria.[18]

Responding to the JCS instruction to defend successive fallback positions in preparation for a withdrawal to Japan, MacArthur reversed the argument. Unless the United States took military action directly against China, he insisted, the defending forces would be overrun. The war must be widened or his mission terminated at once. The general proposed to "blockade the coast of China, destroy through naval gun fire and air bombardment China's industrial capacity to wage war," secure Chinese Nationalist troops to fight in Korea, and "release existing restrictions upon the Formosan garrison . . . possibly leading to counter invasion against vulnerable areas of the Chinese Mainland." Besides halting the current offensive, MacArthur argued, these measures would "neutralize" China's war-making ability "and thus save Asia from the engulfment otherwise facing it."

Since he considered China already fully committed to the war, Mac-

Arthur minimized the risk of it reacting to such measures. As for the Soviets, he doubted their decision to "precipitate a general war" would depend on any factor save their "own estimate of relative strengths and capabilities, with little regard for other factors."

The general predicted that a withdrawal from Korea would free Chinese forces for aggression against Southeast Asia and Japan. Despite the "demand for European security," he felt that "accepting defeat anywhere else" would "insure later defeat in Europe itself." The battle should be joined here and now, not later, somewhere else.

Although not quite stated openly in this recommendation, General Ridgway was convinced that MacArthur "wanted to go to war, full war, with Communist China and he could not be convinced by all the contrary arguments." Ridgway himself felt torn between support of some aspects of MacArthur's proposal and his own agreement with the JCS on the need to prevent a major war with China. The UN commander told Ridgway the time had come to support Chiang's return to the mainland. China was "wide open in the south" and even a moderately successful Nationalist invasion would "relieve the pressure in Korea." In a cable to Army Chief Collins on December 29, Ridgway described this idea as "brilliant" and offered his "full concurrence at once." [19]

The Joint Chiefs waited a week before responding to MacArhtur. Politely, they noted that while his proposals were "under consideration," very "little possibility" of a policy change existed. The chiefs stressed that the NATO allies opposed operations against China and that the administration would consider it only if the enemy launched attacks outside Korea (i.e., Japan or Taiwan). Dismissing the value of Nationalist troops, the Joint Chiefs repeated their plea for the general to move south in successive stages, inflict as much damage as possible on Chinese forces, and prepare for an evacuation to Japan. [20]

Aware that some members of the JCS planning staff sympathized with his call for widening of the war, MacArthur continued to press his case. He informed the Joint Chiefs that their "political" strategy confused him. Should he hold on in Korea "indefinitely, for a limited time, or to minimize losses by evacuation as soon as it can be accomplished?" He complained about "extraordinary limitations" that made his "military position untenable." But his forces would, "if overriding political considerations so dictated," hold as long as ordered up to their "complete destruction." [21]

This message astounded Truman, Acheson, Marshall, and the Joint Chiefs. MacArthur's complaint that dubious political considerations threatened his destruction echoed a theme running back to the siege of the Philippines. It struck the recipients as a "posterity paper" issued by a "basically disloyal" subordinate eager to pass blame up to his superiors. General Bradley surmised that the UN commander's call for a wider war had less to do with the battlefield situation than with his impulse for revenge upon the "Red Chinese generals who had made a fool out of him" and his "infallible 'military genius.' " In response to MacArthur's search for "clarification," the president sent a long, conciliatory message on the importance of limiting the war to Korea. The Joint Chiefs repeated their earlier instructions and dispatched Generals Collins and Vandenberg on a fact-finding mission.[22]

Meanwhile, since a Chinese breakthrough remained a real possibility, the JCS prepared a plan (submitted on January 12 to the National Security Council for later consideration) that resembled the general's recommendations of December 29. In the event Chinese forces pushed UN troops out of Korea or attacked Japan, the JCS proposed a naval blockade of China, naval and air bombardment of China, and assistance to Nationalist military operations against the mainland. These actions were envisioned as a way to punish Peking, tie down its armies in Korea, and protect Taiwan and Southeast Asia. Even though the document included a reference to supporting the "establishment of a government in China friendly to the United States," it was not part of a plan to wage open-ended war on the mainland. While MacArthur had urged undertaking his program at once, the JCS would take more aggressive steps only if Peking widened the war or pushed American forces out of Korea.[23]

When Collins and Vandenberg visited Tokyo in mid-January, they showed the contingency proposal to MacArthur. He interpreted it as confirmation that the Joint Chiefs supported his strategy but political interference from the White House and State Department blocked its implementation. During testimony in May 1951, following his recall, the general insisted that the Joint Chiefs favored operations against China. In fact, both in January and afterward the chiefs emphasized the conditional nature of their January 12 proposal. More to the point, Collins and Vandenberg found the situation in Korea so much *less* critical than MacArthur's reports had led Washington to believe that the Joint Chiefs quickly withdrew the document from NSC consideration.[24]

During their mid-January visit to Tokyo and Korea, the army and air force commanders met several times with MacArthur. Despite his gloomy reports, they discovered a revitalized Eighth Army under Ridgway. The field commander had rallied his forces and, on January 15, began a series of counteroffensives directed at the Chinese. By January 17, Collins and Vandenberg reported (to both Washington and Tokyo) the hopeful signs they had seen in Korea. Privately, the general still complained about unfair combat restrictions, inadequate forces, and the impossibility of both fighting in Korea and protecting Japan. Publicly, however, MacArthur took his cue from Collins and Vandenberg. At a press conference he condemned the "loose talk about the Chinese driving us into the sea" and announced this "command intends to maintain a military position in Korea" just as long as ordered.

This unexpected good news, General Bradley recalled, changed not only the military assumptions held in Washington, but altered completely the credence given to MacArthur's reports. Ridgway obviously deserved the real credit for turning the tide of battle and demonstrating that the Chinese could be stopped in Korea with limited, conventional means. In the judicious words of the official JCS history, after January 17, the "imposing figure in Tokyo no longer towered quite so impressively." MacArthur's "counsels no longer commanded the respect they had once enjoyed." Less formally, Bradley wrote that even though the Joint Chiefs continued to address messages to MacArthur, they no longer considered him the decisive factor in Korea. In their minds, at least, he was "kicked up stairs to chairman of the board and was, insofar as military operations were concerned, mainly a prima donna figurehead who had to be tolerated."[25]

During the next two months, Ridgway's Eighth Army repulsed China's "third phase offensive" and advanced UN lines back to the 38th parallel. Communist forces, confronted by severe logistic problems, declining morale, and immense casualties, were compelled to go on the defensive. Ridgway made clear that under current circumstances he could hold most of South Korea. The administration revealed that while it continued to support the principle of a "unified, independent and democratic Korea," it would not seek forcible unification nor conduct extensive ground operations north of the 38th parallel. Washington was ready for an armistice that would salvage American honor and lives as well as an independent South Korea.[26]

Chinese accounts of the war during early 1951 echo the praise heaped by Bradley and others on Ridgway. They credit him with identifying their logistic weakness and for employing air and artillery attacks in limited counteroffensives that devastated the ranks of the People's Volunteers. By avoiding the grandiloquent rhetoric and strategy of MacArthur, Chinese military historians admit, Ridgway broke their offensive power. By late February, Mao and Peng seemed trapped in a war of attrition along the 38th parallel. In April and May, Chinese forces would make a desperate push to regain momentum. But, in the words of a retired Chinese official, they gambled that sooner or later the "Truman-Acheson-Marshall faction" would triumph over the "MacArthur faction," setting the stage for negotiations.[27]

Domestic Infighting

Even as the pressure on American combat troops ebbed a bit, the administration came under renewed domestic political attack. In December, old isolationists like Colonel Robert McCormick, former ambassador Joseph P. Kennedy, and former president Herbert Hoover (most of them allies of Douglas MacArthur) began demanding a radical revision of the nation's foreign policy. Hoover, for example, spoke of America as a new Gibraltar of western civilization. He proposed to remove American armed forces from continental Europe and Asia. Naval and air power, based in the Atlantic and Pacific ocenas, must become the main line of defense against the Sino-Soviet menace.

As 1951 began, Senator Robert Taft, the most prominent Senate Republican, issued his call for a "great debate" on whether American ground forces should be dispatched to Europe under NATO command. On January 5, Taft charged that Truman had "simply usurped authority—in violation of the laws and Constitution" by dispatching troops to fight an undeclared war in Korea. (The previous summer, the senator had approved the purposes of Truman's actions in Korea but had protested his unilateral action in bypassing Congress.) Taft's new objections were directed really more against the current plan to send several divisions to NATO and did not challenge seriously the Korean venture.

Echoing this complaint, on January 8, Senator Kenneth Wherry (R–Nebraska) introduced a resolution stating that no American troops should be sent to Europe without congressional authorization. Wherry's proposal went down to defeat in April, following a bitter debate and a compromise in which the Senate went on record as opposing the dispatch of more than four divisions to Europe without legislative approval.[28]

In a formal sense, congressional opposition to foreign military adventures offered little comfort to MacArthur's effort to widen the Korean War. Still, the Republican attempt to limit commitments to NATO would have removed one of the administration's strongest arguments against building up forces in Korea. The subtext of debate, of course, revealed Republican efforts to humiliate the administration because of its emphasis on European and Japanese security and unwillingness to confront China more aggressively.

House and Senate Republicans demanded Dean Acheson's resignation followed by a "house cleaning" in the State Department. Senator Joe McCarthy muttered about impeaching the president. Although a minority view, anger at China (and the administration's restraint) boiled over in mid-January when both houses of Congress passed resolutions demanding that the United Nations condemn China as an aggressor.[29]

Partisan wrangling prompted the administration, over the vigorous opposition of its allies, to push such a resolution through the UN. France, Britain, and the British Commonwealth nations all worried that branding China an outcast would only reduce the likelihood of a cease-fire and increase the chances of a wider war. Acheson complained that delaying UN action risked "destruction domestically," while General Bradley told a British military official that the "strength of public and Congressional feeling" was so strong that Washington must force a resolution through. Accordingly, the administration twisted the arms of its allies (while consenting to compromise wording that did not require specific retaliation against Peking) and forced a reluctant UN majority on February 1 to declare China an aggressor in Korea. This gratuitous slap, even if designed to appease domestic critics, complicated the search for an armistice and alienated China from the United Nations for over twenty years.[30]

Stalemate and Recall

As General Ridgway succeeded in pushing the badly bruised Chinese back beyond the 38th parallel, MacArthur played a diminishing role in events. Eager to stay in the limelight (or, in Ridgway's words, "to fire the starting gun"), the general insisted on flying from Tokyo to Korea on the eve of major operations. Fearing this would tip off the enemy, Ridgewy urged him to stay away, at least until after the attack began.[31]

The military achievements of February and March compelled policymakers to confront the question of whether or not again to send American forces north of the 38th parallel. While the State Department frowned upon large scale incursions for fear of increasing Chinese resistance, military officials worried that Communist forces would utilize this sanctuary to build up offensive strength. Without adopting a formal policy, Washignton decided that Ridgway could make limited strikes north of the boundary in order to destroy enemy concentrations and, the administration hoped, set the stage for negotiations. However, both civilian and military leaders expressed little interest in forced unification and eagerly anticipated a settlement allowing the withdrawal of most American troops.

As Ridgway's success became more evident, MacArthur searched for ways to retake the initiative. During February he pressed the Joint Chiefs for authority to bomb the port of Rashin (near the Soviet border) and additional hydroelectric plants on the Yalu. The Joint Chiefs and the State Department rejected these requests, noting both the danger of violating Russian and Chinese territory and the fact that neither target played much of a role in enemy strength.[32]

In Tokyo, diplomat William Sebald found MacArthur "tired and depressed." He complained about the "lack of instructions regarding his mission in Korea" and that he "no longer knows what he is supposed to do." The general at least knew what he opposed, for on February 13 he issued a statement denouncing as "wholly unrealistic and illusory" the idea of "some" officials that he should hold a line across Korea and enter into "positional warfare." Four days later MacArthur told Sebald that he had conceived of a new "concept to bring China to her feet by" means of a naval blockade and air power. The task would take less than a year, although he "was not clear about whether the Soviets would come in or not."[33]

About this time MacArthur may have suggested an even more radical approach to break the Korean stalemate. According to his memoirs (and inteviews granted in later years), during February he conceived a plan to win the war "in a maximum of ten days"—a remarkable shortening of the time he mentioned to Sebald. The general's memoirs assert that in the spring of 1951 he recommended "massive air attacks" against North Korea coupled with a move to "sever Korea from Manchuria by laying a field of radioactive waste . . . across the lines of enemy supply." (He made no mention of atomic bombs at this time.)

Late in 1952, the general urged president-elect Dwight Eisenhower to spread a "belt of radioactive cobalt" along the Sino-Korean border and then invade North Korea with Chinese Nationalist troops and American marines. MacArthur spoke even more graphically of atomic warfare in a pair of interviews granted in 1954 but published only after his death a decade later. He claimed to have wanted to drop as many as fifty atomic bombs on Manchuria as a prelude to laying a barrier of radioactive wastes and invading North Korea.

Although MacArthur may well have fantasized about an atomic victory, the documentary record fails to confirm his claims that he unsuccessfully proposed this plan in 1951. As early as July 1950, he had made vague reference to utilizing atomic bombs to isolate or destory North Korea and China. Yet, during the 1951 recall hearings he testified that the Joint Chiefs, not he, took the initiative in requesting a list of potential nuclear targets. The general indicated he thought this a poor idea and made no mention whatever of the radioactive belt.[34]

Nevertheless, some evidence suggests MacArthur did seek authority to utilize atomic weapons beginning early in March 1951. Reacting to evidence (discussed below) of a large concentration of Chinese and Soviet forces, he requested on March 10 that the Joint Chiefs make available to him on an "on call basis" atomic bombs for use on "D-Day" should the Soviets invade Japan. It was "vital," he stressed, to have them available at once in case of a Russian attack. Although authorities in Washington recognized the same military threat, by now they had grown nearly as fearful of their own commander as of the enemy. The chiefs waited two weeks before responding and then refused to do more than "consider" his request. Their delay in response proved a portent of moves to oust him.[35]

MacArthur seldom spelled out to friend or foe exactly what he meant by "victory." Ridgway surmised that he did not mean merely the destruction of enemy forces in the field and unification of Korea under a democratic government. Rather, "what he envisaged was no less than the global defeat of Communism, dealing Communism a blow from which it would never recover and which would mark the historical turning back of the Red Tide." Thus, he wanted to destroy the "air bases and industrial complex in Manchura" as well as the remainder of China's "industrial centers." He wanted to equip and carry Chiang's troops to the mainland to "break the Communist hold." He believed the "Chinese masses were ready to welcome Chiang back." Although MacArthur doubted the Soviets would enter this "preventive war on Red China," if they did, Ridgway guessed the general would proably have "urged the further step of an attack upon the USSR."[36]

MacArthur certainly made no secret of his disgust with the emerging policy of restoring the prewar boundary. On March 7, he complained to journalists about the "savage slaughter" of Americans certain to result from a war of attrition. As the UN line moved north, China would again enjoy a logistic advantage. A week later he issued a statement criticizing those who favored halting at the 38th parallel. "Fundamental decisions" had to be made on ways to avoid a "heavy cost in Allied blood."[37]

Despite MacArthur's pressure to widen the war, by the middle of March most of Truman's civilian and military advisers thought the time had come for the president to issue a direct appeal to China for peace talks. The Joint Chiefs informed MacArthur on March 20 of a tentative approach to China but provided no details. The appeal would probably have failed. Although some officials (like Robert Lovett) favored flexibility on the issues of Taiwan and China's UN seat, the draft proposal condemned Chinese aggression and repeated the standing UN goal of Korean unification. The only hint of flexibility came in a phrase suggesting that a "prompt settlement" in Korea would "open the way for consideration of other problems in that area." The Chinese were massing forces for a major spring offensive and probably awaited its outcome before deciding on negotiations. The chiefs inquired what freedom of action MacArthur required during the following weeks to protect his forces while efforts were made to induce Peking to come to the conference table.

MacArthur responded tartly a few days later, declining to mention the proposed peace initiative at all. Rather, he decried any "further military restrictions" on his command. He already labored under inhibitions that made it "impracticable to attempt to clear North Korea." Additional interference would tie his hands completely. Instead of "backing off" from China even temporarily, the general unsuccessfully requested authority to send the Seventh Fleet closer to the eastern and southern Chinese coasts "to make a show of force." This would enable naval units and pilots to "obtain area familarization" for future operations against China and would improve morale on Taiwan.[38]

The general waited until Mach 24 to let fly his real attack on Peking *and* Washington. In Tokyo, MacArthur announced his own peace plan. Denouncing the pretensions of "Red China" to modern, great power status, he declared it inadequate to the "challenge of modern war." Even under the inhibitions that restricted his actions, China could not achieve its goals in Korea. Therefore, the enemy must "by now be painfully aware that a decision by the United Nations to depart from its tolerant effort to contain the war to Korea, through an expansion of our military operations to its coastal and interior bases, would doom Red China to the risk of imminent military collapse." As soon as Peking recognized this fact, MacArthur saw no problem in resolving the war "without being burdened by extraneous matters not directly related to Korea, such as Taiwan or China's seat in the United Nations."

The UN commander's peace terms—transmitted to the American and Chinese governments by press release—included no change in the status of Taiwan or China's UN seat, no "sacrifice" of the Korean nation and its people (presumably all Korea), and the abject appearance "in the field" of the defeated "commander-in-chief of the enemy forces." MacArthur took sole responsibility for achieving "political objectives of the United Nations in Korea."[39]

The general subsequently tried to downplay the significance of his action, describing it as a "routine communiqué." In fact, his closest associate in Tokyo, General Courtney Whitney, described the maneuver as a desperate act to stop "one of the most disgraceful plots in U.S. history." MacArthur said much the same thing to William Sebald in a subsequent conservation. The general charged Truman and his advisers with orchestrating a "plot" to "hand over . . . Formosa to Red China." While he never explicitly admitted his motive, MacArthur almost certainly hoped that by antagonizing China he would scuttle the administration's peace feeler and increase public pressure for a showdown with Peking.[40]

Administration figures realized at once that MacArthur hoped to preempt American policy in Korea and East Asia. Former Eighth Army commander, Robert Eichelberger, now a civilian employee in the Defense Department, reported that MacArthur's "latest bomb about making peace with China has the [Pentagon] boys hanging on the ropes." The Tokyo press release resulted in a hectic round of meetings in the Pentagon and White House. The Joint Chiefs and Deputy Secretary of Defense Robert Lovett agreed that MacArthur had embarrassed the American government and caused grave confusion among the allies with whom the peace appeal had already been discussed. Not only China, but the thirteen countries participating in the UN command would question who had ultimate negotiating authority. Given this breech of discipline, Lovett and the Joint Chiefs admitted that "it would be perfectly obvious if it were anybody else who had made the statement which MacArthur made . . . he would be relieved of his command at once." Although the chiefs "recognized that the consequences of relieving MacArthur are startling," they realized they could not "just let this slide by." Since MacArthur held four separate commands in Japan, Korea, and the Far East, any summary dismissal might lead to chaos. Consequently, the Joint Chiefs favored an interim reprimand.

Discussing this dilemma with Acheson, Lovett worried about the public and press reaction to firing the general. Given popular frustration with the war, MacArthur's pronouncement "was probably the most popular public statement anyone has ever made." It promised victory, peace, and a quick exit from Korea. "If the President challenged it," Lovett cautioned, "he would be in the position at once of being on the side of sin." For the moment, MacArthur had snared the administration. It needed time to escape the trap, time which only temporary public "silence" could provide.

Acheson favored an interim reprimand while the president and State Department reassured America's allies. However, this would not do much good "if the same thing is apt to happen next week." Truman (whose reaction, Acheson reported, "combined disbelief with controlled fury") considered MacArthur's behavior "open defiance" of presidential authority, a challenge to civilian control of the military, and grounds for relief. Still, he agreed to move with deliberation. Acheson, Lovett, and the Joint Chiefs drafted a message (sent by the JCS) informing MacArthur that the president wished to "direct your attention" to the order of December 6, 1950, that military officials not comment publicly on sensitive military and diplomatic matters. If, by some miracle, the enemy requested an armistice, the general was immediately to notify the Joint Chiefs. Both within and outside the government, many wondered how seriously he would take these orders. A cartoon in the *Washington Post* parodied the president by depicting him waving the message under MacArthur's nose and declaiming "Honest, no fooling this time!" Disdainfully, the general directed an aide to "file this one with the others."[41]

Intelligence indications about a new enemy threat magnified the significance of MacArthur's challenge. Since late March the Joint Chiefs had received evidence of an extensive buildup of Chinese troops and aircraft in Manchuria, coordinated, it seemed, with a large concentration of Soviet submarines northeast of Japan. American analysts thought these forces might be used in a renewed effort to drive the UN command out of Korea while severing the lifeline to Japan or, still worse, invading that nation. Strangely, few in Washington considered the possibility that the Chinese and Soviet buildup might actually be designed to deter a renewed American push into North Korea or an attack upon China.

While stalling MacArthur's request for access to atomic bombs, the president and his military advisers weighed seriously the need to transfer these weapons from control of the civilian Atomic Energy Commission (AEC) to the Air Force. AEC Chairman Gordon Dean learned from high officials early in April that he might soon be instructed to approve the transfer. These reports troubled him, as he feared that Truman and the JCS were acting prematurely and might use the bomb without sufficient reflection once they gained possession. Dean opposed strongly giving any field commander, especially MacArthur, "who had little knowledge concerning effects," this decision-making power.[42]

With the administration still reeling from the March 24 fiasco and worried increasingly about the Communist military buildup, Acheson's warning about MacArthur provoking a new crisis "next week" proved prophetic. On April 5, House Minority Leader Joseph W. Martin (R–Massachusetts) read to the Congress a telegram from MacArthur. Martin's recent speeches had demanded that Truman "open a second front in Asia" by helping Chinese Nationalist troops to invade the mainland. "Without a single GI being forced to place foot on the soil of the Chinese mainland," the congressman asserted, this would punish the aggressor, relieve pressure on Korea, and save American lives. The Republican leader charged that unless Truman planned to win the war, he bore responsibility for the "murder" of thousands of GIs.

Martin sought MacArthur's reaction to his ideas, promising, if asked, to keep the response confidential. The general composed a reply on March

20 (before the peace initative flap) without requesting confidentiality. He assured the Republican leader that his ideas were in the grand American tradition of "meeting force with counterforce" and went on to offer a synopsis of his official recommendations of the previous six months. In Asia the "Communist conspirators . . . elected to make their play for global conquest." American soldiers in Korea fought "Europe's war with arms while the diplomats there still fought it with words." Defeat in Asia made the fall of Europe "inevitable." Victory would prevent war elsewhere. "As you point out," he wrote to Martin, "we must win. There is no substitute for victory."[43]

As with his March 24 ultimatum, MacArthur pleaded innocent. The letter to Martin, he told William Sebald in Tokyo, was "personal and insignificant." He "didn't even remember it" until he located a copy in his files. In his memoirs, he described the message as a "polite response." Yet, the general issued at least two other statements through journalists early in April that broached the same themes. He decried the lack of a "definite objective" in Korea. The administration's policy would be "ludicrous if men's lives were not involved." He also accused Washington of refusing to expand the South Korean army—even though since January he had opposed any further buildup of ROK forces.[44]

For Truman, this latest incident of "rank insubordination" proved "the last straw." The president wrote on his desk calendar that the "situation with regard to the Far Eastern General has become a political one." (He added a nasty comment about the genreal's two marriages and divorce.) Although he wanted his inner circle to make a recommendation to him on their own, he had decided that the "Big General in the Far East must be recalled."[45]

The same day that Martin released MacArthur's latest blast, April 5, AEC head Gordon Dean learned that Joint Chiefs intended to request the transfer of atomic bombs to military control and their delivery to the Pacific. Also they drafted orders authorizing MacArthur to launch air attacks against China in case of major Communist escalation.[46]

On April 6, General Bradley discussed with Truman the latest intelligence on the Chinese and Soviet air and naval concentration. Although the Joint Chiefs were extremely wary of MacArthur's judgment, they felt the theater commander must have authority to retaliate (with conventional weapons) against Chinese bases in case of a "major attack" upon UN forces launched from Chinese territory. Bradley also stated that the time had come to deploy several atomic bombs to Guam and Okinawa.

Truman conferred with AEC chairman Dean later that day. The president related in detail the report of large troop and aircraft concentrations in Manchuria and the massing of seventy Soviet submarines near Vladivostok and Sakhalin. He feared the Chinese would attempt to drive the UN command off the Korean peninsula while the Soviets moved to "take the Japanese islands and with the submarines cut our supply lines to Japan and Korea." Since such a move might well trigger global war, Truman instructed Dean to authorize transfer of nine atomic bombs to military control.

Dean worried that the small atomic arsenal might be squandered in tactical engagements. The president specified that if they were used, atomic weapons would be employed against major Chinese or Soviet targets. In any case, the terms of the transfer assuaged the concern of Dean and others that MacArthur might use the bombs improperly or prematurely. The air force and Strategic Air Command on Guam and Okinawa, would retain operational control. Dean signed the order and B-29s began transporting the weapons across the Pacific on April 10.

Despite this safeguard, neither Truman nor his military and civilian advisers were comfortable with MacArthur's imminent proximity to the atomic trigger. Their anxiety stemmed in part from the dilemma posed by their own decision to grant the UN commander authority to launch retaliatory air strikes against China if, in his opinion, planes flying from Chinese bases carried out major attacks on UN forces. Given MacArthur's subjective interpretation, such discretion would almost certainly lead to one result. In fact, General Bradley (who had pushed to give the theater commander this power) admitted he "was now so wary of MacArthur" that he "deliberately withheld the message [regarding authority to attack China and details about the transfer of atomic weapons] and all knowledge of its existence from him, fearing that [MacArthur] might make a premature decision in carrying it out." It was not difficult to imagine even a small incident escalating to the point where MacArthur insisted on striking China with conventional or atomic weapons.[47]

The president, like Acheson, understood the importance of moving deliberately. Observers at home and abroad must see the administration acting in a firm, responsible and nonpartisan way. Otherwise, in Lovett's earlier phrase, they might seem "on the side of sin." The political and diplomatic damage of a misstep, Acheson noted, could be incalculable. Accordingly, Truman pressed Acheson, Marshall, Harriman, Bradley (the Big Four, as he called them); the rest of the Joint Chiefs; and several other officials to confer numerous times over the following days before recommending a course of action.

Acheson spoke accurately when he referred to the anxiety in London and other allied capitals at the prospect of giving MacArthur any sort of leeway to attack China. When American officials broached the idea, it elicited fear and disbelief among most British leaders. On April 9, Sir William Slim, Chief of the British Imperial General Staff, told his colleagues that "in his opinion, General MacArthur personally wanted war with China." Permitting him to define and respond to a major enemy attack would prove disastrous, as he would "color both intelligence and operational reports to suit his own ends." Nor should British authorities defer to the judgment of the American Joint Chiefs, since "they were scared of MacArthur" and were easily manipulated by him.

Similarly, the British foreign secretary cabled Ambassador Sir Oliver

Franks in Washington that London's "principal difficulty" with America was General MacArthur. "He seems to want war with China. We do not." The general's "public utterances" had weakened British and European confidence in the "quality of American political judgment and leadership." MacArthur publicly declared that "his policy is not the stated policy of his government" yet that government "is, nevertheless, unwilling and unable to discipline" him.

Sharing many of these same fears, Truman resolved to remove the Korean commander in whom he had lost all confidence. By replacing MacArthur with a reliable officer, the president hoped to solve several problems simultaneously. He would: (1) rid himself of a ruthless, perhaps imbalanced, critic in league with the Republicans; (2) cover his domestic flanks, and assure support for the recall from the Joint Chiefs, by giving a new commander authority to retaliate against Chinese escalation; and (3) possibly deter escalation by China or the Soviet Union (or be in a position to retaliate effectively) by leaking word that the United States would use conventional and nuclear weapons should the major Communist powers expand the war within or outside Korea.[48]

Although MacArthur heard nothing official between April 5 and 10, he and his entourage knew his fate was being weighed. He told General Ned Almond on April 9 that he "had become politically involved and may be removed by the President." Sid Huff, a personal aide exceptionally close to the general, told William Sebald on April 8 that "he wished the General would be recalled in order to have an opportunity to clarify the thinking in Washington" by swaying the public and Congress. Although usually a strong supporter of MacArthur, Sebald disagreed. He felt the general would lose a "showdown," as "he could not stand the searchlight and shafts of public criticism." In his diary, the diplomat quoted with approval Senator Robert Kerr's (D–Oklahoma) comment on the Martin incident:

> MacArthur's prolonged one man act is wearing the patience of the rest of the team mighty thin. . . . In fact, it is getting about as threadbare as the General's much touted Oriental prestige. The time may be nearer than we think when the overall cost of keeping MacArthur as theater commander will be greater than the value of his "position" with Asiatics.[49]

MacArthur, too, may have sensed the waning of his support and may also have wanted to get out of his Korean command before his position among members of Congress and the American public eroded further. Although he may no longer have aspired seriously to the presidency, the general was determined not to end his military career abandoned in what he deemed a hopeless situation, a symbolic Corregidor. The administration would neither fight the kind of war he demanded nor seek the kind of victory he clamored for. Whether or not a draw in Korea advanced the national interest, it clearly presented no opportunity for personal glory or vindication. By forcing Truman to relieve him, MacArthur could at least be assured of a martyr's mantle, of sullying the reputation of a president he disliked, and an opportunity—once again—to grandstand before the American public.

Truman conferred with the Big Four several times on April 6, 7, and 8. Although accounts prepared by the participants vary in some details, all confirm the substantive issues. Harriman and Acheson wanted to relieve MacArthur of all his duties at once. The former argued he should have been fired long before, when he interfered with the 1947–48 change in policy in Japan. Bradley and Marshall agreed in principle but worried that Congress might punish the administration by cutting military appropriations. Bradley also hesitated to act without consulting his fellow chiefs, one of whom was traveling. Some thought was given to sending Marshall to Tokyo to speak with MacArthur or to bringing the general home for consultation. Marshall and Bradley even drafted a letter (unsent) ordering MacArthur to shut up. Harriman and Acheson, fearing political mischief if the general either stayed in place or returned for consultations, rejected all these ideas.

On April 8, Generals Collins and Vandenberg and Admiral Sherman informed Marshall that MacArthur should be relieved. (Bradley expressed no formal opinion.) Even if not technically guilty of insubordination under military law, they noted the general's lack of sympathy with the policy of limiting the war to Korea and felt it was "necessary to have a commander more responsive to control from Washington." Also, he had clearly violated the president's directive against making unauthorized public statements on national policy. Finally, the three commanders felt MacArthur's actions "were continuing to jeopardize the civilian control over the military authorities."

On Monday, April 9, Bradley communicated these views to the president and indicated that he and Marshall concurred with the recommendation of Harriman and Acheson. Although they must have guessed it, only then did Truman reveal his determination to fire MacArthur. The president told members of his staff he "was sure MacArthur wanted to be relieved." The group agreed that General Matthew Ridgway should assume all MacArthur's commands (with General James Van Fleet replacing Ridgway in Korea) and then discussed the best way to inform the general and nation of the decision.

Also on April 9, AEC chairman Gordon Dean briefed members of the House and Senate Joint Atomic Energy Committee on the transfer of atomic bombs to military control. Administration officials hoped this sign of resolve against possible enemy escalation would mute the expected Republican outcry over the imminent recall of MacArthur.[50]

Truman signed the necessary orders on April 10 (the same day as the shipment of atomic weapons to the Pacific began, although the two events were never publicly connected) with the understanding that Army Secretary Frank Pace, then on an inspection tour in Korea, would fly to Tokyo and deliver the news personally to MacArthur. However, as one careful student of the event remarked, "Murphey's law did not fail." Communications with Pusan broke down and Pace did not receive instructions. While White House officials worked on a press release, they learned that a reporter from the *Chicago Tribune* had inquired whether an "important resignation was expected in Tokyo." Truman's staff feared the general knew of the decision and planned to "get the jump on the President" by announcing his own departure along with an attack on administration policies. If that happened, they "would never catch up with the General." Late that evening, as additional rumors of a resignation circulated in Washington, Truman and his staff decided to act.

At a hastily called press briefing (at 1 A.M. on April 11), Press Secretary Joe Short handed reporters copies of Truman's order that MacArthur turn over all his commands to Ridgway. An accompanying statement detailed some of the reasons behind the decision. The president declared that MacArthur was "unable to give his wholehearted support to the policies of the United States Government and of the United Nations." Even though a "full and vigorous debate" over national policy was essential to democracy, by law and tradition "military commanders must be governed by the policies and directives issued to them in the manner provided by our laws and Constitution." In a nod toward the general, Truman affirmed that "MacArthur's place in history as one of our greatest commanders is fully established." The nation owed him a "debt of gratitude" for "distinguished and exceptional service . . . in posts of great responsibility."[51]

MacArthur learned of his recall during lunch on April 11, from an aide who heard the news on a commercial radio broadcast. The general displayed little emotion and ordered preparations for a quick departure. In a private discussion a few hours later with William Sebald, his anger surfaced. If Truman had only asked, he would have retired "without difficulty." Instead, the president subjected him to this "public humiliation." The general denied violating orders and refused to believe the Martin letter played any role in the recall. He "intimated that his removal was [part of] a plot in Washington" to "hand over" Taiwan to "Red China." The Philippines would fall next, followed by Japan. Then "our whole position in the Far East would crumble."

MacArthur worried that while Ridgway might be a "good soldier," he "doubted" his capacity to command a theater or act as supreme commander in Japan. The general hoped his replacement had not "sold his soul" for the job. Predicting a purge of Americans in Tokyo, MacArthur urged Sebald to get out quickly. (Most of MacArthur's closest subordinates left with him or departed soon afterward.)[52]

Ridgway flew to Tokyo on April 12. His written account of meeting MacArthur represented their discussion as cordial and the general as "composed, quiet, temperate, friendly and helpful to the man who was to succeed him." His "apparent lack of rancor or resentment" seemed amazing. Yet in a confidential memorandum Ridgway described a bitter man who questioned the president's "mental stability." MacArthur stated that an "eminent medical man" in contact with Truman's physician had told him "that the President was suffering from malignant hypertension; that his affliction was characterized by bewilderment and confusion of thought." This caused him to threaten to assault a music critic who had disparaged Margaret Truman's singing. He predicted the president "wouldn't live six months." The general indicated he would move to New York and alluded to several lucrative offers to make speeches and "raise hell." [53]

The president explained his decision to the public in a radio address on April 11. By most accounts, his performance fell flat. Instead of highlighting his differences with MacArthur, Truman discussed "what we are doing in Korea and our policy in the Far East." He appealed for moderation on all sides. Just as America would be "tragically wrong" to widen the war, the enemy would be equally "foolhardy" to do so.

As "telegrams and letters of abuse" began pouring in, the president wrote in his diary that while the recall caused "quite an explosion," he had "had to act." In an explanation sent to NATO commander Eisenhower, Truman said he was "sorry to have to reach a parting of the way with the big man in Asia." But, "he asked for it and I had to give it to him." [54]

14
Fading Away...

MacArthur spent his final days in Japan arranging a return to a homeland he had not seen in fourteen years. He accepted an invitation to address Congress, made provisions to move into the Waldorf Astoria Hotel in New York City, and learned of Republican demands for a Senate inquiry into his dismissal. By law, he remained on the army's active list for life and retained his five-star rank and salary (over $18,000 in 1951). His perquisites included an airplane, paid staff, and office in New York. Besides the $500,000 nest egg acquired from Manuel Quezon in 1942, MacArthur had a standing offer from his friend James Rand of the Remington Rand Corporation to serve as board chairman. Concerned that publicity about this lucrative position would tarnish the general's image as a selfless patriot, Herbert Hoover cabled a plea to delay acceptance until the dust had settled. On April 16, the general ordered his personal plane (called *SCAP* since 1945) renamed the *Bataan* and departed Tokyo with his family and closest aides.[1]

MacArthur's recall resulted in harsh attacks upon President Truman. Some angry citizens burned Truman in effigy while a dozen state legislatures passed resolutions supporting the general or condemning the president. Conservative newspapers and politicians severely criticized the administration, the Democratic party, and others judged soft on communism. The *Chicago Tribune* declared Truman "unfit to be President" while the New York *Journal American* asserted that the chief executive must have acted under the influence of "some kind of mental or neural anodyne" administered by a Communist agent in the State Department.

A number of U.S. senators also believed Truman had erred. Senator Richard Nixon (R–California) remarked that the "happiest group in the country are the Communists and their stooges," as Truman had given them what they wanted, "MacArthur's scalp." Senator Joe McCarthy (who had become a fan of the general's after helping to defeat him in the 1948 Wisconsin primary) accused the president of being drunk on "bourbon and benedictine" and suggested "the son of a bitch ought to be impeached." Senator William Jenner (R–Indiana) lamented that a "secret inner coterie . . . directed by agents of the Soviet Union" now controlled American diplomacy, and Senator William Knowland spoke of a "Far Eastern Munich." Other Republican leaders, such as Taft, and Wherry, and House minority leader Martin, discussed possible impeachment.

These politicians were perhaps responding to the immediate popular outcry in support of MacArthur. A Gallup Poll found two-thirds of the public opposed to the recall and only a quarter supporting Truman. Exceptionally large, exuberant crowds turned out to see MacArthur as he left Tokyo and during his journey home. When his plane landed in San Francisco, 500,000 people lined the route from the airport to his hotel. Responding to a reporter's query about his political plans, he remarked: "I do not intend to run for any political office . . . the only politics I have is contained in a simple phrase known to all of you—God Bless America." At the Washington airport, a mob broke through police lines and practically trampled the general and his family.[2]

The White House began a damage control operation as soon as MacArthur returned to America, eager to make a case before the general spoke to both houses of Congress on April 19. Truman wisely refrained from public comment, allowing his staff to devise a strategy of undermining MacArthur's appeal. On April 17, George Elsey and other presidential assistants set to work "buttoning up loose ends." Members of the White House staff were assigned to work with friendly legislators and with the Defense and State departments in devising rules and questions for the upcoming Senate hearings. They initiated a congressional investigation (later aborted) of the so-called China lobby, presumably to show the network of pro-Nationalist agents cooperating with the Republican party and MacArthur. Leading citizens opposed to MacArthur were asked to line up with the president. The staff assembled material on the past record of Taft, Wherry, and McCarthy to show the inconsistency of their current claims. Concerned that the debate might "reach the dirty tricks stage," Elsey's office sought "miscellaneous MacArthur materials" that exposed unflattering past events. These included the general's controversial World War II communiqués and (at Senator Robert Kerr's suggestion) references to General Arthur MacArthur's difficulties in the Philippines fifty years before.[3]

Still, a kind of seige mentality permeated the White House as the time for MacArthur's speech to Congress approached. To break the tension, the president's staff circulated a mock "Schedule for Welcoming General MacArthur," which read:

12:30	Wades ashore from Snorkel submarine
12:31	Navy Band plays "Sparrow in the Treetop" and "I'll Be Glad When You're Dead You Rascal You"
12:40	Parade to the Capitol with General MacArthur riding an elephant
12:47	Beheading of General Vaughan at the rotunda

1:00	General MacArthur addresses members of Congress
1:30–1:49	Applause for General MacArthur
1:50	Burning of the Constitution
1:55	Lynching of Secretary Acheson
2:00	21-atomic bomb salute
2:30	nude D.A.R.s leap from Washington Monument
3:00	Basket lunch, Monument Grounds

This gallow humor revealed the anxiety within Truman's inner circle.[4]

MacArthur's much anticipated speech more than fulfilled the hopes of his supporters. He held Congress and millions of radio listeners enraptured for over thirty minutes. In a strong, resonant voice, he pounded his enemies with a barrage of sweeping generalizations and clichés. Although full of bathos, the speech proved remarkably effective. Borrowing phrases from earlier reports and press releases, the general decried the "defeatism" that claimed that America lacked the power to defend both Europe and Asia simultaneously. "You cannot," he argued, "appease or otherwise surrender to communism in Asia without simultaneously undermining our efforts to halt its advance in Europe." MacArthur focused on defending the anti-Communist outposts of Asia, especially Taiwan. The administration, he charged, had imposed impossible military restrictions (China's "privileged sanctuaries") upon his command and had given him no clear goals.

The general called for an economic and naval blockade of China, aerial reconnaissance of the mainland, and support for Nationalist military operations against the People's Republic. Backing away from his confidential recommendations for direct American intervention, he declared that "no man in his right mind would advocate sending our ground forces into continental China." "Every military leader concerned with Korean campaigns, including our own Joint Chiefs of Staff," he insisted, agreed with his strategy. Although he hated war, MacArthur declared that once a conflict began "there can be no substitute for victory." His strategy posed far less of a danger of igniting global conflict than did that of "those who would appease Red China." The Korean people knew this through bitter experience and begged him to carry a simple message to America: "Don't scuttle the Pacific."

MacArthur closed with a flourish that touched the emotions of all but his most callous listeners. After "fifty-two years of military service" he was ending his career. A life in the army had been his dream as a boy and a West Point cadet. Although these youthful hopes and dreams had "long since vanished," he still remembered the popular barracks ballad that proclaimed, " 'Old soldiers never die; they just fade away.' And like the old soldier of that ballad, I now close my military career and just fade away— an old soldier who tried to do his duty as God gave him the light to see that duty. Good Bye."

Journalist Robert Donovan, who witnessed the scene, quipped that by the end of the speech there was not a "dry eye" among the Democrats or a "dry seat" among the Republicans. (Presumably, no one in Congress knew he had used the old soldiers line and the plea not to "scuttle the Pacific" on several occasions since 1938.) Representative Dewey Short believed he had "heard the voice of God." In a similar theological vein, Herbert Hoover described the general as the "reincarnation of St. Paul into a great General of the Army who came out of the East." After receiving the keys to the District of Columbia, MacArthur left for New York City, where an estimated 7.5 million people cheered his ticker-tape parade from the Bowery to the mayor's office. Adding to the honors, the Waldorf offered him a luxury suite at nominal rent for the rest of his life.[5]

The impressive outpouring of affection that greeted MacArthur's appearances did not imply mass support for his war strategy or Asian policy. Probably, very few who stood to glimpse his return or cheered his speeches understood much about the underlying military or diplomatic debate. For many, MacArthur was the last great hero of the Second World War to return home. Regardless of events in Korea, for nearly a decade politicians and journalists had built him up to superhuman proportions. Since the heady victory of 1945, the cold war, fear of domestic subversion, the "loss of China," and the nuclear threat had called into question the meaning of that great triumph. MacArthur appealed to traditional patriotic virtues and spoke in vivid cadence about honor, duty, and country. Truman's inability to explain the meaning of limited war and his inadvertent crudeness in relieving MacArthur increased the contrast between their styles. As D. Clayton James has written, developments since World War II had "left many Americans disillusioned and ready to cheer a respected commander who promised victory and to disparage an unpopular President who appeared to have led a once powerful and purposeful nation into a state of impotence and drift."[6]

During the next two weeks, before the Senate inquiry began, MacArthur gave speeches to large gatherings in Chicago and Milwaukee criticizing the administration. Declining direct interchanges with the press, he relied on his associate, Courtney Whitney, as his spokesman. This imperious style of handling public relations did not work as well as it had in Tokyo. Soon journalists complained about the new "Dai Ichi Building" in New York where Whitney passed out press releases and expected reporters to "record the pearls that drop from his lips and like it." As one wag noted, in their "idle hours the correspondents covering the Waldorf Front" composed the "Battle Hymn of the Waldorf."

> Here is the Waldorf Astoria,
> The home of the rich and the odd,
> Where the press speaks only to Whitney
> And Whitney speaks only to God.[7]

Meanwhile, the administration mustered more fortitude than its critics expected. On April 21, *The New York Times* carried an account of the Wake Island Conference (leaked by the White House), which revealed that MacArthur had minimized the danger of China entering the war and assured Truman he could easily repulse its forces should it dare to do so. The general disputed the accuracy of the report and then protested that a presidential eavesdropper had secretly recorded his words. Neither charge was valid. General Bradley and others (including a stenographer who sat outside the meeting room) took notes during and after the conference. A short time later, Bradley collated these accounts and circulated the transcript among the participants, including MacArthur. This revelation only tarnished a bit the general's reputation as an oriental sage. But it revealed the administration's determination to defend itself.[8]

On April 25, the Senate voted formally to hold an inquiry into MacArthur's dismissal and the "Military Situation in the Far East." The Foreign Relations and Armed Services committees selected Senator Richard Russell (D–Georgia) to chair their joint sessions. Hoping to use the hearings as a platform to attack the administration's overall foreign policies, many conservative Republicans evinced little sustained interest in the grounds for his recall or the strategic dilemmas in Korea. They demanded open meetings with live radio and television coverage. Fearing a carnival atmosphere and the disclosure of sensitive military secrets, the Democratic majority insisted on closed hearings. In a compromise, State and Defense department security officers culled each day's testimony and released sanitized transcripts within hours of the event. However, the complete and unedited account remained classified until released by Senator J. William Fulbright in 1973.

Despite allegations of an administration cover-up, the bulk of the censored material related to current force levels and military capabilities on both sides, not political questions embarrassing to the administration. However, the executive testimony did contain some important elaborations of the positions held by the general and his critics. The most careful student of the full hearings, John E. Wiltz, notes that "many of the passages . . . which the censor excised from the public record bore upon the General's arguments and nearly all were destructive of the arguments." Ironically, out of excessive zeal to safeguard national security, the administration withheld information that undermined many of MacArthur's claims and bolstered Truman's case for relieving him.[9]

Chairman Russell and his fellow Democrats, including Lyndon B. Johnson (D–Texas), received careful coaching by staff attorneys before the hearings began on May 3. They viewed the inquiry as the "critical turning point of the entire debate" over foreign policy. "Properly handled," it could "turn the tide," resulting in "public willingness to judge MacArthur's views on the basis of merit, rather than emotion." A misstep could "build up MacArthur to the point where *anything* is his for the asking."

The hearings presented a "make or break situation" for both sides. The Democrats felt MacArthur knew that the "first hot flash of public indignation had died down" and only a new, dramatic forum could restore his momentum. His critics could defeat him either by refuting his assertions or stalling until the public lost interest in his cause.

"Every question," the senators decided, should be asked in the light of "two major dangers and two major opportunities." The Democrats had to avoid a "personality argument" or partisan wrangling while still preventing the Republicans from turning the hearings into a vehicle for attacking the administration as "corrupt and infiltrated by Communists." Truman's allies needed to expose the half-truths and contradictions voiced by the general and then move to "separate MacArthur from the majority of his followers" who did not "really share his views but who see him as a baseball bat with which to beat the administration."

Senator Lyndon Johnson played a critical behind-the-scenes role in developing a strategy for conducting the inquiry. (The young and talented Texan was close to Chairman Russell and had loaned key members of his own staff, such as George Reedy, to the investigative committee.) Johnson and his colleagues developed a list of questions for the general. They would permit MacArthur to speak at great length, without interruption. In effect, the Democrats planned to give him enough rope to hang himself. Then they would begin a polite but coordinated assault on his position. How could he know with certainty how Moscow and Peking would respond to bombing and blockading China? As a theater commander, did he claim to possess better intelligence information than the rest of the government? Did the general have any idea how America's allies and the United Nations would react to a wider war with China? Could the United States defend Asia and Europe without their support? Just how wide a war did he propose? Were Chinese cities as well as military bases proper targets? Were American forces capable of defending Japan against enemy retaliation? Given the dismal record of Chiang Kai-shek's forces, why did he assume they would perform effectively in Korea or the Chinese mainland? How much logistical support did he contemplate giving Chiang's troops for their operations against China? The Democrats guessed that MacArthur would dodge these questions (as, in fact, he did), arguing that as a theater commander they were outside his area of responsibility. That of course, was their point.[10]

The MacArthur Hearings

The heralded investigation into MacArthur's recall began dramatically and, as Truman's supporters had hoped, ended inconclusively. During May and June, the Senate committee called some thirteen witnesses (MacArthur; Marshall; Bradley; the three Joint Chiefs—Collins, Vandenberg, and Sherman; Acheson; Louis Johnson; General Albert Wedemeyer; Admiral Oscar Badger; Patrick Hurley; General David Barr; and General Emmett O'Donnell) and took several thousand pages of testimony. At least four of the witnesses (Wedemeyer, Badger, Hurley, and Barr) had almost nothing to say about the Korean War, but were paraded by the Republican members to testify about softness toward Chinese communism endemic to the Roosevelt–Truman administrations. In fact, nearly half the testimony taken at the hearings focused on World War II and early postwar China policy. The Republicans declined to call a single witness prepared to back up the general's strategic arguments.

As the leadoff witness, MacArthur testified for three days. He admitted no error of any kind on any issue military, political, or diplomatic. He blamed the initial Chinese advance on faulty intelligence from Washington and argued that granting China "privileged sanctuaries" in Manchuria prevented him from repelling the attack and completing his job of unifying Korea. The general insisted that his plan to strike at China would have won the war and assured Korean unification, without much risk of global conflict. He repeated his claim that the Joint Chiefs supported his strategy without reservation. Not only had he never usurped authority or defied the president, but, he declared, "no more subordinate soldier has ever worn the American uniform."

MacArthur frequently tried to change the subject from Korean strategy to Asian politics and psychology. He made a series of racist statements (quoted in the earlier discussion of the occupation) about the immature and malleable Japanese brain, which he had tutored toward maturity. He also bemoaned the alleged transformation of Chinese character, from docile to warlike, since 1900. Addressing the senators, he described the rise and fall of Chinese regimes and warlords since the collapse of the Ch'ing Dynasty. Much of this was entertaining fiction that bore little relation to any known chronology. (For example, he had the Manchurian warlord, Chang Tso-lin, leading a nonexistent Chinese national government years before he was even a figure of regional importance.) None of the senators knew better or cared to correct him.

As the Democrats hoped, he soon fell into a carefully laid trap. Senator Brien MacMahon (D–Connecticut) pressed MacArthur on the implications of his strategy. What if he had received authority to escalate the war and the Soviets intervened or initiated a global conflict? The general responded "that doesn't happen to be my responsibility." As a theater commander, he left that concern to the president and Joint Chiefs. MacMahon seized the opportunity to respond that "by your own statement you have not made that kind of study, and yet you advise us to push forward with a course of action that may involve us in that global conflict." MacArthur tried later to amend his statement, but the damage had been done. As the Democratic committee staff members noted in their daily analysis of his testimony, "after all, we have only the return of a general not the Second Coming." Despite ample opportunity to do so, most careful observers believed that MacArthur contributed little that went beyond his public assertions.[11]

Defense Secretary Marshall and the members of the Joint Chiefs of Staff provided the major counterpoint to the general's claims. (Republican senators grilled Dean Acheson for nine days but focused their questions on the "loss" of China.) Testifying in succession, Marshall, Bradley, Vandenberg, Collins, and Sherman refuted nearly all MacArthur's arguments.

The general maintained that if the United States extended the war to China, the Soviets were not likely to intervene. At most, Moscow would increase military assistance to China and North Korea. He doubted the Russians' ability to bring much power to bear in northeast Asia and considered it virtually unthinkable they would initiate a global war. Similarly, China lacked the capacity to counter an American escalation.

The Joint Chiefs and Defense Secretary Marshall disagreed, unanimously. Unfortunately, many of their most telling points had to be censored so as not to reveal American capabilities or what Washington knew about enemy forces. (MacArthur's testimony, on the other hand, relied on such generalities and subjective information that it posed no such problem.) If the United States attacked China by sea and air, they expected the Russians to establish air superiority in northeast Asia. Soviet and Chinese air and submarine forces could drive the UN command out of Korea and, probably, Japan. General Vandenberg noted that Manchurian airfields were heavily defended and that any assault would lead to huge losses of American aircraft. In censored testimony, he called the "U.S. air force really a shoe string" operation incapable of carrying out a major offensive. Despite MacArthur's talk of an easy air war, Vandenberg warned that in the "vast expanse of China," American air power would be a "drop in the bucket." [12]

General Bradley quickly disposed of the privileged sanctuary argument. China, like the United States, was fighting a "limited war," holding back its air force from attacking American lines of communication, airfields, and ports in South Korea and Japan. Thanks to the enemy's restraint, Bradley stressed, "we are fighting under rather favorable rules for ourselves." The "sanctuary business," as Vandenberg called it, "is operating on both sides." Marshall joined all the chiefs in predicting that an expansion of the air war would hurt, not help, the UN effort. [13]

In his testimony and public assertions, MacArthur had claimed that the January 12, 1951, contingency plan devised by the Joint Chiefs proved they really agreed with his proposal to hit China and assist a Kuomintang invasion of the mainland. Defense Secretary Marshall and each of the chiefs took issue with this assertion. Their plan had evolved in response to MacArthur's prediction of doom and defeat. When they learned that Chinese forces were not about to overrun South Korea and destroy American forces, they shelved the idea.

True, under highly specific circumstances, they would recommend attacking China. "We just couldn't take" expulsion from Korea "as a final solution" to the war, Bradley testified. If American forces were driven off the peninsula by China, or if Peking launched attacks outside of Korea (for example, against Japan or Taiwan) they would urge retaliation. In fact, they revealed that Ridgway now possessed authority (denied MacArthur) to respond to a major Communist escalation by bombing Manchuria. If the fighting dragged on without a sign of settlement, Bradley explained, the Joint Chiefs might press for a wider war as a means of forcing China to accept a Korean armistice. (Unlike MacArthur, none of the chiefs suggested a campaign to defeat and topple the People's Republic.) However, he pointed out, Chinese forces were taking such terrible losses in their current spring offensive that Peking might soon be willing to begin negotiations. Bradley's prediction proved quite accurate.[14]

The Joint Chiefs dismissed as illusory MacArthur's call for using Chinese Nationalist troops in Korea or on the China mainland. They knew of no large anti-Communist guerrilla movement ready to link up with Chiang's forces and paralyze or topple the Peking regime. Pulling Kuomintang troops out of Taiwan for such an operation would probably insure a Communist invasion of the island, not its survival. "Even if we could get him ashore," Bradley argued, few Chinese were eager to rally around Chiang. After all, General Collins remarked, "these were the same people that were run off China in the first place." They were unlikely to fight more effectively in Korea than they had for their own country.

MacArthur's idea of alleviating Chinese pressure on Korea by helping Chiang to invade south China, Marshall warned, had a deceptive appeal. But once the United States landed Chiang's soldiers they would face probable annihilation. At that point, to save them, we would be "drawn into their support" and a war on the Chinese mainland. Bradley got off the most memorable line by predicting that MacArthur would get the United States "in the wrong war, at the wrong place, at the wrong time and with the wrong enemy."[15]

These points were actually made in successive testimony. Bradley had testified first and his unexpected vigor so disturbed Republican senators that when he finished (on May 24) they suggested canceling further appearances by the service chiefs. The Democratic majority, of course, insisted on hearing them all. When the three completed their testimony, the Republicans did not call even a single rebuttal witness to defend MacArthur's strategy. They decided, instead, to beat a dead horse—China policy since 1944. The last half dozen witnesses sat through a barrage of questions by Republicans about the alleged betrayal of Chiang Kai-shek by FDR and Truman.

Not surprisingly, public interest in the hearings faded soon after MacArthur left the stand. Newspaper coverage declined while White House and congressional mail shifted notably away from support of the general. The hearings adjourned late in June leaving members of the Senate committee unsure of what to do with the massive transcript in their possession. In mid-August they decided against issuing any formal report, although individual members could file their views, and sent the entire record of testimony and related materials to the full Senate. Eight Republicans filed their opinions as a de facto minority report. They acknowledged Truman's authority to remove MacArthur (just as the general had predicted Congress would do should FDR remove him in November 1942) but lamented the circumstances as a "shock to national pride" and his transgressions as "inadequate" to warrant his relief. But the Republicans directed the bulk of their criticism at the administration's past China policy.[16]

Truman must have been as pleased with this outcome as MacArthur was disappointed. The hearings cushioned the political hysteria of the moment and allowed Truman to weather the storm. Republican critics had an opportunity to vent their wrath and soon moved on to other concerns. Although the president's overall popularity never recovered from the impact of the war and the recall, this reflected a variety of problems. More to the point, Truman's loss never proved MacArthur's gain.

Unfortunately for the general's serious critics, the Senate hearings revealed only the broad outlines of the administration's case. If released publicly, the uncensored transcripts might have revealed the manipulative talk about privileged sanctuaries, military restrictions, unleashing Chiang Kai-shek, and lack of danger in attacking China. Even for those people who paid attention, the hearings only showed that MacArthur had no infallible plan for winning the war and that the administration had not turned its back on an easy victory. Given the intensity of the time, perhaps this sufficed.

The episode affected administration policy in another, more troublesome, way. To impress both the Chinese and domestic critics, Truman adopted a noticeably tougher policy toward Asian communism. The administration dropped almost all flexibility on the questions of Taiwan's future or China's UN membership. The United States pressed its allies to tighten the economic blockade against China and increased, dramatically, military aid to the French fighting in Indochina. Abandoning plans for an unarmed Japan, Washington retained extensive bases and compelled the Japanese to rebuild their armed forces.

A Last Hurrah

Soon after completing his own testimony, MacArthur set off on a national speaking tour, which lasted until the 1952 Republican convention. Still a dashing figure in his uniform, he began the circuit in Texas by attacking appeasement in Asia, New Deal liberalism, Harry Truman, and high federal taxes. The last theme, critics noted, may have reflected the politics of H. L. Hunt and Clint Murcheson, two Texas oil men who financed his tour.

The speeches sometimes addressed large policy issues but often contained inappropriate political observations directed against the general's critics. As the blush of excitement faded from his appearances, MacArthur found himself addressing smaller crowds. In Paris, NATO commander Eisenhower (who had avoided public comment on the recall) denounced his former boss to journalist C. L. Sulzberger. Politically, the NATO chief likened the general to Senator Joe McCarthy. Ike felt "MacArthur now, as always, was an opportunist seeking to ride the crest of the wave."[17]

The national speaking kept MacArthur before the public and may have indicated his continued interest in the presidency. Yet, like in 1944 and 1948, he shied away from any organized effort to gain the nomination. As in those abortive campaigns he awaited a draft from a deadlocked convention that never came. The general received a boost when Republican party leaders selected him to make the keynote address at the Chicago gathering in July. But, to everyone's surprise, he delivered what even friendly commentators described as an ineffective speech that lost the delegates' attention. Sensing failure, he returned to New York immediately. Shortly afterward, the convention nominated Dwight D. Eisenhower on the first ballot.[18]

In contrast to his preconvention activity, MacArthur ceased all public appearances for many months after Ike's selection. With his military career over and his political aspirations frustrated, the time had come to seek other opportunities. On August 1, he accepted the position as board chairman with Remington Rand, closing a remarkable half century of public life.

Despite MacArthur's predictions of appeasement and surrender, America's Far Eastern policy muddled through. A Japanese peace treaty, nearly drafted by John Foster Dulles at the time of the general's recall, was signed in September 1951. Its economic and military provisions bound Japan tightly to the Western alliance. The occupation ended the following April.

Soon after MacArthur left Tokyo, Ridgway and his field commander, General James Van Fleet, repulsed China's great spring offensive without resorting to nuclear weapons or a wider war. Chinese military commanders, hoping to capitalize on possible disarray following MacArthur's dismissal, made the mistake of underestimating American strength. In late April and mid-May, efforts to break through American lines stalled, and hungry Chinese troops began to surrender en masse. The next month American troops launched their own offensive, which halted for political reasons at the 38th parallel. Then, as hoped, Soviet diplomats proposed armistice talks, to which China agreed. Negotiations began in July, although the talking—and fighting—dragged on for two years until the summer of 1953. Despite continued hostilities, neither belligerent intended to fight for Korean reunification.

MacArthur's supporters have claimed, in justification of the general's position, that only threats by Eisenhower to escalate the war yielded a settlement. It remains uncertain whether Eisenhower actually threatened— or intended—to use nuclear weapons against China. (In December 1952, MacArthur urged the president-elect to invade North Korea and seal it off from China by laying down a belt of radioactive waste.) Allusions by Eisenhower and Dulles to a wider war might have been a bluff or a serious contingency. But even while the new administration rattled both nuclear and conventional sabers, its negotiators reached compromises with the

Chinese on such sticky issues as prisoner repatriation and redrawing the Korean boundary.

Following Stalin's death in March 1953, major political changes occurred in both the Soviet Union and China. Leaders in Moscow and Peking had a renewed interest in improving relations with the new American president. All these factors—not just the threat of a wider war—contributed to the July armistice, which restored a north–south division near the 38th parallel. These terms, it should be remembered, represented the original war aims of *both* the United States and China. In the end, American losses included some 33,000 dead and 100,000 wounded. Chinese and North Korean military losses have been estimated at 1.5 million killed and wounded. About one million Korean civilians also became casualties.

Far from turning its back on Asia, the Truman administration and its four successors paid increasing attention to the region's turmoil. Taiwan, firmly under Nationalist control and with an American security guarantee, prospered in the aftermath of the Korean settlement. Indochina, never of prime concern to the general, emerged as the real flash point between the United States and communism. During the Korean conflict, America aid sustained France's colonial war. Following the French defeat in 1954, the United States created a new anti-Communist regime in Saigon. For most of the next two decades, America's commitment to South Vietnam dominated American foreign policy in Asia.

Although MacArthur's dream of destroying the Chinese Communist regime failed to materialize, he must have taken some solace in having barred the door to any improvement in Sino-American relations. From 1951 through 1971, the United States and China engaged in a shadow war. Beginning under President Truman and continuing until the Nixon administration, the two countries remained totally estranged and hostile. Meanwhile, the CIA and other government organs sponsored covert guerrilla operations aimed at destabilizing the People's Republic. American leaders justified growing involvement in Vietnam as a way of containing China.

MacArthur, during his final decade of life, played no part in and rarely commented on foreign policy. (An exception came in a bizarre exchange of letters with former Vice President Henry A. Wallace in which the two men praised each other's ideas and shared their dislike for Truman.) He seemed to enjoy his role as the genteel board chairman of Remington (later Sperry) Rand. Early in 1961, President John F. Kennedy spoke with him about the growing crisis in Southeast Asia. Because of the earlier failure to suppress Asian communism, MacArthur observed, "all the chickens are coming home to roost and you [Kennedy] are living in the coop." At the same time, he advised against fighting a land war on the Asia mainland. When President Lyndon Johnson visited the general on his deathbed in March 1964, he repeated the advice, in vain. Yet never did MacArthur indicate regret or assume responsibility for his own actions, which contributed so much to America's Vietnam crusade.[19]

Ultimately, MacArthur's legacy must be judged a failure. From his succession of headquarters in the Philippines, Australia, and Japan, during sixteen years of uninterrupted service in Asia and the Pacific, he distorted information and manipulated events to serve selfish, often political, ends. He used the press, public opinion, and political threats to bludgeon bureaucratic rivals and two presidents. In Japan, arguably the most successful of his commands, MacArthur supervised the demilitarization and democratization of a great power. Yet even there he deserved only partial credit for implementing the reform program and frequently subverted it to suit his own presidential ambitions. During the late 1940s, he fought the administration's efforts to disengage from China's civil war and provided secret military assistance to Taiwan. As UN commander in Korea, he willfully risked a war with China or even the Soviet Union, to achieve personal vindication. Following his removal, MacArthur encouraged political action against a sitting president that foreclosed any chance of improving relations with China.

America's "greatest expert on oriental psychology," it is safe to say, knew little about Asian realities and not much more about American politics. For twenty years after he returned home the United States pursued a program of military containment that MacArthur's actions had helped make inevitable. Ironically, American policy in Asia finally achieved a peaceful balance only by coming to terms with the Chinese regime he had hoped to destroy.

In a curious footnote, a new China—MacArthur controversy erupted in 1987, twenty-three years after the general's death. In a proposed cultural exchange, the National Portrait Gallery planned to exhibit a collection of portraits of prominent Americans at a Peking museum. The Chinese government insisted that for reasons of "political sensitivity," a picture of Douglas MacArthur be removed. The United States Information Agency decried the censorship and canceled the exchange. The general, no doubt, would have been pleased."

NOTES

Chapter 1

1. Douglas MacArthur, *Reminiscences* (New York, 1964), 2–21. Much controversy surrounds this autobiography, which focuses on the 1940s and early 1950s. Besides the debate over its interpretive biases, MacArthur drew much of the text from earlier biographers with whom he had collaborated.

2. The document is quoted extensively in Carol M. Petillo, *Douglas MacArthur: The Philippine Years* (Bloomington, Ind., 1981), 45–47.

3. MacArthur, *Reminiscences*, 12–13; Petillo, *Douglas MacArthur*, 2, 4, 252 n 5; D. Clayton James, *The Years of MacArthur* (Boston, 1970–85), 1. 28–29.

4. Petillo, *Douglas MacArthur*, 49–56; James, *Years of MacArthur*, 33–40.

5. MacArthur, *Reminiscences*, 30–32; James, *Years of MacArthur*, 1: 91–94; Petillo, *Douglas MacArthur*, 84–95.

6. MacArthur, *Reminiscences*, 32.

7. The improved circumstances at Fort Leavenworth convinced Douglas to defy Pinkie's efforts to get him to enter private business. Fearful that her son might suffer her husband's disappointments, she induced railroad magnate E. H. Harriman to offer him a job. Douglas rejected the proposal. See Petillo, *Douglas MacArthur*, 99–100.

8. Ibid.; James, *Years of MacArthur*, 1: 95–109.

9. Petillo, *Douglas MacArthur*, 126–34.

10. Ibid., 137; James, *Years of MacArthur*, 1: 305.

11. Petillo, *Douglas MacArthur*, 141; James, *Years of MacArthur*, 1: 323.

12. For an example of the gossip, see diary entry of September 25, 1943, Harold L. Ickes Papers, Library of Congress; the general made a successful marriage a few years later and never publicly uttered Louise's name following the divorce. See MacArthur, *Reminiscences*, 91.

13. *New York Times*, Apr. 21, 1929.

14. Petillo, *Douglas MacArthur*, 147–48; James, *Years of MacArthur*, 1: 334.

15. James, *Years of MacArthur*, 1: 342–44; Theodore Friend, *Between Two Empires: The Ordeal of the Philippines, 1929–46* (New Haven, Conn., 1965), 78. Just before his selection, MacArthur had requested permission to tour China and Japan preliminary to a requested reassignment in New York. Franklin Roosevelt

later claimed that MacArthur was actually weighing an offer "to go to China to build up a strong army" for the Nationalist regime. See diary entry of Oct. 9, 1937, Ickes Papers.

16. Petillo, *Douglas MacArthur*, 151–53.

17. MacArthur, *Reminiscences*, 89.

18. MacArthur correspondence with Isabel Cooper, multiple letters, September–October, 1930, box 111, Morris Ernst Papers, Humanities Research Center, University of Texas, Austin. The letters became part of the Ernst Papers when he represented Drew Pearson in a suit brought by MacArthur in 1934. See the following chapter for an explanation.

19. James, *Years of MacArthur*, 1: 376–77, 390; *New York Times*, June 3, 1931. Many newspapers reprinted MacArthur's denunciation of pacifism and radicals.

20. MacArthur, *Reminiscences*, 92–93.

21. Dwight D. Eisenhower, *At Ease* (New York, 1967), 216–17; James, *Years of MacArthur*, 1: 394–404; *New York Times*, July 29, 1932. MacArthur made several accusations about revolutionary plots but offered no evidence of any.

22. Rexford G. Tugwell, *The Democratic Roosevelt* (Garden City, N.Y., 1957), 348–51; Eisenhower, *At Ease*, 316–17.

Chapter 2

1. MacArthur kept a low profile during the election campaign by touring Europe. In November, he reported to Secretary of State Henry Stimson his impressions of the military situation in practically every nation on the continent. He found the Soviet threat diminished because, as he told Stimson, Stalin had "killed off all . . . the able men from the educated classes." "Jewish commissars without brains" now controlled the army which he characterized as "sunk." See diary entry of Nov. 9, 1932, Henry Stimson Papers, Yale University.

2. James, *Years of MacArthur*, 1: 417–26.

3. MacArthur, *Reminiscences*, 101. Whether true or not, MacArthur recounted similar episodes of facing down FDR and other authority figures in subsequent years. The accounts all end with the other party pleading with the general not to resign. MacArthur told an almost identical story about confronting FDR in Honolulu in 1944. See Chapter Six.

4. Ibid.

5. Diary entry of July 27, 1933, Ickes Papers.

6. Petillo, *Douglas MacArthur*, 164–66; William Manchester, *American Caesar.* (Boston, 1978), 144–45, 156.

7. *New York Times*, May 17, 1934; James, *Years of MacArthur*, 1: 685 n.24; Oliver Pilat, *Drew Pearson: An Unauthorized Biography* (New York, 1973), 140.

8. Tyler Abell, ed., *Drew Pearson Diaries* (New York, 1974), 28–29.

9. The MacArthur–Cooper love letters and the legal documentation for the case are located in box 111, Ernst Papers, Cooper to Ernst, Dec. 6, 1934, and Ernst to MacArthur, Dec. 6, 1934.

10. Pearson to Ernst, Dec. 13, 1934, ibid.

11. Admiral William Leahy later commented that MacArthur, a bachelor in 1934, could have faced down moral criticism for having a mistress. He could have looked his accusers in the eye and said "So what? Cunt can make you look silly at times."

He caved in, Leahy claimed, for fear of shaming Pinkie. "He didn't want his mother to learn about the Eurasian girl!" See Pilat, *Drew Pearson*, 142–46; receipt from MacArthur to Pearson and Allen, Dec. 19, 1934, box 111, Ernst Papers; diary entry of May 24, 1942, Ickes Papers; Petillo, *Douglas MacArthur*, 166; Manchester, *American Caesar*, 156. The sensationalist *National Enquirer* got wind of the MacArthur–Cooper letters shortly after the Ernst Papers became available. The tabloid ran an essentially accurate feature on the love affair. The MacArthur estate got an injunction preventing the publication of additional letters. See *National Enquirer*, Sept. 21, 1976, 1.

12. Joseph C. Goulden, *Korea: The Untold Story of the War* (New York, 1982), xxii. Goulden claims that former State Department official Charles Burton Marshall told him that Davis prepared a written account of these incidents for the Joint Chiefs during the Korean War. Davis, like Eisenhower, fell out with MacArthur in the late 1930s. As no JCS document has been found on the subject, the evidence remains hearsay. However, Davis's verifiable role in setting up and terminating the affair with Isabel Cooper lends it some credence. Also see Robert Ferrell, ed., *The Eisenhower Diaries* (New York, 1981), 18–22.

13. Goulden, *Korea*, xxii–xxiii; also see C. B. Marshall, "The Very Image of a General," *Washington Post*, Oct. 11, 1970.

14. The George Van Horne Moseley Papers at the Library of Congress and the Hoover Institution at Stanford University contain a lively correspondence between Moseley and MacArthur. The latter, as always, put little of his inner thoughts down on paper.

15. For a discussion of the decision to extend MacArthur's service and regarding his accomplishments during the final year as Chief of Staff, see James, *Years of MacArthur*, 1: 442–47; 448–70.

16. David J. Steinberg, *Philippine Collaboration in World War II* (Ann Arbor, Mich., 1967), 4.

17. Sidney Fine, *Frank Murphy: The New Deal Years* (Chicago, 1979), 15.

18. Friend, 83–85.

19. Fine, *Frank Murphy: The New Deal Years*, 7–18; Friend, *Between Two Empires*, 85–92.

20. Fine, *Frank Murphy: The New Deal Years*, 44–45; Friend, *Between Two Empires*, 137–41.

21. Friend, *Between Two Empires*, 141.

22. Fine, *Frank Murphy: The New Deal Years*, 47–65.

23. Friend, *Between Two Empires*, 165; MacArthur, *Reminiscences*, 140.

24. James, *Years of MacArthur*, 1: 480.

25. MacArthur, *Reminiscences*, 102; Manuel L. Quezon, *The Good Fight* (New York, 1946), 153–55; John Gunther, "Manuel Quezon," *Atlantic Monthly*, Jan. 1939, 61.

26. Dwight D. Eisenhower diary, undated draft, 1935–38, box 1, Dwight D. Eisenhower Papers, Dwight D. Eisenhower Presidential Library.

27. Fine, *Frank Murphy: The New Deal Years*, 186–87.

28. Dwight D. Eisenhower diary, undated draft, 1935–38, box 1, Eisenhower Papers; James, *Years of MacArthur*, 1: 482–85.

29. James, *Years of MacArthur*, 1: 487–91; Fine, *Frank Murphy: The New Deal Years*, 155–65.

30. James Farley, *Jim Farley's Story: The Roosevelt Years* (New York, 1948), 55; James, *Years of MacArthur*, 1: 593–94.

31. Louis Morton, "War Plan ORANGE: Evolution of a Strategy," *World Politics* 2 (1959), 221–50; Friend, *Between Two Empires*, 161–63; Steinberg, *Philippine Collaboration*, 20.

32. Eisenhower, *At Ease*, 219–21; Ferrell, ed., *Eisenhower Diaries*, 9–10.

33. Eisenhower, *At Ease*, 222–23.

34. Ibid., 223.

Chapter 3

1. Fine, *Frank Murphy: The New Deal Years*, 162–67.

2. Ibid., 170–80.

3. Report on National Defense in the Philippines, April 27, 1936, box 74, Dwight D. Eisenhower Pre-Presidential Papers, Eisenhower Library.

4. Gen. Embick to Chief of Staff, Dec. 2, 1935, WPD 3389-29, RG 165, Chief of Staff Malin Craig and his staff endorsed this report; Fine, *Frank Murphy: The New Deal Years*, 191.

5. Ferrell, ed., *Eisenhower Diaries*, 10–15.

6. Ibid., 14–15.

7. Ibid., 19–20.

8. Ibid., 21; James, *Years of MacArthur*, 1: 505; Stephen Ambrose, *Eisenhower: Soldier, General of the Army, President Elect, 1890–1952* (New York, 1983), 107.

9. Ferrell, ed., *Eisenhower Diaries*, 22–23; MacArthur to O'Laughlin, Dec. 12, 1936, box 54, John C. O'Laughlin Papers, Library of Congress.

10. Roosevelt memoranda for McIntyre, Sept. 12, 1936 and to Woodring, Sept. 17, 1936, OF 178, Franklin D. Roosevelt Papers, Franklin D. Roosevelt Presidential Library.

11. Sayre memorandum of conversation with Roosevelt, Nov. 16, 1936, cited in Fine, *Frank Murphy: The New Deal Years*, 196; on neutrality ideas, see Dorothy Borg, *The United States and the Far Eastern Crisis of 1933–38* (Cambridge, Mass., 1964), 244–46; diary entry of Nov. 20, 1936, Ickes Papers.

12. Sayre memorandum of discussion, Dec. 28, 1936, cited in Fine, *Frank Murphy: The New Deal Years*, 197; diary entry of Jan. 20, 1937, Ickes Papers.

13. MacArthur, *Reminiscences*, 106.

14. Eisenhower, *At Ease*, 225; James, *Years of MacArthur*, 1: 515–18.

15. Ferrell, ed., *Eisenhower Diaries*, 26.

16. MacArthur to Craig, Sept. 16, 1937, PPF 4914, Roosevelt Papers; Eichelberger, "Dictations," Robert Eichelberger Papers, Duke University. These postwar recollections were dictated orally by Eichelberger as part of his plan to write an exposé of MacArthur. They supplement his diary and letters of the 1930s and 1940s. James, *Years of MacArthur*, 1: 522–23.

17. James, *Years of MacArthur*, I, 524–25; Harold Ickes, as numerous references in his diary between 1932 and 1945 attest, sought to prove that MacArthur had significant, hidden financial interests in the Philippines. The evidence remains inconclusive. As mentioned below, Clare Boothe Luce also heard stories about the general's investments during her late 1941 visit to Manila. D. Clayton James and Carol Petillo researched the charge but found no conclusive evidence.

18. Eisenhower, *At Ease*, 225–28.

19. See James, *Years of MacArthur*, 1: 535–38; Eisenhower, *At Ease*, 228.

20. James *Years of MacArthur*, 1: 526–31; Fine, *Frank Murphy: The New Deal Years*, 197.

21. Friend, *Between Two Empires*, 193–94; James, *Years of MacArthur*, 1: 537–38; Fine, *Frank Murphy: The New Deal Years*, 199; Stephen Shadegg, *Clare Boothe Luce: A Biography* (New York, 1970), 129.

22. Eisenhower, *At Ease*, 231.

23. See Eichelberger Dictations, Eichelberger Papers; although Eisenhower's diaries and letters from the early 1930s on reveal continual astonishment with MacArthur's selfish behavior, he tried hard to separate his personal judgment from professional decisions. During the Second World War, for example, MacArthur continually denounced Ike's campaigns in Europe and lobbied against giving more resources to him. Eisenhower never worked against MacArthur in this way; diary entry of Nov. 1, 1942, Gerald Wilkerson papers, Churchill College, Cambridge University (copy at MacArthur Memorial Bureau of Archives).

24. MacArthur to O'Laughlin, July 20, 1938, box 54, O'Laughlin Papers.

25. MacArthur to Early, July 14, 1939, OF 4771, Roosevelt Papers.

26. James Leutze, *A Different Kind of Victory: A Biography of Admiral Thomas C. Hart* (Annapolis, Md., 1981), 185–88. This study draws upon, and quotes extensively, Hart's diary and letters.

27. Theodore White, *In Search of History* (New York, 1978), 108.

28. MacArthur to Early, Mar. 21, 1941, OF 400, Roosevelt Papers.

29. Watson to MacArthur, Apr. 15, 1941, OF 4771, Roosevelt Papers; Early to MacArthur, Apr. 14, 1941, box 10, Steve Early Papers, Roosevelt Library; MacArthur to Early, May 11, 1941, ibid.

30. James, *Years of MacArthur*, 1: 585–88.

Chapter 4

1. On the development of American policy toward China and Japan during the 1930s, see Robert Dallek, *Franklin D. Roosevelt and American Foreign Policy, 1932–1945* (New York, 1978); Borg, *The U.S. and the Far Eastern Crisis;* Dorothy Borg and Shumpai Okamoto, *Pearl Harbor as History* (New York, 1973); Stephen E. Pelz, *Race to Pearl Harbor* (Cambridge, Mass., 1974); Michael Schaller, *The U.S. Crusade in China, 1938–45* (New York, 1979).

2. James, *Years of MacArthur*, 1: 546–47, 551.

3. Ibid., 579–80.

4. John Hersey, *Men on Bataan* (New York, 1943), 287–91.

5. Diary entry of July 1, 1941, Ickes Papers.

6. See Dallek, *Roosevelt and American Foreign Policy*, 273–75.

7. For a discussion of the role of intelligence and codebreaking, see Ronald Spector, *Eagle Against the Sun* (New York, 1985), 85; Ronald Lewin, *The American Magic* (New York, 1982), 1–75.

8. Diary entry of May 21, 1941, Stimson Papers; James, *Years of MacArthur*, 1: 585–88.

9. On negotiations during this period, see Harold L. Ickes, *The Secret Diary of Harold Ickes.* (New York, 1953–54), 3: 552–60, 583–88; John Morton Blum, *From the Morgenthau Diaries: Years of Urgency, 1938–41* (Boston, 1965), 377–79; Irvine H. Anderson, *The Standard Vacuum Oil Company and United States East Asian Policy, 1933–41* (Princeton, N.J., 1975); R. J. C. Butow, *Tojo and the*

Coming of the War (Princeton, N.J., 1961); R. J. C. Butow, *The John Doe Associates: Backdoor Diplomacy for Peace* (Stanford, Calif., 1974).

10. James, *Years of MacArthur*, 1: 589.

11. For a discussion of MacArthur's rebirth in the Filipino mind, see Petillo, *Douglas MacArthur*, 198.

12. James, *Years of MacArthur*, 1: 592–93; Leutze, *A Different Kind of Victory*, 212–17.

13. MacArthur to Cal O'Laughlin, Oct. 6, 1941, box 54, O'Laughlin Papers.

14. James, *Years of MacArthur*, 1: 609.

15. Louis Morton, *Strategy and Command: The First Two Years* (Washington, D.C., 1962), 90–91; diary entry of Sept. 12, 1941, Stimson Papers; Dallek, *Roosevelt and American Foreign Policy*, 303; Mark S. Watson, *Chief of Staff: Prewar Plan and Preparations* (Washington, D.C., 1950), 445–46; Forrest C. Pogue, *George C. Marshall: Ordeal and Hope* (New York, 1966), 2: 166; James, *Years of MacArthur*, 1: 610–11; Spector, *Eagle Against the Sun*, 74–75. For the secret plan to utilize China as a base for attacking Japan, see Schaller, *The U.S. Crusade in China*, 65–85.

16. Hart's letters and diary entries of Nov. 1941, quoted in Leutze, *A Different Kind of Victory*, 218.

17. Ibid., 219.

18. For an excellent appraisal of Japanese doctrine and capabilities, see Spector, *Eagle Against the Sun*, 35–50.

19. Ibid., 77–88; Dallek, *Roosevelt and American Foreign Policy*, 302.

20. Dallek, *Roosevelt and American Foreign Policy*, 304–5.

21. Spector, *Eagle Against the Sun*, 82–85; Schaller, *The U.S. Crusade in China*, 39–85; Dallek, *Roosevelt and American Foreign Policy*, 306–8.

22. Dallek, 306–8.

23. Francis B. Sayre, *Glad Adventure* (New York, 1957), 221; James, *Years of MacArthur*, 1: 616–17.

24. Luce to Noel Busch, quoted in Shadegg, *Clare Boothe Luce*, 128–30. In her letter, Luce referred to MacArthur's difficulties with Quezon before the general's reactivation, MacArthur's exorbitant salary, the fact that most Americans in the Philippines disliked him, and the reality that the American army, not the Filipino army, had to be credited with most of the effective defense effort in the islands.

Chapter 5

1. Diary entry of December 7, 1941, Stimson Papers. On Pearl Harbor, see Gordon Prange, *At Dawn We Slept* (New York, 1981) and Prange's *Pearl Harbor: The Verdict of History* (New York, 1986).

2. Leutze, *A Different Kind of Victory*, 231; James, *Years of MacArthur*, 1: 618–19.

3. Louis Morton, *The Fall of the Philippines*, 80–87; Lewis H. Brereton, *The Brereton Diaries* (New York, 1946), 35–57; James, *Years of MacArthur*, 2: 3–16; Leutze, *A Different Kind of Victory*, 231–32.

4. C. L. Sulzberger, *A Long Row of Candles* (Toronto, 1969), 672. Eisenhower recounted this story after MacArthur's recall from Korea, stressing his poor record of strategic forecasts. For a lucid explanation of the controversy see James, *Years of MacArthur*, 2: 1–15.

5. Diary entries of Dec. 14 and 21, 1941, Ickes Papers.

6. Diary entries of Dec. 14, 17, 22, and 25, 1941, Stimson Papers.

7. Diary entry of Jan. 2, 1942, Stimson Papers; entries of Jan. 8, 13, 17, and 19, 1942, in Ferrell, ed., *Eisenhower Diaries*, 43–44.

8. Eisenhower quoted in James, *Years of MacArthur*, 2: 84.

9. For an excellent description of the fighting on Bataan and Corregidor during this period, see James, *Years of MacArthur*, 2: 50–87.

10. Ferrell, ed., *Eisenhower Diaries*, entries of Jan. 29 and Feb. 3, 8, and 9, 1942, 46–47; *New York Times*, Feb. 23, 1942; James, *Years of MacArthur*, 2: 50–87.

11. The finest study of the political crisis in the wartime Philippines is Steinberg, *Philippine Collaboration*, 33–48.

12. See Executive Order #1, Jan. 3, 1942, copy in box 4 (57), Richard J. Sutherland Papers, Modern Military Records Branch, National Archives.

13. For the best discussion of this episode see, Carol Petillo, "Douglas MacArthur and Manuel Quezon: A Note on an Imperial Bond," *Pacific Historical Review* 48 (Feb. 1979), 107–17; see also Petillo, *Douglas MacArthur*, 208–12.

14. Quezon to Roosevelt, Jan. 13, 1942, PSF, Safe #7, Roosevelt Papers; MacArthur to Marshall, Feb. 4, 1942, ibid.; MacArthur to Marshall, Feb. 8, 1942, ibid.

15. Diary entries of Feb. 9 and March 23, 1942, Stimson Papers; See draft responses to MacArthur in PSF, Safe #7, Roosevelt Papers; James, *Years of MacArthur*, 2: 96–97. Some in Washington thought that MacArthur's concern over his family's safety led him to consider a deal with the Japanese. Permitting their evacuation would, presumably, strengthen his resolve.

16. "Messages received from General MacArthur during night of Feb. 11–12, 1942," PSF, Safe #7, Roosevelt Papers; MacArthur to Marshall, Feb. 16, 1942, ibid.; Marshall memo for Roosevelt, "Condensed record of plan for evacuation of governmental officials from the Philippine Islands," ibid.

17. MacArthur to War Department, Radiogram #285, Feb. 15, 1942, PSF 64, Roosevelt Papers; cash receipts of Feb. 19 and Feb. 25, 1942, Box 4 (57) Sutherland Papers; Petillo, *Douglas MacArthur*, 208–11. Roosevelt, Stimson, Ickes, Eisenhower, and Marshall all knew some details of the transaction but did not, at first, understand its implications.

18. Ferrell, ed., *Eisenhower Diaries*, entry of Feb. 23, 1942, 49; MacArthur, *Reminiscences*, 140–41; James, *Years of MacArthur*, 2: 98–99.

19. Diary entry of February 15, 1942, Ickes Papers; Gardner Jackson to Marvin H. McIntyre, Mar. 11, 1942, PPF, 3453, Roosevelt Papers; Vandenberg letter of Feb. 1942, in Arthur H. Vandenberg, Jr., ed., *The Private Papers of Senator Vandenberg* (Boston, 1952), 76; for a marvelous description of the mania of naming things after MacArthur, see James, *Years of MacArthur*, 2: 132–39. The *New York Times Index* for 1942 lists an entire column of honors bestowed upon the general during the spring.

20. Spector, *Eagle Against the Sun*, 143–44.

21. On establishment of SWPA, see James, *Years of MacArthur*, 2: 100–124.

22. For Marshall's comments on drafting the award, see Marshall to Roosevelt, Aug. 22, 1944, PSF, F106, Roosevelt Papers; James, *Years of MacArthur*, 2: 129–32.

23. Hart diary, quoted in Leutze, *A Different Kind of Victory*, 287; Murphy

quoted in Sidney Fine, *Frank Murphy: The Washington Years* (Ann Arbor, Mich., 1984), 216; FDR to Marshall, Aug. 22, 1944, PSF, 106, Roosevelt Papers.

24. Diary entries of Mar. 17 and 23, 1942, Stimson Papers.

25. Ferrell, ed., *Eisenhower Diaries*, entry of Mar. 19, 1942, 51–52.

26. Diary entry of Mar. 22, 1942, Ickes Papers.

27. Diary entries of Mar. 30 and Apr. 30, 1942, Ickes Papers.

28. During April Ickes questioned Filipino exiles about the transaction. The interior secretary, not realizing the funds were for MacArthur's personal use, thought the money might be used as a slush fund for Quezon and his cronies. Ickes tried, without much success, to get Roosevelt to investigate the matter. He also renewed his discussion with Drew Pearson about MacArthur's affair with Isabel Cooper. See diary entries of Apr. 5 and May 24, 1942, and Sept. 1943, Ickes Papers; before Eisenhower decided how to handle Quezon's proposition, he discussed the issue with Gen. Marshall. Besides their concern with ethics and regulations, Marshall probably warned Ike not to accept the gift since MacArthur would certainly learn about it and be in a position to use the information against his rival. See memorandum by Eisenhower, June 20, 1942, Ferrell, ed., *Eisenhower Diaries*, 63.

29. Diary entry of May 14, 1942, Stimson Papers; William D. Hassett, *Off the Record with FDR* (New Brunswick, N.J., 1958), 88.

30. See anonymous report, actually by Sherrod, in "Public Activities, Biographical, Information, Navy File," box 73, Subject File LBJA, Lyndon B. Johnson Papers, Lyndon B. Johnson Presidential Library; entries of June 1942, "LBJ's WWII Diary," box 74, ibid.; diary entry of July 26, 1942, Ickes Papers; for the background to Johnson's trip, and more about his adventures in the South Pacific, see Ronnie Dugger, *The Politician: The Life and Times of Lyndon Johnson* (New York, 1982), 242–50.

31. Diary entry of Mar. 23, 1942, Stimson Papers.

32. Diary entry of Apr. 18, 1942, Stimson Papers; James, *Years of MacArthur*, 2: 141–49.

33. Gen. Robert Eichelberger recalled the intense dislike MacArthur expressed toward Wainwright throughout the Pacific campaigns. According to Eichelberger, MacArthur relented only when he realized that the public considered Wainwright a hero upon his release from a prison camp in August 1945. See Eichelberger Dictations, Eichelberger Papers.

34. Ferrell, ed., *Eisenhower Diaries*, entry of May 6, 1942, 52.

Chapter 6

1. Daniel E. Barbey, *MacArthur's Amphibious Navy: Seventh Amphibious Force Operations, 1943–49* (Annapolis, Md., 1969), 100; MacArthur's letters to: Robert Wood, Nov. 17, 1943; George Van Horn Moseley, Jan. 5, 1944; Maj. Gen. George B. Duncan, Mar. 3, 1944, all in RG 10, Douglas MacArthur Papers, MacArthur Memorial Bureau of Archives MacArthur resented the fact that during 1943 and 1944, the European Theater and Pacific Ocean Areas grew much faster than SWPA. Of course, most enemy strength lay in those theaters. On force levels, see James, *Years of MacArthur*, 2: 349–61.

2. Roosevelt quoted in James, *Years of MacArthur*, 2: 214; Marshall to MacArthur, Aug. 10, 1942, RG 4, MacArthur Papers.

3. Eichelberger dictations, Eichelberger Papers. Although one of MacArthur's most successful combat generals, Eichelberger became disenchanted with his commander even before he agreed to serve under him in occupied Japan. By the time he left in 1948, he hated MacArthur, and these sentiments no doubt colored his memory. Still, numerous references in Eichelberger's wartime diary and letters confirm his later commentary.

4. James, *Years of MacArthur,* 2: 168–69; On Midway, see Gordon Prange, *Miracle At Midway* (New York, 1983); Spector, *Eagle Against the Sun,* 168. MacArthur's handling and evaluation of intelligence remained controversial throughout the war. He virtually banned the Office of Strategic Services (OSS) from operating in his theater, relying instead on the sometimes questionable work of his G-2, Gen. Charles Willoughby. Eventually, MacArthur, like other theater commanders, received information derived from the navy cryptoanalysts who read intercepted Japanese messages. Willoughby and MacArthur were uneasy about relying on the navy for intelligence and resented their lack of control over the material. The navy's code-breaking ability also permitted it to read Japanese battle reports, which often contradicted inflated SWPA communiqués. The intelligence material, known both as MAGIC and ULTRA, proved more useful to the navy (which utilized it to ambush Japanese shipping) than to SWPA planners engaged in tactical assaults on Japanese jungle units. See Lewin, *American Magic,* 175; Spector, *Eagle Against the Sun,* 445–57.

5. Diary entries of September 25 and 26, 1942, Henry H. Arnold Papers, Library of Congress; H. H. Arnold, *Global Mission* (New York, 1949), 336–49; James, 2: *Years of MacArthur,* 210–11. Given the frequent description of MacArthur's youthful appearance (and the fact that he lived another two decades), it is surprising that several associates commented on the general's chronic infirmities. Arnold noted how his "hands twitch and tremble." Gen. Eichelberger also spoke of a "decided shake in his hands" so severe that he sometimes used both hands "to get his cup of coffee up to his mouth." See Eichelberger Dictations, Eichelberger Papers.

6. Eichelberger Dictations, Eichelberger Papers; Jay Luvaas, ed., *Dear Miss Em* (Westport, Conn., 1972), 61–65.

7. Richard H. Rovere and Arthur Schlesinger, Jr., *The MacArthur Controversy and American Foreign Policy* (New York, 1965), 71–72, 75–76. This critical and insightful study appeared originally in 1951 as *The General and the President;* For a discussion of SWPA commanders, see William M. Leary, ed., *We Shall Return: MacArthur's Commanders and the Defeat of Japan* (Lexington, Ky., 1988).

8. James, *Years of MacArthur,* 2: 280.

9. This controversy is detailed in ibid., 295–303.

10. Gavin Long, *MacArthur As Military Commander* (London, 1969), 119; Rovere and Schlesinger, *The MacArthur Controversy,* 76.

11. Anonymous informant, quoted in James, *Years of MacArthur,* 2: 281.

12. Stanley L. Falk, "MacArthur and Japan," in Leary, ed., *We Shall Return,* 1–22; journal entries of Nov. 1, 1942 and Jan. 14 and 26, 1943, Wilkinson Papers; for more information on Wilkinson and his relationship with MacArthur, see Christopher Thorne, "MacArthur, Australia and the British, 1942–43: The Secret Journal of MacArthur's British Liaison Officer, Part 1," *Australian Outlook* 29 (April 1975), 53–67, and "Part 2" 29 (August 1975) 197–210.

13. Spector, *Eagle Against the Sun,* 253–55.

14. For his flip-flop, see Eichelberger Dictations, Eichelberger Papers.

15. Spector, *Eagle Against the Sun*, 277–80; James, *Years of MacArthur*, 2: 336–402.

16. Pogue, *George C. Marshall* (New York, 1964–87), 3: 440–41.

17. MacArthur made his case in his *Reminiscences*. See also Manchester, *American Caesar*, 277–373; for a more realistic assessment of SWPA casualties, see Falk, "MacArthur and Japan."

18. Pogue, *George C. Marshall*, 3: 440–41.

19. Journal entries of Nov. 1, 1942, and Jan. 14, Jan. 26, and Feb. 13, 1943, Wilkinson Papers.

20. See Hadley Cantril, ed., *Public Opinion, 1935–46* (Princeton, 1951), 626–35; *New York Times*, Jan. 24, July 14, Sept. 18, 1943; Elmo B. Roper, *You and Your Leaders: Their Actions and Your Reactions, 1936–56* (New York, 1957), 56–57, 90–99, 100–103, 151–54.

21. For a thoughtful discussion of the influence of the old isolationists, see Justus Doenecke, *Not to the Swift: The Old Isolationists in the Cold War Era* (Lewisburg, Pa., 1970). The same coalition supported MacArthur's 1948 presidential effort.

22. For a detailed account of MacArthur's support on the state level during the 1944 and 1948 campaigns, see Carolyn Jane Mattern, "The Man on the Dark Horse: The Presidential Campaigns for General Douglas MacArthur, 1944 & 1948," unpub. Ph.D. dissertation, University of Wisconsin, 1976; Vandenberg, ed., *Private Papers of Senator Vandenberg*, 76.

23. For two of the few critical assessments of MacArthur's wartime publicity blitz, see John McCarten, "General MacArthur: Fact and Legend," *The American Mercury*, 58 (January 1944), 7–18; Irving Brant, "The Truth About MacArthur," *New Republic*, Dec. 28, 1942; see also Brant to Steve Early, Aug. 26, 1942, PPF 7859, Roosevelt Papers.

24. See MacArthur's statement of Oct. 28, 1942, quoted in James, *Years of MacArthur*, 2: 250.

25. Diary entries of May 24 and June 23, 1942, Ickes Papers; diary entry of Oct. 29, 1942, Stimson Papers.

26. Diary entry of June 2, 1943, and Eichelberger Dictations, Eichelberger Dictations.

27. Vandenberg, ed., *Private Papers of Senator Vandenberg*, 77–78.

28. Ibid.

29. Wood to Vandenberg, June 24, 1943; Vandenberg to Wood, June 28, 1943,; Vandenberg to Wood, July 2, 1943; Vandenberg to Joseph Savage, July 2, 1943; Savage to Vandenberg, July 1, 1943; Wood to Vandenberg, July 3, 1943; Wood to Savage, July 6, 1943; Vandenberg to Wood, July 8, 1943; Vandenberg to Savage, July 8, 1943; Wood to Vandenberg, July 30, 1943; Vandenberg to Wood, Aug. 5, 1943; Wood to Vandenberg, Aug. 10, 1943; Vandenberg to Wood, Sept. 15, 1943; Wood to Vandenberg, Sept. 22 and Oct. 5, 1943; Vandenberg to Wood, Oct. 21, 1943; Vandenberg to Wood, Nov. 5, 1943; Wood to Vandenberg, Nov. 6, 8, 1943; Vandenberg to Wood, Dec. 16, 1943; Vandenberg to Gannett, Dec. 16, 1943; Wood to Vandenberg, Dec. 20, 1943; Vandenberg to Wood, Dec. 22, 1943; Vandenberg to Wood, Jan. 18, 1944; Wood to Vandenberg, Jan. 19, 20, 1944; Vandenberg to Wood, Jan. 21, 1944; Wood to Lansing Hoyt, Jan. 21, 1944— all in Vandenberg correspondence, Robert Wood Papers, Herbert Hoover Presidential Library. Vandenberg to Willoughby, Aug. 17, 1943, Charles A. Willoughby

Papers, MacArthur Memorial, Vandenberg, ed. *Private Papers of Senator Vandenberg*, 78–79; James, *Years of MacArthur*, 2: 404–6; MacArthur's backers kept Thomas Dewey informed about their efforts. Dewey saw the general as more a threat to Willkie than himself. See Richard Norton Smith, *Thomas E. Dewey and His Times* (New York, 1982), 385.

30. Luvaas, ed., *Dear Miss Em*, 91; Sid Huff, *My Fifteen Years with General MacArthur* (New York, 1964), 89. See the numerous references to the campaign and Hunt in Eichelberger dictations, Eichelberger Papers; James, *The Years of MacArthur*, 2: 418.

31. James, *The Years of MacArthur*, 2: 412–13, 429. James discovered that in December 1943 Robert Wood alerted MacArthur that an independent group planned to enter his name in the Illinois primary. Wood urged him to prevent this by returning an enclosed certificate asking that his name be struck from the ballot. The general thanked his friend but declined to sign.

32. Diary entries of Oct. 10, 1942 and Sept. 5, 12, 25, 1943, Nov. 27, 1943, Ickes Papers. Ickes suspected that Herbert Hoover planned to make a dramatic statement accepting full blame for the Bonus Incident of 1932 and exonerating MacArthur; Gen. Eichelberger walked on eggshells when MacArthur assigned him to escort the First Lady around Australia while the SWPA chief departed for New Guinea. See Luvaas, ed., *Dear Miss Em*, 90–91.

33. McCarten, "General MacArthur: Fact and Legend," 7–18.

34. Ibid.

35. Luvaas, ed., *Dear Miss Em*, 90–91.

36. James, *Years of MacArthur*, 2: 414–16; Possibly to counter McCarten's charge that he seldom visited the front, on Jan. 26, 1944 (the general's sixty-fourth birthday), he staged a deceptive "photo opportunity." MacArthur traveled to the Australian outback to be photographed with American troops training there. Pictures released to the press bore the caption "General MacArthur at the Front with General Eichelberger in New Guinea." Inadvertently, the nose of a civilian Packard automobile could be seen in the corner of the photograph. See Luvaas, ed., *Dear Miss Em*, 90, 100, 148; Eichelberger Dictations, Eichelberger Papers.

37. Memorandum of conversation, Mar. 10, 1944, Post-Presidential Files, Richard K. Sutherland file, Herbert Hoover Papers, Hoover Library.

38. Vandenberg to Wood, Mar. 18, 1944, Wood Papers, Vandenberg correspondence, Hoover Library; Vandenberg, ed., *Private Papers of Senator Vandenberg*, 82–83; James, *Years of MacArthur*, 2: 432–33.

39. Vandenberg, ed., *Private Papers of Senator Vandenberg*, 83–84; Vandenberg to Wood, Apr. 10, 1944, Vandenberg correspondence, Wood Papers.

40. Vandenberg to Wood, Apr. 15, 1944, Vandenberg correspondence, Wood Papers; Miller to MacArthur, Sept. 18, 1943; Jan. 27, 1944, and Apr. 10, 1944; MacArthur to Miller, Oct. 2, 1943, and Feb. 11, 1944, all in RG 10, VIP File, MacArthur Papers.

41. *New York Times*, Apr. 17, 1944.

42. *New York Times*, Apr. 30, 1944.

43. Vandenberg, ed., *Private Papers of Senator Vandenberg*, 84–86, 87–89; Vandenberg to Wood, May 1, 1944, Vandenberg correspondence, Wood Papers; James, *Years of MacArthur*, 2: 439–40.

44. Pogue, *George C. Marshall: Organizer of Victory, 1943–45* (New York, 1973), 3: 442–43; James, *Years of MacArthur*, 2: 520–42.

45. James, *Years of MacArthur*, 2: 525.

46. MacArthur, *Reminiscences*, 196–99; William D. Leahy, *I Was There* (New York, 1950), 249–52; Robert Sherwood, *Roosevelt and Hopkins* (New York, 1950), 2: 440–41. Samuel I. Rosenman, *Working With Roosevelt* (New York, 1952), 456–58; memorandum of Dec. 20, 1947, included in diary, Eichelberger Dictations, Eichelberger Papers, ibid.; Roosevelt to MacArthur, Aug. 9, 1944, and MacArthur to Roosevelt, Aug. 26, 1944, PPF 4914, Roosevelt Papers; James, *Years of MacArthur*, 2: 529–35.

47. Diary entry of August 13, 1944, Ickes Papers.

48. Diary entry of August 13, 1944, Ickes Papers; Frazier Hunt, *The Untold Story of General MacArthur* (New York, 1954), 332–33; Walter Robertson Oral History, Hoover Library; Eichelberger memorandum in Luvaas, ed., *Dear Miss Em*, 155–56; memorandum of Dec. 20, 1947, in diary, Eichelberger Papers.

49. Memorandum of Dec. 20, 1947, in diary, Eichelberger Papers; Herbert Hoover diary, "Japan, May 4, 5, 6, 1946," in Post-Presidential Subject File, Famine Emergency Committee, World Mission, 1946 Journey, Hoover Papers.

50. For a discussion of these decisions see James, *Years of MacArthur*, 2: 536–42.

51. Clark Lee and Richard Henschel, *Douglas MacArthur* (New York, 1952), 172.

Chapter 7

1. On Australia, see D. M. Horner, "Blamey and MacArthur: The Problem of Coalition Warfare," in Leary, ed., *We Shall Return*, 23–59. Gary Hess, *The United States' Emergence as a Southeast Asian Power, 1940–50* (New York, 1987). This thorough discussion of American policy emphasizes that most Americans considered the Philippines a model for Southeast Asian political and economic development and tried to press the Europeans to place power in the hands of "moderate" Asian leaders in order to block the appeal of communism and revolution. On the evolution of MacArthur's attitude toward colonialism, see Petillo, *Douglas MacArthur*, 89–97; for a discussion of Anglo-American tensions over colonial issues, see William Roger Louis, *Imperialism at Bay* (New York, 1977), Christopher Thorne, *The Issue of War* (New York, 1985), and Thorne's *Allies of a Kind*.

2. James, *Years of MacArthur*, 2: 467–69; For an excellent discussion of American policy toward Indonesian independence, see Robert J. McMahon, *Colonialism and Cold War: The United States and the Struggle for Indonesian Independence, 1945–49* (Ithaca, N.Y., 1981).

3. "Notes on Meeting in the Pacific of Sherman and MacArthur," Nov. 9, 1944, box 5, Operational Archives, Adm. Forrest Sherman Papers, Center for Naval History.

4. Bert Andrews, of the *New York Herald Tribune*, gave Navy Secretary James Forrestal his notes on MacArthur's talk. See diary entry of Nov. 22, 1944, James Forrestal Papers, Princeton University.

5. Turner Catledge, *My Life and the Times* (New York, 1971), 155–59.

6. Record of conversation with MacArthur, Feb. 1945, U.S. Department of Defense, *The Entry of the Soviet Union into the War Against Japan: Military Plans, 1941–45* (Washington, D.C., 1955), 51–52; diary entry of Feb. 28, 1945, Forrestal Papers; James, *Years of MacArthur*, 2: 763–65. After Japan's surrender

MacArthur denied he had been consulted regarding Soviet entry into the war and insisting he had opposed it as politically unwise and militarily unnecessary. See the *New York Times*, Oct. 21, 1955, and MacArthur, *Reminiscences*, 261–62. In August 1945, he expressed relief at both the Soviet entry into the war and the use of the atomic bomb. Later, he condemned both events. Gen. Eichelberger voiced a common thought during the spring of 1945 when he wrote of his hope that "the Japanese will quit when Stalin begins to push down along the Manchurian Railway." See Eichelberger letter to his wife, Feb. 16, 1945, Luvaas, ed., *Dear Miss Em*, 217.

7. Sherwood, *Roosevelt and Hopkins*, 2: 525–26.

8. Ibid.

9. Among the best studies of the growth of revolutionary nationalism in wartime Asia are Thorne's *Allies of a Kind* and *The Issue of War;* Steinberg's *Philippine Collaboration;* Chalmers Johnson, *Peasant Nationalism and Communist Power: The Emergence of Revolutionary China, 1937–45* (Stanford, 1963); Joyce Lebra, *Japanese Trained Armies in Southeast Asia* (Hong Kong, 1977).

10. Steinberg, *Philippine Collaboration*, 33.

11. Ibid., 73–77.

12. Ibid., 100–108.

13. For the best study of the Huks, see Benedict John Kerkvliet, *The Huk Rebellion: A Study of Peasant Revolt in the Philippines* (Berkeley, 1977).

14. Diary entries of Sept. 5 and 12, 1943, and July 9, 1944, Ickes Papers. MacArthur's appointment of Col. Andres Soriano, a Spaniard formerly close to Franco, upset Ickes terribly. On Osmena, see Steinberg, *Philippine Collaboration*, 100–108.

15. The most concise and informative treatment of MacArthur's plans for this period are found in Ronald K. Edgerton, "General Douglas MacArthur and the American Military Impact in the Philippines," *Philippine Studies* 25 (1977), 420–40.

16. MacArthur quoted ibid.

17. Diary entries of Oct. 6 and Nov. 15, 1944, and Jan. 27, 1945, Ickes Papers.

18. Edgerton, "MacArthur and the American Military Impact in the Philippines."

19. Ibid.; diary entry of Jan. 27, 1945, Ickes Papers.

20. James, *Years of MacArthur*, 2: 554–55.

21. See Eichelberger's commments in Luvaas, ed., *Dear Miss Em*, 184–97.

22. James, *Years of MacArthur*, 2: 613–50; Eichelberger's comments in Luvaas, ed., *Dear Miss Em*, 230–31.

23. James, *Years of MacArthur*, 2: 670–90; Friend, *Between Two Empires*, 267; Falk, "MacArthur and Japan."

24. Diary entries of Feb. 4, and 10, 1945, Ickes Papers; diary entries of Sept. 26 and Oct. 26, 1944, Feb. 8, 1945, Stimson Papers; Friend, *Between Two Empires*, 246–47.

25. Steintorf memorandum of Mar. 20, 1945, in Joseph C. Grew to Truman, Apr. 30, 1945, PSF, Harry S. Truman Papers, Harry S. Truman Library.

26. Steinberg, *Philippine Collaboration*, 108–9.

27. Ibid., 109–11.

28. James, *Years of MacArthur*, 2: 692–93; Steinberg, *Philippine Collaboration*, 115; MacArthur, *Reminiscences*, 205, 236–37. Most students of the subject question Roxas's patriotic credentials.

29. Willoughby reports of October 1944, in Edgerton, "MacArthur and American Military Impact in the Philippines."

30. Ibid.; Kerkvliet, *The Huk Rebellion.*

31. Edgerton, "MacArthur and American Military Impact in the Philippines."

32. Diary entries of Apr. 29, June 9, and July 21, 1945, Ickes Papers; Robert Ferrell, ed., *Off the Record: The Private Papers of Harry S. Truman* (New York, 1980), 47.

33. Mar. 8, 1946, excerpts from Drew Pearson Papers, Johnson Library.

34. Steinberg, *Philippine Collaboration*, 116–27, 141–46; Friend, *Between Two Empires*, 251–55, 262–63; *The New York Times*, Mar. 17 and May 16, 1946; Kerkvliet, *The Huk Rebellion.*

35. Friend, *Between Two Empires*, 262–63; Hess, *The United States' Emergence as a Southeast Asian Power.*

Chapter 8

1. James, *The Years of MacArthur*, 2: 724.

2. Eichelberger letters of Mar. 4, Apr. 8, June 6, and July 13, 1945, in Luvaas, ed., *Dear Miss Em*, 229, 247, 279, 292–93; diary entry of Apr. 10, 1945, Eichelberger Papers. In spite of MacArthur's near paranoia, Marshall frequently defended him against navy criticism. At one particularly acrimonious meeting of the Joint Chiefs in 1944, Marshall snapped at Adm. King that he would not tolerate the navy's "policy of hatred" toward MacArthur and with that, Henry Stimson reported, he "shut King up." See diary entry of Nov. 22, 1944, Stimson Papers.

3. Diary entry of Mar. 30, 1945, Stimson Papers.

4. Diary entries of Apr. 10 and 28, 1945, Eichelberger Papers; Eichelberger letter of Apr. 28, 1945, in Luvaas, ed., *Dear Miss Em*, 260; James, *Years of MacArthur*, 2: 729–30.

5. See Ronald Schaffer, *Wings of Judgement: American Bombing in World War II* (New York, 1985), and Michael Sherry, *The Rise of American Air Power* (New Haven, 1987). Both studies examine the ethical as well as technical issues surrounding the strategic air war; Spector, *Eagle Against the Sun*, 487–97.

6. Spector, *Eagle Against the Sun*, 487–505.

7. James, *Years of MacArthur*, 2: 737–50; 832–33.

8. Sherwood, *Roosevelt and Hopkins*, 2: 526.

9. The British embassy in Washington kept a file on American opinion of the Japanese. See FO 371/41793, Public Record Office, London; for details of the Roosevelt–Hrdlicka correspondence, see Thorne, *Allies of a Kind*, 158–59, 167–68; John Morton Blum, ed., *The Price of Vision: The Diary of Henry A. Wallace, 1942–46* (Boston, 1973), 448; Roger Buckley, *Occupation Diplomacy: Britain, the United States and Japan, 1945–1949* (Cambridge, Eng., 1982), 15; Arnold's proposal is discussed in a memorandum by Joseph C. Grew, July 31, 1945, V 7, 46, Joseph C. Grew Papers, Harvard University; Bilbo to MacArthur, Sept. 14, 1945, RG 10, VIP File, MacArthur Papers. For the most recent and complete analysis of the racial hatred inherent in the Pacific War, see John W. Dower, *A War Without Mercy* (New York, 1986).

10. Akira Iriye, *Power and Culture: The Japanese–American War, 1941–45* (Cambridge, Mass., 1981). This powerful study of a group Iriye calls "Japanese

Wilsonians" shows how the peace faction grappled with the problem of planning for a postwar settlement.

11. John W. Dower, *Empire and Aftermath: Yoshida Shigeru and the Japanese Experience, 1878–1954* (Cambridge, Mass., 1981), 258–59, 264. Dower not only examines the career of Yoshida, but explains the continuities between pre- and postwar Japan. The initial failure of the peace advocates seemed dismal for Yoshida, as the secret police detained him. However, this brief arrest proved his salvation when the Americans later interpreted it as proof of his antimilitarism. It opened the way for a new political career.

12. Hugh Borton, *American Presurrender Planning for Postwar Japan* (New York, 1967), 12; Hilldring and Pence to Dunn, Feb. 18, 1944, Department of State, *Foreign Relations of the United States* (Washington, D.C., 1965), 5: 1190–94 (hereafter abbreviated as *FRUS*, with volume and page).

13. PWC 111, Mar. 13, 1944, *FRUS 1944*, 5: 1212–15; PWC 116d, May 9, 1944, ibid., 1250–55; Borton, *Presurrender Planning*, 17; see also Theodore Cohen, *Remaking Japan: The American Occupation as New Deal* (New York, 1987), 1–48. Marlene Mayo, "American Wartime Planning for Occupied Japan: The Role of the Experts," in Robert Wolf, ed., *Americans as Proconsuls: The United States Military Government in Germany and Japan* (Carbondale, Ill., 1984), 3–51.

14. Borton, *Presurrender Planning*, 20.

15. The deterioration of Soviet–American relations during the spring of 1945 is covered fully in Dallek, *Roosevelt and American Foreign Policy*, 442–528.

16. Diary entry of Apr. 23, 1945, William Leahy Papers, Library of Congress.

17. Memorandum by Forrestal, May 12, 1945, in Walter Millis, ed., *The Forrestal Diaries* (New York, 1951), 56; Herbert Feis, *Contest Over Japan* (New York, 1967), 27–28; Grew to Stimson, May 12, 1945, ABC 336, "Russia in the Far East," Department of the Army Records, RG 336; James Matray, *The Reluctant Crusade: American Foreign Policy in Korea, 1941–1950* (Honolulu, 1985), 34; on the development of the atomic bomb, see Martin Sherwin, *A World Destroyed: The Atomic Bomb and the Grand Alliance* (New York, 1977), and Gregg Herken, *The Winning Weapon: The Atomic Bomb and the Cold War, 1945–50* (New York, 1982).

18. Memorandum by Grew, May 19, 1945, quoted in Waldo Heinrichs, *American Ambassador: Joseph C. Grew and the Development of the United States Diplomatic Tradition* (Boston, 1966), 365–73, 374. Some influential Japanese knew of Grew's position and hoped to cooperate with him in negotiating a settlement. Intercepts by the OSS of communications between Tokyo and its representatives in Europe revealed that some Japanese elements wanted to arrange talks with certain British and American officials before the Soviets struck. The ambassador to Switzerland, Kase Shunichi, referred specifically to Grew as the "best U.S. authority on Japan" and believed he shared the moderates' view that the sole condition for surrender should be the "retention of the emperor as the only safeguard against Japan's conversion to communism." See Donovan to Truman, May 12, 1945, PSF, Harry S. Truman Papers, Truman Library; Japanese peace feelers are discussed at length in *FRUS 1945*, 6: 478, passim; memorandum by Grew of conversation with Truman, May 28, 1945, *FRUS 1945*, 6: 545–47; Heinrichs, *American Ambassador*, 374–78.

19. Memorandum by Stimson of conversation with Truman, June 6, 1945, folder 18, box 172, Stimson Papers; for more detailed information on the role played by the atomic bomb in decisions on China and Japan, see Schaller, *The U.S. Crusade*

in China; Sherwin, *A World Destroyed;* Herken, *Winning Weapon.* At about the same time as Grew and Stimson lobbied for their position, Herbert Hoover entered the debate on the same side. Partly back in favor with Truman after FDR's death, the former president warned that unless the United States arranged an immediate peace with Japan, the Soviets would take control of mainland Asia. He urged permitting Japan to maintain some of its foreign empire and influence in China as the means of ensuring its "return to Western civilization" and assistance as an anti-Communist bulwark. See memoranda by Hoover, June 4 and 9, 1945, "Stimson Top Secret Safe File: Japan after Dec. 7, 1941," Secretary of War Files, Department of the Army Records, RG 107.

20. Stimson to Truman, "Proposed Program for Japan," June 26, 1945, second draft, Japan box, Stimson Safe File, July 1940–September 1945, Department of the Army Records, RG 107.

21. Stimson to Grew, May 21, 1945, ABC 336, "Russia in the Far East," Department of the Army Records, RG 107; Marshall and Handy to Stimson, July 7, 1945, "Stimson Top Secret Safe File," Japan after Dec. 7, 1941, ibid.; *FRUS,* "Conference in Berlin," 1: 905.

22. James, *Years of MacArthur,* 2: 771–72. Plans for deploying Allied forces in East and Southeast Asia took on added importance when nationalist and revolutionary guerrillas demanded to accept the Japanese surrender themselves, gaining important control over weapons, territory, and political authority.

23. Henry Stimson and McGeorge Bundy, *On Active Service in Peace and War* (New York, 1947), 637; Daniel Yergin, *Shattered Peace: The Origins of the Cold War and the National Security State* (Boston, 1977), 113; W. Averell Harriman and Ellie Abel, *Special Envoy to Churchill and Stalin, 1941–1946* (New York, 1975), 488; Stimson diary, July 14, 1945, Stimson Papers; on the frictions within the Truman administration, see Robert Messer, *The End of an Alliance: James F. Byrnes, Roosevelt, Truman and the Origins of the Cold War* (Chapel Hill, N.C., 1982), passim.

24. Walter Brown diary, entry of July 16–24, 1945, James F. Byrnes Papers, Clemson University. Brown, a journalist and close friend of Byrnes, kept a detailed record of the events at Potsdam.

25. Diary entry of July 23, 1945, Stimson Papers; Walter Brown diary, entries of July 18, 20, and 24, 1945, Byrnes Papers. Previously, American officials had encouraged the Chinese Nationalist government to accept Soviet demands for enhanced influence in Manchuria. Hoping that a delay in reaching agreement might block any Russian move into Manchuria or against Japan, Washington now urged Chinese negotiators to stall a treaty.

26. Notes, July 24, 1945, folder 18, box 172, Stimson Papers; undated notes on Potsdam Conference, folder 19, ibid.; see also Sherwin, *A World Destroyed,* 234–37, and Herken, *Winning Weapon,* 20; Stimson and Bundy, *On Active Service,* 612–56.

27. CCS Report, July 24, 1945, *FRUS Berlin,* 2: 1462–69; Tripartite Military Meeting, July 24, 1945, ibid., 344–53; Marshall to MacArthur, July 25, 1945, Department of the Army Records, OPD [Plans and Operations Division] 014.1, TS, Sec. III, RG 319; H. A. Craig to John E. Hull, July 25, 1945, ibid.; Matray, *Reluctant Crusade,* 41–42.

28. Brown notes, July 24, 1945, box 54, Byrnes Papers; Harriman and Abel, *Special Envoy,* 486; diary entry of July 30, 1945, Stimson Papers; Walter Brown diary, entry of August 3, 1945, Byrnes Papers.

29. For reasons that remain unclear, Washington thought that Chinese intransigence would keep the Soviets out of Manchuria. On August 9, Stalin informed the Chinese foreign minister, T. V. Soong, that if he stalled further, the Red Army would not only enter China without a treaty, but would see that the Chinese Communists also "get into Manchuria." This panicked the Kuomintang and the Truman administration. From Moscow, Ambassador Harriman urged Truman to land American forces in Korea, north China, and Manchuria in order to block Soviet forces. The crisis abated by August 14 when the Chinese and Soviets announced agreement on a Friendship Treaty and Moscow accepted an American demand for an occupation zone in southern Korea. See Harriman to Byrnes, Aug. 7 and 8, 1945, *FRUS 1945*, 7: 957–58; memorandum by Harriman of conversation with Stalin, Aug. 8, 1945, ibid., 960–65; Byrnes to Harriman, Aug. 9, 1945, ibid., 965–66; Harriman to Truman and Byrnes, Aug. 10, 1945, ibid., 967; Harriman and Abel, *Special Envoy*, 495–97; Matray, *Reluctant Crusade*, 42–45; Bruce Cumings, *The Origins of the Korean War: Liberation and the Emergence of Separate Regimes, 1945–1947* (Princeton, 1981), 114–20; Schaller, *The U.S. Crusade in China*, 259–61; Blum, *The Price of Vision*, 470–75. During the days before and after Tokyo's offer to surrender, Truman received SWNCC documents outlining plans to utilize the Japanese government, to maintain an American monopoly of occupation forces, and to pursue policies assuring the "regeneration of Japan in the minimum amount of time." See Stimson–Grew discussions of proposed "Post Hostilities Policy Toward Japan," Aug. 4, 6, 1945, *FRUS 1945*, 6: 584–89; "National Composition of Forces," SWNCC 70/5, Aug. 11, 1945, ibid., 603; "U.S. Initial Post Defeat Policy Relating to Japan," SWNCC 150/2, ibid., 609; Paper on Composition of Forces, signed by Truman, Aug. 19, 1945, PSF, Truman Papers.

30. James, *Years of MacArthur*, 2: 773–76; diary entry of Aug. 7, 1945, Eichelberger Papers; White, *In Search of History*, 224.

31. Harriman and Abel, *Special Envoy*, 499–501; *FRUS 1945*, 6: 628–31. Assistant Secretary of War John J. McCloy believed Harriman had not only saved Japan, but was one of the few Americans thinking about the "future of Japan." Everyone else, he complained, thought only about "the end of the war."

32. Hirohito's letter, dated Sept. 9, 1945, is quoted in the *New York Times*, Apr. 28, 1986; Aiichiro quoted in John D. Montgomery, *Forced to be Free: The Artificial Revolution in Germany and Japan* (Chicago, 1957), 106–7.

33. Diary entry of Aug. 11, 1945, Eichelberger Papers; James, *Years of MacArthur*, 2: 780; MacArthur's claims regarding FDR and Truman are found in a "Memorandum by Hoover of discussions with MacArthur," May 4, 5, 6, 1946, Post-Presidential Subject File—Famine Emergency Committee—World Mission—HH Diary 1946 Journey, Hoover Papers, Hoover Library.

34. Eichelberger letter of Aug. 19, Luvaas, ed., *Dear Miss Em*, 300; James, *Years of MacArthur* 2: 779.

35. Faubion Bowers, Oral History, Columbia University. Another oral history by Bowers, recorded by Cineworld, is deposited at the MacArthur Memorial.

36. Ferrell, ed., *Off the Record*, 61; diary entries of Aug. 18 and 26, 1945, Ickes Papers.

Chapter 9

1. Otis Cary, ed., *War Wasted Asia: Letters, 1945–46* (Tokyo, 1975), 48. These letters were written by a group of army and navy interpreters, trained in Japanese and Chinese language. Many became leading postwar scholars of East Asia. Cohen, *Remaking Japan*, 54. This posthumous memoir is filled with revealing anecdotes and insights based on the author's service as a labor adviser during the occupation.

2. Cohen, *Remaking Japan*, 54, 63. Cohen noted that Japanese newspapers courted MacArthur's good will by declining to publish documents or news releases concerning Truman administration initiatives regarding Japan.

3. On SCAP personalities, see Cohen, *Remaking Japan*, 84–96.

4. Faubion Bowers, Oral History; Charles Kades, Oral History, Columbia University; for a description of the peculiar recruiting techniques utilized by MacArthur, see Harry Emerson Wildes, *Typhoon in Tokyo: The Occupation and its Aftermath* (New York, 1954), 8–9; on the chronic inability of SCAP to oversee the Japanese, see John R. Stewart, *Notes on the Economic Aspects of the Allied Occupation of Japan* (New York, 1947). Stewart served with SCAP, as had Wildes; George Atcheson to Dean Acheson, Nov. 7, 1945, *FRUS 1945*, 6: 838–41.

5. MacArthur, *Reminiscences*, 293; William MacMahon Ball, *Japan: Enemy or Ally?* (New York, 1949), 27–35; Faubion Bowers, Oral History; for a more detailed description of the negotiations leading to the creation of the control councils in Japan, see Michael Schaller, *The American Occupation of Japan: The Origins of the Cold War in Asia* (New York, 1985), 57–62. The British and Chinese, as much as the Soviets, resented exclusion by the United States. Not surprisingly, the Soviets frequently compared Washington's stance in Japan to their own domination of Eastern Europe.

6. MacArthur, *Reminiscences*, 322–23; Courtney Whitney, *MacArthur: His Rendezvous with History* (New York, 1955), 213; Charles A. Willoughby and John Chamberlain, *MacArthur, 1941–51* (New York, 1954); Harry S. Truman, *Memoirs* (Garden City, N.Y., 1955–56); Dean Acheson, *Present at the Creation: My Years in the State Department* (New York, 1969); in *American Caesar*, William Manchester describes MacArthur's years in Japan as a virtual golden age. D. Clayton James's authoritative *The Years of MacArthur* (Vol. 3) presents a balanced and generally positive view of the occupation. For a discussion of the evolution of the literature on MacArthur and occupied Japan, see Carol Gluck, "Entangling Illusions—Japanese and American Views of the Occupation," in Warren Cohen, ed., *New Frontiers in American–East Asian Relations* (New York, 1983), 169–236.

7. JCS to MacArthur, JCS 1380/5, in SCAP, *The Political Reorientation of Japan, September 1945–September 1948. Report of the Governmental Section of SCAP* (Washington, 1948) 2: 428 (hereafter cited as *PRJ*).

8. Americans debated for some time the economic and political role of the *zaibatsu*. During the war journalists and scholars published numerous articles and books about the historic role of combines and their links to aggression. These included William Johnston's *The Future of Japan*, Owen Lattimore's *Solution in Asia*, Andrew Roth's *Dilemma in Japan*, and Thomas A. Bisson's *Japan's War Economy*. Such journals as *Pacific Affairs*, the *Nation*, and *Amerasia* also carried many articles proclaiming that a real transformation of Japan depended on replacing the *zaibatsu* with a "democratic economy." The most thorough study of the

internal struggle between conservative Japan specialists in the American government and New Deal economic reformers can be found in two essays by Marlene Mayo. See "American Economic Planning for Occupied Japan: The Issue of *Zaibatsu* Dissolution, 1942–1945," in *Proceedings of a Symposium by the MacArthur Memorial* (Norfolk, Va., 1978), and her "American Wartime Planning for Occupied Japan: the Role of the Experts," in Wolf, ed., *Americans as Proconsuls*, 3–51; for a personal account of the role played by New Deal bureaucrats in shaping the occupation program see Cohen, *Remaking Japan*, 3–48.

9. MacArthur to Robert E. Wood, January 12, 1948, MacArthur correspondence, Wood Papers.

10. Cohen, *Remaking Japan*, 64, 69.

11. Perry Miller quoted in John C. Perry, *Beneath the Eagle's Wing* (Boston, 1981), 167; MacArthur's testimony in U.S. Senate, Committee on Armed Services and Committee on Foreign Relations, 82d Cong., 1st sess., *Hearings to Conduct an Inquiry into the Military Situation in the Far East and the Facts Surrounding the Relief of General of the Army Douglas MacArthur from His Assignments in that Area*, 1951, pt. 1: 312–313 (hereinafter abbreviated *MSFE*); Arthur MacArthur suggested that the Philippines be given "tuitionary," rather than colonial, status; not surprisingly, hate and paternalism were the flip sides of anti-Japanese racism. During the war, American official and popular treatment of the enemy portrayed Japanese as merciless beasts and simple, childlike idiots. For a masterful discussion of this dichotomy, see Dower, *War Without Mercy*.

12. Edwin A. Locke to Truman, Oct. 19, 1945, PSF, Truman Papers; report by D. R. Jenkins to Treasury Department, February 21, 1946, box 23, Office of the Asssistant Secretary for International Affairs File, Records of the Department of the Treasury; W. W. Butterworth to Stuart and Byrnes, Aug. 20, 1946, *FRUS 1946*, 7: 303–4.

13. Diary entry of July 10, 1946, Forrestal Papers.

14. MacArthur statement of Sept. 2, 1946, *PRJ*, 2: 756–57; MacArthur message to Congress, Feb. 20, 1947, ibid., 763 64.

15. William P. Woodard, *The Allied Occupation of Japan, 1945–52, and Japanese Religions* (Leiden, Netherlands, 1972), 14–18, 243–46; Lawrence S. Wittner, "MacArthur and the Missionaries: God and Man in Occupied Japan," *Pacific Historical Review* 40 (Feb. 1971), 77–97; John Gunther, *The Riddle of MacArthur* (New York, 1950), 75; James, *Years of MacArthur*, 3: 287–93; Jones quoted in Perry, *Beneath the Eagle's Wing*, 76.

16. MacArthur statement on selection of Katayama Tetsu as prime minister, May 24, 1947, *PRJ*, 2: 770.

17. Memorandum of conversation between Kennan and MacArthur, Mar. 1, 1948, *FRUS 1948*, 6: 697–99.

18. Faubion Bowers, Oral History, Columbia University and Cineworld Transcripts.

19. William J. Coughlin, *Conquered Press: The MacArthur Era in Japanese Journalism* (Palo Alto, Calif., 1952), 111–16; James, *Years of MacArthur*, 3: 301–7; among American periodicals, *Newsweek*, *The New Republic*, *The Nation*, and the *Chicago Sun-Times* had frequent disputes with SCAP over censorship and the denial of entry visas for their reporters.

20. *Time*, July 10, 1950.

21. The SCAP reform program is discussed more thoroughly in Dower, *Empire and Aftermath*; Schaller, *The American Occupation of Japan*; and Cohen, *Remak-*

ing Japan. Kades's reflections appear in his Oral History, Cineworld Transcript; the comment on the emperor's divinity comes from January 10, 1946, excerpts from the Pearson Papers; Chiang's lament is recounted by Gen. Edward M. Almond, Oral History, U.S. Military History Institute. In a candid letter to Robert E. Wood, MacArthur defended his reform program as a middle ground between socialism and reactionary capitalism. His explanation mirrored Roosevelt's defense of New Deal reforms. See MacArthur to Robert E. Wood, January 12, 1948, MacArthur File, Wood Papers.

22. I. F. Stone, "Behind the MacArthur Row," *Nation* 161 (Sept. 29, 1945), 297–99; John Maki, "Japan: Political Reconstruction," *Far Eastern Survey* 1 (April 9, 1947), 73–77.

23. On medical experiments, see Peter Williams and David Wallace, *Unit 731: The Japanese Army's Secret of Secrets* (London, 1988); on the trial of war criminals in general, see Richard H. Minear, *Victor's Justice: The Tokyo War Crimes Trial* (Princeton, N.J., 1971), and Arnold Brackman, *The Other Nuremberg* (New York, 1987).

24. Miriam Farley, *Aspects of Japan's Labor Problems* (New York, 1950), 44–50. Mark Gayn, *Japan Diary* (New York, 1948), 231–32, 263–65; Joe Moore, *Japanese Workers and the Struggle for Power* (Madison, Wisc., 1983), 3–31; MacArthur Statement of May 20, 1946, *PRJ*, 2: 750; MacArthur Statement of September 2, 1946, ibid., 756–57; MacArthur Statement Calling off General Strike, Jan. 31, 1947, ibid., 762. MacArthur worked to break the hold of the left on Japanese labor by fostering the creation of anti-Communist labor federations. He enlisted the help of the AFL-CIO to fracture the leftist unions and weaken the union-backed socialist movement while remaining on good terms with American labor leaders. On this subject see Howard Schonberger, "American Labor's Cold War in Occupied Japan," *Diplomatic History* 3 (Summer 1979), 249–72; see also Schaller, *The American Occupation of Japan*, 135.

25. Mark Gayn, *Japan Diary*, 344–45.

Chapter 10

1. Statement by SCAP, Sept. 17, 1945, *FRUS 1945*, 6: 715; diary entires of Sept. 14 and 16, 1945, Eichelberger Papers.

2. Robert E. Wood to MacArthur, Sept. 4, 1945, MacArthur file, Wood Papers; Philip LaFollette to MacArthur, Sept. 3, 1945, RG 10, MacArthur Papers. LaFollette's letters to the general during 1945–46 discussed reactions among various political groups in America to SCAP policies. He assured MacArthur that people at home saw the "contrast between what is happening in Europe [under Eisenhower] and what is happening in Japan." Policies in Europe opened the "whole continent . . . to communism." In Japan, the actions of the Supreme Commander were "exactly opposite." See LaFollette to MacArthur, Oct. 15, Nov. 2, 1945, ibid.; U.S. Congress, Senate, *Congressional Record* 91 (Sept. 18, 1945), 816.

3. Truman, *Memoirs*, 1, 520–21; MacArthur to Marshall, Sept. 19, 1945, box 7, Radios, Richard K. Sutherland Papers, RG 30, MacArthur Memorial.

4. Diary entries of Sept. 18, 19, and 21, Oct. 1 and 20, 1945, Eichelberger Papers; Gen. Bonner Fellers to Robert E. Wood, Oct. 1, 1945, MacArthur File, Wood Papers. MacArthur complained that Eisenhower was permitted to testify secretly as to his views of postwar troop requirements, but the administration wanted him (MacArthur) to testify publicly. One of MacArthur's happiest moments after

Japan's surrender came when he learned that Truman sent the European commander a letter critical of the army's treatment of displaced persons in Europe. MacArthur gloated that it "was the worst letter he had ever seen a superior send an officer" and he "now believed Ike would not be made chief of staff." See diary entry of Oct. 1, Eichelberger Papers.

5. The reactions to the troop level and related statements are covered in *FRUS 1945*, 6: 716–21; Truman, *Memoirs*, 1: 520–21; diary entry of Sept. 18, 1945, Harold Smith Papers, Truman Library; diary entry of Sept. 18, 1945, Eben Ayers Papers, Truman Library; MacArthur to Marshall, Sept. 18, 1945, box 7, Sutherland Papers, RG 30.

6. Diary entries of Oct. 7 and Nov. 14, 1945, Ickes Papers. McCloy disliked the Japanese as people, complaining that they all "look alike in more ways than one." At the same time, the envoy feared liberal reforms, whether pushed by SCAP or Washington, would drive Japan "further left," and that the Tokyo war crimes trial was a "total fiasco." Only the "incredible discipline of the Japanese people," he wrote, prevented chaos. McCloy soon advocated the reconstruction of both German and Japanese industry and the abandonment of most reforms. See McCloy Memorandum to Gen. George Lincoln, "Visit to Japan," Feb. 17, 1946, box 222, Records of the Far Eastern Commission, RG 43.

7. MacArthur and Kramer quoted in Edwin A. Locke to Truman, Oct. 19, 1945, PSF, Truman Papers; SCAPIN 244, *PRJ*, 2. 565; *Report of the Mission on Japanese Combines* (Washington, D.C., 1946). MacArthur seemed to agree with then Foreign Minister Yoshida Shigeru who, on October 19, told American journalists that the old, established *zaibatsu* bore little responsibility for aggression. Only the "new *zaibatsu*," linked to the military, had benefited from the war. The tart-tongued Yoshida suggested that Washington prosecute the du Pont company for war profiteering. Yoshida's statement is quoted in Thomas A. Bisson, *Zaibatsu Dissolution in Japan* (Berkeley, Calif., 1954), 70.

8. *Report of the Mission on Japanese Combines* (Washington, D.C., 1946), Publication 2628, Far Eastern Series. The Pauley mission initially devised plans for German reparations and after August shifted its focus to Japan. An independent oil entrepreneur (and Democratic fund-raiser), Pauley had an aversion to both domestic and foreign monopolies.

9. Some of MacArthur's more conservative subordinates, like Gen. Willoughby, objected to "radicals" such as Owen Lattimore, who served on Pauley's staff. Lattimore and his colleagues proposed speeding Asian development through the transfer of industrial facilities from Japan. This would compensate Japan's victims while creating a more equal, competitive environment for the future. Other Asian states would, for the first time, coexist as equals with Japan. Pauley's staff argued that accelerated development in Asia would also enhance world economic recovery. See Edwin Pauley, "U.S. Reparations Policy—Preliminary Statement," Oct. 31, 1945, box 21, records of the U.S. Mission on Reparations, RG 59; Owen Lattimore to H. D. Maxwell, Nov. 6, 1945, box 24, ibid.; Lattimore to Maxwell, Nov. 10, 1945, box 22, ibid.; draft report by Lattimore, Nov. 12, 1945, box 21, ibid.; notes of staff discussion by Lattimore, for Pauley, Nov. 15, 1945, box 21, ibid.; Lattimore diary, Dec. 4, 1945, ibid.; Pauley to MacArthur, Dec. 5, 6, 1945, ibid.; Pauley report for President Truman, "Reparations from Japan—Immediate Program," December 8, 1945, ibid.; Pauley Report, May 10, 1946, ibid.

10. Corwin Edwards, "The Dissolution of Japanese Combines," *Pacific Affairs* 19 (Sept. 1946), 227–40.

11. The Edwards report remained in limbo during 1946. In May 1947, SWNCC forwarded it to the moribund Far Eastern Commission for further consideration. There is was renamed FEC 230. See Schaller, *The American Occupation of Japan*, 38–41.

12. When Eisenhower visited Tokyo in May 1946, MacArthur greeted him cordially, but denounced him in private. Ike returned the compliment, telling Eichelberger that MacArthur had always denigrated his talented subordinates by taking credit for their work and recommending promotions solely on the basis of loyalty. Ike observed that he himself gained his best publicity by "the way he glorified his generals." Eisenhower told MacArthur he had no presidential ambitions and urged the occupation commander to run, though he offered no endorsement. For his part, MacArthur claimed "it was evident Ike was running for President." See diary entries of May 10, 12, and 13, 1946, June 26, 1946, Eichelberger Papers; Stephen Ambrose, *Eisenhower: Soldier, General of the Army, President Elect, 1890–1952* (New York, 1983), 440–41; For MacArthur's views on Marshall's alleged political goals, see diary entries of Mar. 12 and 13, 1946, Eichelberger papers; memorandum of Interview with MacArthur, by E. A. Bayne, for T. V. Soong, Dec. 21, 1946, folder 7, box 1, W. Walton Butterworth Papers, George C. Marshall Research Library; diary entries of July 10 and 14, 1946, in Millis, ed., *The Forrestal Diaries*, 177–79; memorandum of conversations with MacArthur, May 4, 5 and 6, 1946, by Herbert Hoover, "Famine Emergency Committee–World Mission, Gen., H. H. diary 1946 Journey," Hoover post-presidential subject file, Hoover Papers. MacArthur, in discussions with Hoover, accused FDR of precipitating the Pearl Harbor attack despite Japanese efforts to prevent war, of spurning Japanese surrender efforts after MacArthur had retaken the Philippines, and of plotting to keep him out of postwar Japan. Truman, he suggested, succumbed to similar folly in using the atomic bomb. Hoover urged MacArthur to make a speaking tour of the U.S. in 1947 as a way of kindling interest in his presidential candidacy. The former president followed his suggestion with letters to MacArthur urging him to perform a "transcendent service" to America by undertaking a speaking tour. The world was faced "with Asiatic versus Western Civilization," Hoover wrote in October, and the general could help the country "shake off the various domestic red tinges." See Hoover to MacArthur, Oct. 17, 1946, "Mil. Sec. Corres.," RG 5, MacArthur Papers.

13. Diary entry of Oct. 6, 1947, Forrestal Papers.

14. *Newsweek* 29 (Jan. 27, 1947), 40. On Harry Kern and the American Council for Japan, see Howard Schonberger, "The Japan Lobby in American Politics, 1947–52," *Pacific Historical Review* 46 (Aug. 1977), 327–59.

15. Diary entries of Mar. 3 and 13, Apr. 16 and 18, 1947, Forrestal Papers; minutes of the meetings of the secretaries of State, War, and Navy, Apr. 16, 1947, ibid.; Acheson to Patterson, Apr. 14, 1947, filed with 740.00119 Control (Japan) 9-1347, *Department of State* (DOS), RG 59; Patterson to Acheson, Apr. 21, 1947, ibid.; note to E. A. Locke, attached to ibid.

16. The idea of cranking up Japanese industry through a program resembling the Marshall Plan developed simultaneously among areas specialists in the State, Defense, and Army departments and in the office of the Joint Chiefs of Staff. All posited a four- or five-year program costing about $500 million. Planners expected Japan to trade mostly with Asian nations, exchanging finished products for raw materials. In effect, Japan would become a modest, but critical, hub of the East and Southeast Asian economy. See SWNCC Series #360, 381, 384, RG 353; the

most thoughtful discussion of the evolution of Kennan's strategic thinking is found in John L. Gaddis, *Strategies of Containment* (New York, 1982).

17. Record of conversation with MacArthur, Gascoigne to Foreign Office, Mar. 4, 1947, FO 371/63766, PRO. MacArthur may have expected Gascoigne to pass his warnings on to London and Washington, thereby pressuring the administration.

18. MacArthur interview with press correspondents, Mar. 17, 1947, *PRJ*, 2: 765–67. Needless to say, the general neglected to clear his policy declaration with Washington. MacArthur knew unofficially that a group within the State Department, coordinated by Hugh Borton, had already drafted a peace treaty. It provided for a prolonged period of supervision by an international agency including the Soviets. It also imposed many restrictions on industry and armaments. Although the general did not particularly like these provisions, he would accept almost anything that led to a quick settlement. The so-called Borton draft treaty found few supporters. The British, Soviets, and Chinese all considered it too lenient, while most American officials judged it far too harsh. When virtually all members of the Far Eastern Commission raised objections to a preliminary call for a peace conference, the administration abandoned the idea. With typical lack of coordination, the State Department had neglected to clear the calls for a preliminary peace conference with the military services, Congress, or the White House. By September 1947, virtually all agencies in the government were pleased to see the trial balloon fall flat. See Hugh Borton, Oral History, Columbia University; Schaller, *The American Occupation of Japan*, 98–106.

19. Diary entry of July 1, 1948, Eichelberger Papers.

20. Dean Acheson, "The Requirements of Reconstruction," speech of May 8, 1947, Department of State *Bulletin* 16 (May 18, 1947), 991–94.

21. For the best analysis of the evolution of FEC-230, see Eleanor Hadley, *Anti-Trust in Japan* (Princeton, N.J., 1974), 495–515; see also Schaller, *The American Occupation of Japan*, 109.

22. Minutes of meeting of Aug. 7, 1947, Records of the Policy Planning Staff, RG 59; John P. Davies to Kennan, Aug. 11, 1947, *FRUS 1947*, 6: 485–86; Kennan to Robert Lovett and Lovett to Kennan, Aug. 12, 1947, ibid., 486–87; Col. S. F. Giffen to PPS, Aug. 14, 1947, PPS Records, Freedom of Information Act (FOIA) Request.

23. Minutes of discussion of draft treaty, Meeting 48, Aug. 25, 1947, box 32, PPS Records, RG 59; Meeting 54, Sept. 4, 1947, ibid.; Meeting 65, Sept. 22, 1947, ibid. While the PPS concentrated on "strategic issues," economic specialists in the State and Army departments hammered out an economic rescue plan that closely resembled the European Recovery Program. SWNCC 381 (the State Department plan) and SWNCC 384 (the army plan) both proposed a multiyear aid package aimed at providing raw material and capital for the renewal of heavy industry. Under the terms of these proposals, the United States would encourage Japanese trade with Southeast Asia to assure Tokyo a future market and inexpensive source of vital raw materials. These ideas formed the basis of the Economic Recovery in Occupied Areas program (EROA), passed by Congress in June, 1948. See SWNCC 381, 384, and 360 files, SWNCC Records, RG 353.

24. Kauffman's report was printed in part in *Newsweek*, 30, (Dec. 1, 1947). The lawyer consulted with Kennan, Forrestal and other SCAP critics during 1947–48.

25. Kennan to Lovett and Marshall, Oct. 14, 1947, *FRUS 1947*, 6: 536–43;

memorandum by Carlisle Humelsine to Willard Thorp, et al., Oct. 29, 1947, PPS Records, FOIA; Humelsine to Gen. Marshall Carter, Oct. 16, 1947, ibid.; diary entries of Oct. 31 and Nov. 7, 1947, Forrestal papers; Kenneth C. Royall, Oral History, Columbia University.

26. Memorandum, Dec. 12, 1947, William H. Draper to Gordon Gray, Dec. 14, 1947, Under Secretary of the Army, General Correspondence–Security Classified, August 1947–January 1949, SAOUS 004, Japan, records of the Office of the Secretary of the Army, RG 335.

27. Diary entries of Sept. 11, Dec. 20, 1947, May 1, 1948, Eichelberger Papers. At MacArthur's urging, Eichelberger spoke with Col. Robert McCormick, Henry Luce, Dwight Eisenhower, Omar Bradley, and Philip LaFollette, among others, during his visit home. The general's supporters urged him to come home to campaign and to smear Ike with stories about his alleged wartime love affair. Eisenhower's friends warned that if MacArthur threw mud, they would publicize his tempestuous affair with Isabel.

28. MacArthur to Robert E. Wood, Nov. 16, 1947, MacArthur file, Wood Papers; MacArthur to Hanford MacNider, Oct. 14, 1947, MacArthur Presidential Candidacy File, Hanford MacNider Papers, Hoover Library; diary entry of Dec. 20, 1947, Eichelberger Papers.

29. See the extensive MacNider–Eastwood correspondence file, MacArthur Presidential Candidacy File, MacNider Papers, Hoover Presidential Library. For a detailed description of the MacArthur campaign effort, see Howard Schonberger, "The General and the Presidency: Douglas MacArthur and the Election of 1948," *Wisconsin Magazine of History* 57 (Spring 1974), 201–19; Mattern, "The Man on the Dark Horse."

30. Kenneth Royall, Oral History; see below, Chap. 13.

31. Department of the Army to MacArthur, Oct. 20 and 21, 1947; SCAP to Department of the Army, Oct. 25, 1947, all in memorandum of Dec. 9, 1947, by Robert Blum, CD 3-1-9, records of the Office of the Secretary of Defense, RG 330. Blum, an aide to Forrestal, compiled a file on the disputes between MacArthur and the administration. Royall to MacArthur, Dec. 6, 1947, ibid.; telephone transcript of Forrestal–John Biggers conversation, Dec. 5, 1947, ibid. Several members of Congress followed Knowland in threatening an investigation of SCAP policy. Someone, they charged, was "bringing socialism to the former enemy nation." Between December 1947 and January 1948, numerous conservative newspapers, including the *Chicago Tribune, Washington Times-Herald*, and the *San Francisco Chronicle* accused MacArthur of standing by while (according to the *Times-Herald*) a "bunch of New Deal Socialists or semi-socialists . . . framed up a socialist future for Japan with everybody sharing the poverty." Several newspapers demanded that journalists be given a freer hand to cover Japanese affairs and not be required to parrot "merely the official account of what goes on" as decreed by "General MacArthur's command." See *Congressional Record*, 11 (Dec. 19, 1947), 686–88; Knowland to Royall, Dec. 29, 1947, and Royall to Knowland, Dec. 31, 1947, and Jan. 10, 1948—all in Office of the Secretary of the Army, Unclassified General Correspondence, July 1947–Dec. 1950, "Japan" Foreign Country File, Records of the Office of the Secretary of the Army, RG 335. For the record of press reaction to MacArthur's plans compiled by the State Department, see "U.S. Opinion on Japan and Korea," Dec. 1947–Jan. 1948, Records of the Office of Public Opinion Studies, Department of State Records, RG 59.

32. Gascoigne to British Prime Minister Clement Attlee, Dec. 26, 1947, FO 371/

63830, PRO. As in his references to "Franklin Rosenfeld," MacArthur used the term "Jew" both as an epithet and to categorize a slew of enemies. Faubion Bowers, an aide during the occupation, listened to his boss's frequent derision of rivals as Jews. At one time he even called Truman "that Jew in the White House." To Bower's astonishment, the general insisted that Truman, like Ike, really was Jewish. MacArthur counted among his close friends some virulent anti-Semites, including his deputy chief of staff in the early 1930s, Gen. George Van Horn Moseley who praised Nazi policy toward Jews. On the other hand, MacArthur worked comfortably with several Jewish officers in SCAP, including Theodore Cohen and Charles Kades. Faubion Bowers, Oral History, Columbia University and Cineworld Transcript; MacArthur correspondence, Moseley Papers, Library of Congress; Cohen, *Remaking Japan.*

33. MacArthur to Robert E. Wood, Jan. 12, 1948, MacArthur file, Wood Papers.

34. Speech by Kenneth Royall, January 5, 1948, printed in Jon Livingston, et al., eds. *Postwar Japan: 1945 to the Present* (New York, 1973), 116–19; letter from Royall to Speaker of the House of Representatives, Committee on International Relations, *Selective Executive Session Hearings of the Committee, 1943–50, U.S. Policy in the Far East, pt. 1* (Washington, D.C., 1976), 6: 277–78; MacArthur to Draper for House and Senate Appropriations Committee, Jan. 18, 1948, *PRJ*, 2: 778–79; Overseas Consultants, Inc., *Report on Industrial Reparations Survey of Japan to the United States of America* (New York, 1948); MacArthur to Sen. Brien McMahon, Feb. 1, 1948, *PRJ*, 2: 783; MacArthur to J. H. Gipson, Feb. 1, 1948, ibid., 780–81.

35. Kenneth Royall Oral History; Eisenhower quoted in Sulzberger, *A Long Row of Candles*, 683–85.

36. Diary entry of Mar. 22, 1948, Eichelberger Papers; Gascoigne to Foreign Office, Apr. 6, 1948, FO 371/69886, PRO.

37. Report by Kennan, *FRUS 1948*, 6: 697–706; George F. Kennan, *Memoirs, 1925–50* (Boston, 1967), 381, 383–84; Kennan to MacArthur, March 5, 1948, box 19, PPS Records, FOIA.

38. Kennan to W. W. Butterworth, March 9, 14, and 16, 1948, box 19, PPS Records, FOIA.

39. PPS 28, Mar. 25, 1948, written primarily by George Kennan, PPS Records, RG 59; William H. Draper, Oral History, Truman Library; MacArthur, Draper, Kennan conversation transcript, Mar. 21, 1948, *FRUS 1948*, 6: 706–12; *Pacific Stars and Stripes*, Apr. 12, 1948; *New York Times*, Apr. 20, 1948. The formal findings of the Johnston committee were released on April 26. The group endorsed a plan to achieve economic self-sufficiency by promoting Japanese access to Asian raw materials and markets. It urged ending most restrictions on Japanese industry, terminating the reparations and deconcentration programs, placing strict curbs on labor unions, controlling inflation, fixing the yen's foreign exchange value, cutting the domestic budget, and lowering the domestic living standard to favor export production. See Percy Johnston, et al., "Report on the Economic Position and Prospects of Japan and Korea and the Measures Required to Improve Them," Joseph Dodge Papers, Detroit Public Library.

40. David M. Oshinsky, *A Conspiracy So Immense—The World of Joe McCarthy* (New York, 1983), 199; Cohen, *Remaking Japan*, 66; Schonberger, "The General and the Presidency"; Mattern, "The Man on the Dark Horse," 180–241; Gunther, *The Riddle of MacArthur*, 60–62. Also see three perceptive newspapers

reports on the sources of the MacArthur boom: Robert Riggs, "Strange Bedfellows Back MacArthur," *Louisville Courier Journal,* Apr. 4, 1948; Marquis Childs, "Washington Calling," *Washington Post,* Apr. 2, 1948; Joseph Driscoll, "Hearst and McCormick Papers Whooping it up for MacArthur," *St. Louis Post–Dispatch,* Mar. 18, 1948; *New York Times,* Apr. 5, 1948.

41. "Unfading Old Soldier," *Time* 39 (Feb. 11, 1952), 19.

42. The Economic Reconstruction in Occupied Areas Program appropriated about $125 million. It formed part of the much larger GARIOA program funded at $422 million. Congress permitted the army some discretion in shifting GARIOA funds to EROA. In addition, Draper convinced Congress to approve PL 820, creating a revolving fund for the purchase of American cotton by Japanese mills. For a discussion of these programs, see Schaller, *The American Occupation of Japan,* 130–33.

43. Memorandum for the record by Ralph W. E. Reid, Nov. 19, 1948, Under Secretary of the Army, General Correspondence–Security Classified, Aug 1947–Jan. 1949, SAOUS 091, Japan, Records of the Office of the Secretary of the Army, RG 335; Reid memorandum for the record, December 9, 1948, ibid.; "Informal memorandum of understanding between State and Army departments concerning implementation of NSC 13," December 7, 1948, Under Secretary of the Army, Draper/Voorhees Project Decimal File, 1947–50, 091 Japan, Records of the Department of the Army, RG 335; Draper–Lovett exchange, Dec. 13, 1948, *FRUS 1948,* 6: 1060; Statement on economic stabilization of Japan, Dec. 10, 1948, ibid., 1059–60. See also Schaller, *The American Occupation of Japan,* 132–38.

44. Diary entries of July 1 and 30, Aug. 2, Sept. 18, 19, and 20, 1948, multiple entries, Feb.–Mar. 1949, Eichelberger Papers. Eichelberger was quickly hired as a civilian employee in the Office of the Under Secretary of the Army, specializing on Japan. He served as a link between the Defense Department and a private anti-MacArthur lobbying group, the American Council for Japan. The most complete study of this organization is found in Howard Schonberger, "The Japan Lobby in American Politics," 327–59.

45. Guy Swope to Justin Williams, letters of Mar. 17, May 21, and Sept. 13, 1948, file 106, Justin Williams Papers, University of Maryland.

46. Pink to Foreign Office, August 26, 1948, FO 371/69823, PRO.

47. Willoughby began his secret investigation, entitled "Leftist Infiltration of SCAP," early in the occupation. A draft dated June 7, 1947, was implemented during 1948. See box 18, Willoughby Papers, RG 23, MacArthur Memorial. Theodore Cohen recalled that Willoughby conducted routine wiretaps and surveillance of SCAP staff. See Cohen, *Remaking Japan,* 92–96. Willoughby's crusade against Sorge and leftist American journalists familiar with Asia is documented in Steve and Jan MacKinnon, *Agnes Smedley* (Berkeley, 1987), 321–46. During the 1980s, a historian who had served in SCAP, Gordon Prange, published posthumously some of the material he gathered on Sorge under Willoughby's direction.

48. MacArthur to Prime Minister Ashida, July 22, 1948, *PRJ,* 2, 581; Alvary Gascoigne to Foreign Office, FO 371/69823, PRO; William Sebald to Secretary of State, Dec. 9, 1948, *FRUS 1948,* 6: 916–21; Sebald to Lovett, Jan. 3, 1949, *FRUS 1949,* 7, 601–3; Farley, *Aspects of Japan's Labor Problem,* 189–207. SCAP pressed the cabinet and Diet to pass restrictive labor laws. Gradually these provisions were extended to private sector unions. The reversal in labor policy caused several resignations in SCAP's labor section. They were replaced by representatives of the

AFL-CIO interested in organizing militant anti-Communist unions. See Schonberger, "Labor's Cold War in Japan."

49. Sebald to John Allison, Nov. 29, 1948, 820.02, records of the Foreign Service Posts of the Department of State, 59 A 543, pt. 9, RG 84, Washington National Records Center (WNRC); MacArthur to Army Department, Dec. 4, 1948, Draper/Voorhees Project Decimal File, RG 335; Draper to SCAP, Dec. 10, 1948 and SCAP to Draper, Dec. 12, 1948, both in "Japan—1947–48: MacArthur Communications," file 42, Adm. William Leahy Files, JCS Records, RG 218; MacArthur to Draper, Dec. 18, 1948, P&O, 091 Japan TS, sec. 1B, pt. 2, Records of the Department of the Army, RG 319. Despite his protests to Washington about the new policies, MacArthur brooked no delays from the Japanese. He informed Prime Minister Yoshida that the government should not permit unions or left-wing agitators to impede the reverse course. Japan had no choice but to accept "increased austerity in every phase of . . . life" and to "surrender some of the privileges and immunities inherent in a free society." SCAP would "curb" any attempt to delay the "acceleration of production." Unless the stabilization program were successful, he warned, "Japan may perish." See MacArthur to Yoshida, Dec. 19, 1948, *FRUS 1948*, 6: 1066–67.

50. Max Bishop to Butterworth, Dec. 28, 1948, PPS Records, FOIA; Lovett to National Security Council, Dec. 29, 1948, ibid. Most Japanese doubted Southeast Asia could supply adequate raw materials and markets. They preferred to expand trade with America, Europe, and even Communist China. For a fuller discussion of Japan's projected role as regional power, see Schaller, *The American Occupation of Japan*. Theodore Cohen argues that recovery was imminent by the time Dodge arrived. His program only insured a recovery controlled by and helpful to the *zaibatsu*. See Cohen, *Remaking Japan*.

Chapter 11

1. *Time*, July 10, 1950, 14–15.

2. For a discussion of Marshall's changed perspective and shift in priorities, see Schaller, *The American Occupation of Japan*, 72–73. On the Marshall mission, see Steven I. Levine, "A New Look at American Mediation in the Chinese Civil War: The Marshall Mission in Manchuria," *Diplomatic History* 3 (Fall 1979), 349–75.

3. Diary entry of July 10, 1946, Forrestal Papers; memorandum by E. A. Bayne of conversation with MacArthur, Dec. 21, 1946, folder 7, box 1, Butterworth Papers, Marshall Library; diary entry of Mar. 22, 1948, Eichelberger Papers.

4. MacArthur Statements to Eaton Committee, March 3, 1948, CCS 383.21, Japan (3-13-45), sec. 20, Joint Chiefs of Staff (JCS) Records, RG 218; Gascoigne to Prime Minister Clement Attlee, Dec. 26, 1947, FO 371/63830, PRO; Gascoigne to M. E. Dening, Feb. 28, 1948, FO 371/69818, ibid.

5. The most comprehensive analysis of Kennan's strategic thinking appears in Gaddis, *Strategies of Containment*, 41. For examples of Kennan's thoughts on containment in China and East Asia, see National War College Lecture, Oct. 6, 1947, "Soviet Diplomacy," box 17, George F. Kennan Papers, Princeton University; "Talk on Russian–American Relations," to Board of Governors of the Federal Reserve System and the Secretary of the Navy's Council, Dec. 3, 1947, ibid.; "Contempo-

rary Problems of Foreign Policy," National War College Lecture, Sept. 17, 1948, box 17, ibid.; PPS 13, Nov. 6, 1947, *FRUS 1947*, 1, 771; "Policy with Respect to American Aid to Western Europe," May 23, 1947, ibid., 3, 225; Kennan to R. A. Winnacker, Dec. 8, 1948, box 13, PPS Records, RG 59. Also see Schaller, *The American Occupation of Japan*, 87–91.

6. The two best treatments of Korean politics and the early cold war are found in Cumings, *The Origins of the Korean War*, and Matray, *The Reluctant Crusade*. Also see the excellent systhesis of postwar events contained in Peter Lowe, *The Origins of the Korean War* (New York, 1986).

7. Memorandum by Forrestal, September 26, 1947, *FRUS 1948*, 6: 817.

8. See Souers to Truman, April 2, 1948, *FRUS 1948*, 6: 1167.

9. *New York Times*, Aug. 15, 1948; *Time*, Aug. 23, 1948; Matray, *The Reluctant Crusade*, 133, 150, 179, 180–84; Lowe, *The Origins of the Korean War*, 30–50.

10. Faubion Bowers, Oral History, Columbia University.

11. Memorandum of MacArthur's Briefing to Huber Congressional Committee, Sept. 5, 1949, *FRUS 1949*, 9, 545–46; *New York Times*, Mar. 2, 1949, 22; see Acheson's statement as reported in the Department of State *Bulletin* 22 (Jan. 22, 1950), 114–15; James, *Years of MacArthur*, 3: 410–12; MacArthur had traced a similar defense perimeter in a March 1948 discussion with George F. Kennan.

12. For a dramatic portrait of this private policy network, see Bruce Cumings, *The Origins of the Korean War*, vol. 2, forthcoming.

13. NSC 13/3, May 6, 1949, *FRUS 1949*, 7: pt. 2, 730–36; Acheson to Certain Diplomatic Officials, May 8, 1949, ibid., 736–37.

14. Joint Chiefs of Staff, "Strategic Evaluation of U.S. Security Needs in Japan," June 9, 1949, enclosed in NSC 49, June 15, 1949, *FRUS 1949*, 7: pt. 2, 773–77.

15. MacArthur to Acheson, June 16, 1949, ibid., 778–81. The general stood almost alone in trusting the Soviets to stay out of Japan. George Kennan came to agree on this point, but Acheson, Truman, and the entire military establishment thought otherwise.

16. For a discussion of the change in MacArthur's attitude, see memorandum by Max Bishop of conversation with MacArthur, Feb. 16, 1949, ibid., 656–57; memorandum by Cloyce Huston, July 16, 1949, ibid., 806; William Sebald to W. W. Butterworth, July 26, 1949, ibid., 808–12; Sebald to Acheson, Aug. 20, 1949, ibid., 830–40; Sebald memorandum of conversation with MacArthur, Sept. 21, 1949, ibid., 862–64.

17. Memorandum of conversation by Robert Feary, "MacArthur's Views on a Japanese Peace Treaty," Nov. 2, 1949, ibid., 890–94; transcript of a meeting with Col. Stanton Babcock, November 10, 1949, JSSC 388.1 Japan, sec. 1 (9-1-47), JCS Records, RG 218. Babcock served as a SCAP liaison with the JCS.

18. The most complete discussion of the response to the Chinese revolution is found in Robert M. Blum, *Drawing the Line: The Origins of the American Containment Policy in East Asia* (New York, 1982), and Nancy Tucker, *Patterns in the Dust: Chinese–American Relations and the Recognition Controversy, 1949–50* (New York, 1983). Truman's more moderate advisers feared that American intervention on the island would drive the Chinese Communists closer to Moscow and offend Asian nationalism.

19. Transcript of a meeting with Col. Babcock, Nov. 10, 1949, JSSC 388.1 Japan, sec. 1, (9-1-47), JCS Records, RG 218; Report by the JSSC to the JCS on

"Impact of an Early Peace Treaty with Japan on U.S. Strategic Requirements," Nov. 30, 1949, JCS 1380/75, ibid.; notes on a memorandum of conversation with Goldthwaite Dorr by John Howard, Dec. 8, 1949, *FRUS 1949*, 6: 1128; notes by Under Secretary of the Army Tracy Voorhees of MacArthur's opinions on a Japanese Peace Treaty, Dec. 14, 1949, CJCS 092.2, Japanese Peace Treaty 1950, JCS Records, RG 218; Gen. Carter B. Magruder to JCS, Dec. 3, 1949, enclosed in JCS 1380/76, CCS 388.1, Japan, sec. 1 (9-1-47), ibid.; JCS 1380/77, Dec. 10, 1949, ibid.; Johnson to Acheson, Dec. 23, 1949, with enclosure of memorandum by the JCS to the Secretary of Defense, Dec. 22, 1949, *FRUS 1949*, 7: pt. 2, 922–23; memorandum of conversation by Maxwell Hamilton of meeting with Gens. Bradley and Burns, Dec. 24, 1949, ibid., 924–26.

20. Memorandum of conversation with MacArthur, February 16, 1949, *FRUS 1949*, 7: pt. 2, 656–57; report of a meeting between Gen. MacArthur and Chennault, Nov. 21, 1949, NLT-12, CIA, FOIA; memorandum of conversation: MacArthur briefing to Huber Subcommittee, Sept. 5, 1949, CCS 452, China (4-3-43), sec. 7, pt. 5, JCS Records, RG 218.

21. NSC 41, "U.S. Policy Regarding Trade With China," Feb. 28, 1949, NSC Files, Modern Military Branch, National Archives. Approved by the NSC and Truman on Mar. 3, 1949. See also, Nancy B. Tucker, "American Policy Toward Sino-Japanese Trade in the Postwar Years: Politics and Prosperity," *Diplomatic History* (Summer 1984), 183–208.

22. For evidence of MacArthur's duplicity, see memorandum by R. Magill to Philip Sprouose, May 20, 1949, box 15, Chinese Affairs Lot File, DOS, RG 59; MacArthur to Huber Subcommittee, memorandum, Sept. 5, 1949, P&O, 091 Formosa TS, Department of the Army Records, RG 319; MacArthur's statements to Sir Alvary Gascoigne endorsing a blockade are quoted in Tucker, "American Policy Toward Sino-Japanese Trade." In May 1950, the general told a visiting British representative that he foresaw a large increase in Japan's trade with China during the next few years. See Lowe, *The Origins of the Korean War*, 76; nearly all Japanese favored economic links with China. Peking shared this enthusiasm, as much of China's existing industry required Japanese parts and the regime had no desire to depend exclusively on the Soviet Union.

23. The origins and debate over this program are detailed in two sets of congressional hearings. See U.S. Senate, Committee on Foreign Relations, *Military Assistance Program: 1949, Joint Hearings Held in Executive Session before the Committee on Foreign Relations and the Committee on Armed Services*, 81st Cong., 1st sess. (Washington, 1974); U.S. Congress, House of Representatives, Committee on International Relations, *Selected Executive Session Hearings of the Committee, 1943–50, Vol. 5, Military Assistance Programs, pt. 1, Mutual Defense Assistance Act of 1949* (Washington, D.C., 1976); the most thorough analysis of the program and Section 303 is found in Blum, *Drawing the Line*. For Pearson's assertion that MacArthur persuaded Truman to support the covert Asia program see Sept. 17, 1949, excerpts from Pearson Papers.

24. Acheson memorandum for Jessup, July 18, 1949, box 14, Chinese Affairs Lot File, DOS, RG 59; for the genesis and impact of the "White Paper," see Tucker, *Patterns in the Dust;* and Blum, *Drawing the Line*. Acheson assigned a committee to prepare the cover letter accompanying the massive report. Insulting several of its intended audiences, the letter called Mao a Soviet stooge, the Kuomintang corrupt and incompetent, and Chiang's American supporters naive. All responded by denouncing the "White Paper."

25. The drafts of NSC 48, Aug.–Oct. 1949, are contained in a file obtained under the FOIA.

26. See Schaller, *The American Occupation of Japan*, 203–4.

27. Ibid., 204; Jessup's statement of consultant's views on area approach to the Far East, Nov. 3, 1949, box 846, 890.00, DOS, RG 59; outline of Far Eastern and Asian policy for review with the president, Nov. 14, 1949, *FRUS 1949*, 7: 1210–14; memorandum by Acheson of conversation with Truman, Nov. 17, 1949, box 13, PPS Records, RG 59.

28. Schaller, *The American Occupation of Japan*, 206–7.

29. The military's plans for covert activities under MDAP are discussed in memorandum by Gen. L. L. Lemnitzer to JCS, Sept. 13, 1949, JCS 1868/107, P&O, 091 China, sec 2a, case 27, RG 319; Bolte to Collins, Nov. 10, 1949, 091 China TS, ibid.; report by JSSC on military aid to China, Oct. 6, 1949, JCS 11721/37, CCS 45d, China (4-3-45), sec 7, pt. 5, JCS Records, RG 218; memorandum of MacArthur's briefing of Huber Subcommittee, September 5, 1949, P&O, 091 Formosa TS, RG 319; Paper 21, GHQ Seminar for JCS representatives, "Relation of World-Wide Political Situation to CINFE Mission," Oct. 1, 1949, ibid.; Consul General at Taipei to Acheson, Sept. 7 and 8, 1949, *FRUS 1949*, 9: 385–86; Jan. 18, 1950, excerpts from Pearson Papers; in his forthcoming second volume on the origins of the Korean War, Bruce Cumings details the shadowy activity of Chennault, Pawley, Donovan, and others during this period; multiple diary entries, Sept.–Oct., 1949, H. Alexander Smith Papers, Princeton University.

30. Oral History, vol. 6, part J, Wellington Koo Papers, Columbia University. Koo noted that MacArthur's efforts had begun in late 1949 and continued into early 1950.

31. Voorhees notes of meeting with MacArthur, Dec. 14, 1949, P&O, Formosa 091 TS, RG 319.

32. JSSC Report of Dec. 16, 1949, CCS 381, Formosa (11-3-48), sec. 2, JCS Records, RG 218; Gruenther to JCS, Dec. 22, 1949, P&O, 091 Formosa TS, RG 319; Joint Chiefs to Johnson, Dec. 23, 1949, *FRUS 1949*, 9: 460–61; Johnson's deputy, Paul Griffith, coordinated Pentagon efforts with Wellington Koo. The ambassador received intelligence information and other materials from friendly Pentagon sources to help his effort to reverse Truman's policy toward Taiwan. See memoranda in box 130, Koo Papers, and Schaller, *The American Occupation of Japan*, 210–11.

33. Memorandum by Acheson of conversation, Dec. 28, 1949, *FRUS 1949*, 9: 463–67; NSC 48/2, Dec. 30, 1949, NSC File, Modern Military Branch, National Archives. Because of JCS and Defense opposition, the policy paper fudged on the question of a Japan treaty. It simply calling for the administration to "improve" the situation in that country. The discussion also revealed the growing anxiety about Indochina. The document called upon France to transfer power to Vietnamese such as the Emperor Bao Dai in an effort to undercut support for the Communist movement led by Ho Chi Minh.

34. For Truman's statement, see Department of State *Bulletin* 22 (Jan. 16, 1950), 79; Acheson's press club speech in ibid., 22 (Jan. 22, 1950), 114–15. Acheson's testimony of Jan. 10, 1950, in U.S. Senate, Committee on Foreign Relations, *Reviews of the World Situation: 1949–50, Hearings Held in Executive Session on the World Situation*, 81st Congress, 1st and 2d sess. (Washington, D.C., 1974), 113–17, 134–35, 154–70.

35. Memoranda by Jessup of conversations with MacArthur, Jan. 5, 8, and 9,

1950, 694.001/1-1050, DOS, RG 59. A sanitized version of this appears in *FRUS 1950*, 6: 1114–15; see also Lucius Battle to William McWilliams, Feb. 10, 1950, ibid.

36. Memorandum and summary of discussion with Truman, in John Howard to W. W. Butterworth, "Japanese Peace and Security Settlement," March 9, 1950, *FRUS 1950*, 6: 1138–49; the views of the JCS are contained in Howard to Butterworth, ibid., 1133.

37. Memorandum by Butterworth of conversation with MacArthur, Feb. 5, 1950, ibid., 1133–35.

38. Chinese motives remain obscure. Some observers speculated that the Communist regime acted tough to impress the Soviets. During January and February, Mao Tse-tung was in Moscow, negotiating a treaty with Stalin. Acheson, apparently, did not want to terminate diplomatic representation but feared that Americans might become targets of Communist zealots. Removal would allow tempers to cool. However, sending diplomats back to China would represent a major act by the administration, one sure to be criticized by Taiwan's supporters. The Chinese, or at least one faction, were dismayed by the American decision to leave China. In a counterproductive effort, the Communist government actually blocked Consul General Edmund Clubb's departure. See Schaller, *The American Occupation of Japan*, 221, 251.

39. McCarthy's impact is analyzed in Blum, *Drawing the Line*, 187–91, and William W. Stueck, *The Road to Confrontation: American Policy Toward China and Korea, 1947–50* (Chapel Hill, N.C., 1981), 143–46; for more detailed information on McCarthy's anti-Communist crusade, see Thomas C. Reeves, *The Life and Times of Joe McCarthy* (New York, 1982); and Oshinsky, *A Conspiracy So Immense*. Acheson and his staff laughed at the senator's bizarre conspiracy tales until they found an audience. See minutes of Secretary's Daily Meetings, February–June 1950, records of the Executive Secretariat, DOS, RG 59; note on Sino-Soviet Treaty, *FRUS 1950*, 6: 311.

40. William M. Leary, *Perilous Missions: Civil Air Transport and CIA Covert Operations in Asia* (University of Alabama Press, 1984); Slessor to P. Reilly, Dec. 30, 1949, FO 371/83012, PRO; Fergusson to Gascoigne, Feb. 11, 1950, FO 371/83013, PRO. See also Schaller, *The American Occupation of Japan*, 251. Eventually, the aircraft went to the "private investors," not to China.

41. Oral History, vol. 6, pt. J, Koo Papers.

42. Acheson speech of March 15, 1950, Department of State *Bulletin* 22 (Mar. 27, 1950), 467–72; America's wartime support for Indochinese nationalism changed after 1945 as fears of communism in Europe and Asia replaced opposition to colonialism. Still, the United States pressed France to share power with moderate, anti-Communist Vietnamese and provided Paris only indirect military aid for its war effort between 1946 and 1949. To placate the Americans, France proclaimed Bao Dai as the emperor of Vietnam. Given his shady past and lack of local support, Washington balked at the French ruse. However, by early 1950, as the Vietminh threat increased, the Truman administration relented and recognized Bao Dai as the legitimate ruler of an "independent" Vietnam. This fiction turned a colonial war into a struggle against Communist imperialism. During March and April, the State and Defense departments agreed to send an American military-economic advisory mission to Saigon to oversee the distribution of MDAP funds and weapons. See Schaller, *The American Occupation of Japan*, 234–45.

43. *New York Times*, March 28, 1950; Vandenberg to Acheson, Mar. 31, 1950,

and Acheson memorandum of conversation with Dulles, Apr. 5, 1950, box 65, Dean Acheson Papers, Truman Presidential Library; "Appointment of JFD as Consultant to Acheson, Apr. 6, 1950, box 47, John Foster Dulles Papers, Princeton University; *FRUS 1950*, 6: 1160–61; Dean Acheson, *Present At the Creation: My Years in the State Department* (New York, 1969), 337, 432–33; on Rusk's career, see Warren Cohen, *Dean Rusk* (Totawa, N.J., 1980).

44. Memorandum by Rusk for Acheson, Apr. 26, 1950, *FRUS 1950*, 6: 333–35; memorandum of conversation by John Howard, "Japanese Peace Settlement," Apr. 7, 1950, ibid., 1161–66; memorandum by John Howard of conversation, "Japanese Peace Treaty," Apr. 24, 1950, ibid., 1175–82; diary entries of May 14 and 24, 1950, Sen. H. Alexander Smith Papers, Princeton University.

45. Butterworth to Acheson, May 3, 1950, 694.001/5-350, DOS, RG 59; memorandum by Breen for Allison, August 2, 1950, *FRUS 1950*, 6: 1262–63; Ikeda's statement of May 2, 1950, in Reid to Butterworth, May 10, 1950, ibid., 1194–98; memorandum by Butterworth for Acting Secretary Webb, May 12, 1950, ibid., 1198.

46. Memorandum by Dulles for Dean Rusk, Paul Nitze, and Under Secretary Webb, May 18, 1950, *FRUS 1950*, 1: 314–16; many of Dulles's arguments about defending territory regardless of its objective importance echoed a major policy paper, NSC 68, being prepared under the guidance of Paul Nitze, Kennan's successor as head of the Policy Planning Staff. Going beyond the earlier containment doctrine, NSC 68 rejected geographic, economic, political, or military boundaries to the cold war. It called for a huge military buildup and actively confronting revolutionary movements in Africa, Asia, and the Middle East. Instead of containing Soviet influence, the new program championed "rollback" and "liberation." Most of the new program was implemented during the Korean War. See NSC 68, "United States Objectives and Programs for National Security," Apr. 14, 1950, ibid., 237–92.

47. Memorandum by Burns to Rusk, May 29, 1950, ibid., 6: 346–47; memorandum, Fisher Howe to W. Park Armstrong, May 31, 1950, ibid., 347–49; Rusk's report of May 31, 1950, box 18, Chinese Affairs Lot File, DOS, RG 59. Rusk hoped that if Taiwan's leaders abandoned their claim to the Chinese mainland, the two regimes might coexist. For a fuller discussion of Rusk's efforts, see Schaller, *The American Occupation of Japan*, 260–71.

48. Vol. 6, pts. A and J, Koo Oral History, Koo Papers. Koo's diary and official files confirm many of his later recollections and deserve careful scrutiny by scholars. What remains unclear is how much of these covert military plans were simply talk and how much action.

49. Cooke's recommendations and estimates are in a memorandum from the Chief of Naval Operations to the Joint Chiefs, May 1, 1950, JCS 1966/27, CCS 381 Formosa (11-8-48), sec. 3, JCS Records, RG 218; Cooke to MacArthur, May 2, 1950, MacArthur Papers; Cooke to Sen. William F. Knowland, May 23, 1950, box 273, William F. Knowland Papers, University of California, Berkeley; Chiang Kai-shek to MacArthur, undated, box 2, Charles M. Cooke Papers, Hoover Institution, Stanford University; memorandum for Adm. Cooke by K. C. Shah, Apr. 7, 1950, ibid.; memorandum for the generalissimo, by Cooke, May 1950, ibid.; Bradley to Johnson, May 31, 1950, ibid.; MacArthur to Department of the Army, C 56410, May 20, 1950, Formosa folder, box 2, RG 6, MacArthur Papers.

50. Diary entries of May 12 and June 6, 1950, Eichelberger Papers.

51. Schaller, *The American Occupation of Japan*, 267–68. Defense Secretary

Johnson and his assistant secretary, Paul Griffith, met regularly with Koo and passed on sensitive information to the Chinese. See memoranda of conversations with Griffith, June 3 and 7, 1950, box 180, Koo Papers; memorandum of Koo conversation with Dulles, June 12, 1950, ibid.

52. Maj. Gen. Carter B. Magruder, Acting Special Assistant for Occupied Area, to Gen. Omar Bradley and Secretary Johnson, June 5, 1950, CJCS 092.2, Japanese Peace Treaty—1950, JCS Records, RG 218; memorandum by Dulles for Acheson, June 6, 1950, *FRUS 1950*, 6: 1207–12; memorandum by Dulles, June 15, 1950, ibid., 1222–23.

53. Dulles later reported some minimal contact with the military delegation. See *FRUS 1950*, 6: 1229, 1260; memorandum of conversation by John Allison, June 19, 1950, ibid., 7: 107–9; William Sebald, *With MacArthur in Japan* (New York, 1965), 252–53.

54. William R. Mathews to Dulles, June 20, 1950, John Foster Dulles Papers, Princeton University.

55. MacArthur presented these views to the visiting delegates and to the administration in a pair of memoranda. See MacArthur Memorandum on Formosa, June 14, 1950, *FRUS 1950*, 7: 161–65; MacArthur Memorandum on Peace Treaty Problem, ibid., 6: 1213–21.

56. Acheson's remarks to Frank as quoted in Lowe, *The Origins of the Korean War*, 53. The effort at linking Chiang's removal to the protection for Taiwan is discussed in *FRUS 1950*, 4: 346–47, 347–49, 351–52; Rusk report of May 31, 1950, with accompanying documents, box 18, Chinese Affairs Lot File, DOS, RG 59; Rusk to Acheson, "Bi-Partisan Policy on China–Formosa Problems," June 9, 1950, PPS Records, FOIA; Harlan Cleveland to Paul Hoffman, William Forster and Richard Bissell, Jr., "Formosa Policy," June 23, 1950, records of the Economic Cooperation Administration, 53-A-405, RG 286, RG 286; Kenneth T. Young to Gen. Burns, "Action on Formosa," June 19, 1950, CD-6-4-6, records of the Office of the Secretary of Defense, RG 330; Livingston Merchant to Dean Rusk, June 23, 1950, 694.001/6-2350, DOS, RG 59. This subject is also discussed in Schaller, *The American Occupation of Japan*, 263–66 and in Cumings, *Origins of the Korean War*, 2 (forthcoming).

57. Memorandum by Gen. Omar Bradley to the JCS, June 26, 1950, CJCS, 092.2 Japanese Peace Treaty, 1950, JCS Records, RG 218; memorandum on the concept governing security in postwar Japan by MacArthur, June 23, 1950, *FRUS 1950*, 6: 1227–29.

Chapter 12

1. Nikita Khrushchev, *Khrushchev Remembers* (Boston, 1970), 367–73; Robert Simmons, *The Strained Alliance* (New York, 1975), argues that the Soviet Union, rather than China, encouraged North Korea. This study suggests an overeager Kim "jumped the gun" by starting the war in June, rather than in August, which the Russians expected. The most recent studies of the war include Stueck, *Road to Confrontation*, Dobbs, *Unwanted Symbol*, Matray, *Reluctant Crusade*, Cumings, *Origins of the Korean War*, Lowe, *The Origins of the Korean War*, Burton I. Kaufman, *The Korean War: Challenges in Crisis, Credibility and Command* (New York, 1986); also see the forthcoming second volume of Bruce Cumings, *Origins of the Korea War*. Cumings argues persuasively that the North Koreans planned

and executed the attack with minimal outside involvement, and that their initial success was due largely to their support by the South Korean population.

2. *Time* (July 10, 1950), 9.

3. Diary entry of July 1, Eben Ayers Papers, Truman Library; diary entry of June 25, 1950, William Sebald Papers, Nimitz Library, United States Naval Academy; John Allison, *Ambassador From the Prairie; or Allison Wonderland* (Boston, 1973), 129–31.

4. James, *Years of MacArthur*, 3: 419–21.

5. *FRUS 1950*, 7: 174–75.

6. Webb's recollection appears in a letter to John Snyder, April 25, 1975, General Correspondence, S, 1973–75, folder 2, James Webb Papers, Truman Library.

7. Memorandum of conversation by Jessup, June 25, 1950, *FRUS 1950*, 7: 157–61; memorandum by Gen. Omar Bradley for the president, June 25, 1950, CJCS 091 China, JCS Records, RG 218; memorandum by Gen. Omar Bradley for the president, June 26, 1950, CCS 381, Formosa (11-8-48), sec. 3, ibid.; Omar N. Bradley and Clay Blair, *A General's Life* (New York, 1983), 534–35; Stephen Pelz, "U.S. Decisions on Korean Policy, 1943–50: Some Hypotheses," in Cumings, ed., *Child of Conflict*, 93–132.

8. Memorandum of conversation by Jessup, June 26, 1950, *FRUS 1950*, 7: 178–83; memorandum by George M. Elsey of conversation with Truman, June 26, 1950, box 71, George M. Elsey Papers, Truman Library. Most American intelligence analysts agreed that the North Korean attack was a Soviet probe designed to intimidate American allies in Asia, the Middle East, and Europe. See, for example, Intelligence Estimate, Office of Intelligence Research, June 25, 1950, *FRUS 1950*, 7: 148–54; Maj. Gen. Charles A. Bolte to Army Secretary Frank Pace, June 28, 1950, 091 Korea TS, JCS Records, RG 218.

9. Memorandum of Truman's meeting with congressional leaders, June 27, 1950, box 71, Elsey Papers; statement issued by the president, June 27, 1950, *FRUS 1950*, 7: 202–3. UN resolution, ibid., 211.

10. Bradley and Blair, *A General's Life*, 538. Eisenhower wrote these comments to Bradley after discussing them with the general's staff.

11. Diary entry of June 26, 1950, Sebald Papers; James, *Years of MacArthur*, 3: 422.

12. Diary entry of June 27, 1950, Sebald Papers; Allison, *Ambassador from the Prairie*, 135–37.

13. MacArthur, *Reminiscences*, 332–34.

14. Memorandum of meeting of June 29, box 71, Elsey Papers.

15. Copy of statement given to George Elsey, June 30, 1950, ibid.

16. JCS to MacArthur, June 29, 1950, *FRUS 1950*, 7: 240–41; Roy E. Appleman, *South to the Naktong, North to the Yalu* (Washington, D.C., 1961), 44.

17. Memorandum of meeting, June 30, 1950, box 71, Elsey Papers. See James F. Schnabel and Robert J. Watson, *The History of the Joint Chiefs of Staff: The Joint Chiefs of Staff and National Policy, 1945–53* (Wilmington, Del., 1979), *The Korean War*, pt. 1, 3: 137–50 (hereafter abbreviated as *The Korean War*); Acheson, *Present at the Creation*, 413. MacArthur told members of the Senate in 1951 that his "connection with the United Nations was largely nominal."

18. Memoranda of conversations with Johnson and Griffith, June 30, 1950, box 180, Koo Papers.

19. Diary entry of July 3, 1950, Ayers Papers. Ayers cited examples of Johnson leaking information to the press and undermining the president.

20. MacArthur to Truman, July 11, 1950, RG 7, MacArthur Papers.

21. Diary entry of July 1, 1950, Ayers Papers; on July 7, Dulles again cautioned the administration about appointing MacArthur field commander. Then, on July 10, he sent the general a message hailing his appointment. While drafting the Japanese peace treaty, Dulles had to stay in the general's favor. See Lowe, *The Origins of the Korean War*, 69–70; James, *Years of MacArthur*, 3: 437.

22. For an analysis of the military side of the war, see Callum A. MacDonald, *Korea: The War Before Vietnam* (London, 1986), Max Hastings, *The Korean War* (New York, 1987), and Clay Blair, *The Forgotten War: America in Korea 1950–53* (New York, 1987).

23. J. Lawton Collins, *War in Peacetime: The History and Lessons of Korea* (Boston, 1969), 81–82, 85; James, *Years of MacArthur*, 3: 445. When Truman recalled MacArthur in April 1951, about 360,000 soldiers were in the theater. During the first few months of the war, Britain, Australia, New Zealand, Turkey, Canada, and the Netherlands sent small units. Eventually, nineteen nations offered some form of assistance. Japan contributed sailors and ships for transportation, mine sweeping, and many other tasks.

24. See Koo Oral History, vol. 6, J, Koo Papers.

25. MacArthur to Truman, July 20, 1950, box 120, PSF, Truman Papers.

26. Collins, *War in Peacetime*, 82–83; July 9, 1950, excerpts from Pearson Papers.

27. Summary of MacArthur's remarks of July 13, 1950, in Schnabel and Watson, *Korean War*, 507; Walter S. Poole, *The History of the Joint Chiefs of Staff: The Joint Chiefs of Staff and National Policy, 1945–53* (Wilmington, Del., 1979), 1950–52, 4: JCS, 391–93; James, *Years of MacArthur*, 3: 452–53; diary entry of July 24, 1950, Sebald Papers.

28. Almond Oral History; Burton Crane Oral History, Columbia University; Koo Oral History, vol. 7, A, Koo Papers.

29. Sebald, *With MacArthur*, 123; JCS to MacArthur, Aug. 3, 1950, RG 9, MacArthur Papers; Johnson to MacArthur, Aug. 4, 1950, ibid.; MacArthur to Johnson, Aug. 5, 1950, ibid.; James, *Years of MacArthur*, 3: 455–56. For evidence of American complicity in plots against Chiang, see Frank Dorn, *Walkout With Stilwell in Burma* (New York, 1971), 75–79; Cumings, *Origins of the Korean War*, 2: (forthcoming).

30. Goulden, *Korea*, xx–xxiv. The author attributes the story to Charles Burton Marshall, a high State Department official during the Korean War. On Lowe's mission, see James, *Years of MacArthur*, 3: 458–59; see, Lowe Reports, Aug. 1950–Apr. 1951, PSF, Truman Papers.

31. Memorandum of conversation between Gen. MacArthur and W. A. Harriman, PSF 129, Truman Papers. Excerpts of Harriman's account appear in *FRUS 1950*, 6: 427–30, and 7: 542–44; Chennault's Civil Air Transport (CAT) served the CIA, Taiwan, and MacArthur in carrying out missions against China. See Leary, *Perilous Missions*, 124. Ambassador Koo's Oral History and diaries, August–December 1950, allude to MacArthur's involvement in guerrilla operations against China, Koo Papers.

32. Notes on conference with MacArthur, Aug. 8, 1950, box 16, Matthew B. Ridgway Papers, United States Military History Institute; Matthew B. Ridgway, *The Korean War* (New York, 1967), 37–38; see also Ridgway's memorandum on his discussion with MacArthur, attached to Harriman memorandum, in note 31 above.

33. Truman press conference of Aug. 10, 1950, Truman, *Public Papers*, 1950, 580; press release, Aug. 10, 1950, RG 6, MacArthur Papers; James, *Years of MacArthur*, 3: 457.

34. See MacArthur to Clyde Lewis, August 20, 1950, RG 7, MacArthur Papers; selections from the message were published in *The New York Times*, Aug. 29, 1950, and *U.S. News and World Report*, 29 (Sept. 1, 1950): 32–34, as well as in other periodicals.

35. Kennan to Acheson, Aug. 21, 1950, box 65, Acheson Papers; memoranda for file by George Elsey, Aug. 26 and Oct. 2, 1950, box 72, Elsey Papers; also see Lucius Battle, "Memorandum for the Record of the Events of Saturday, August 26, 1950," Acheson Papers; Unsigned Report, "Foreign Policy Aspects of the MacArthur Statement," Aug. 26, 1950, ibid.

36. *New York Times*, Aug. 29, 1950; diary entry of September 4, 1950, Sebald Papers; Truman, *Memoirs*, 2: 406–8, 436; see *MSFE*, 989–90, 1217, 2002–3. For a discussion of Marshall's limitations as Defense Secretary, see Stueck, *Road to Confrontation*, 236, 250–57.

37. Edward Almond, Oral History.

38. MacArthur, *Reminiscences*, 349–50; James, *Years of MacArthur*, 464–79.

39. Collins, *War in Peacetime*, 141–42.

40. Acheson, *Present At the Creation*, 418–420; Lowe, *The Origins of the Korean War*, 166–69; Kaufman, *The Korean War*, 52–54; Frank Pace, Oral History, Truman Library.

41. The debate over crossing the 38th parallel and disposing of the Communist regime can be traced in *FRUS 1950*, 7: see especially 272, 346, 346, 386–87, 383–95, 449–54, 458–61, 483–85, 502–10, 600–3, 623–28; for an overall statement of national policy, see NSC 81, Sept. 1, 1950, ibid., 685–693; NSC 81/1, Sept. 9, 1950, ibid., 712–21. Within the State Department, the Office of Northeast Asian Affairs, headed by John Allison, took a hard line and advocated the destruction of the North Korean state. The Policy Planning Staff, headed by Paul Nitze, but still influenced by George Kennan, advocated a more moderate approach.

42. JCS to MacArthur, Sept. 27, 1950, *FRUS 1950*, 7: 781–82, 792–93; Marshall to MacArthur, Sept. 29, 1950, ibid., 826; MacArthur to Marshall, Sept. 30, 1950, RG 6, MacArthur Papers.

43. Gascoigne to Foreign Office, Sept. 28 and Oct. 4, 1950, FO 317/83008, PRO.

44. This discussion of Chinese perceptions and policy draws heavily on an unpublished paper, "The Korean War and Sino-American Relations," by Jonathan Pollack, of the Rand Corporation. Pollack conducted extensive interviews with Chinese officials and historians during 1986 and received access to memoir materials and historical studies available only in China.

45. Ibid.

46. Ibid.

47. *FRUS 1950*, 7: 851; Allen S. Whiting, *China Crosses the Yalu: The Decision to Enter the Korean War* (Stanford, Calif., 1960), 108; James, *Years of MacArthur*, 3: 490–91.

48. Truman, *Memoirs*, 2: 462–63; Acheson, *Present at the Creation*, 452; Gascoigne to Foreign Office, Oct. 3, 1950, quoted in Lowe, *The Origins of the Korean War*, 192.

49. See Cumings, *Origins of the Korean War*, 2: (forthcoming); Rosemary Foot,

The Wrong War: American Policy and the Dimensions of the Korean Conflicts, 1950–53 (Ithaca, N.Y., 1985). Gen. Bradley asserts that the JCS discounted the likelihood of Chinese intervention in Korea, believing it primarily a Soviet sphere. See Bradley and Blair, *A General's Life,* 563–64.

50. Schnabel and Watson, *Korean War,* 260–61; Bradley and Blair, *A General's Life,* 570–71.

51. Charles Ross quoted in Robert J. Donovan, *Tumultuous Years: The Presidency of Harry S. Truman, 1949–53* (New York, 1982), 284.

52. Truman, *Memoirs,* 2: 362–63; Ferrell, ed., *Off The Record,* 196, 199.

53. MacArthur claimed these warnings came from telegrams he received from friends in America. Diary entries of October 11 and 12, 1950, Sebald Papers.

54. Wake Conference transcripts, *FRUS 1950,* 7: 948–62.

55. Ferrell, ed., *Off The Record,* 200; Truman, *Memoirs,* 2: 364–65; MacArthur, *Reminiscences,* 410–13; Whitney, *Rendezvous,* 388–89; Eben Ayers reported Truman saying that MacArthur apologized profusely for the VFW incident. See diary entry of October 19, 1950, Ayers Papers.

56. James, *Years of MacArthur,* 3: 513.

57. Bradley and Blair, *A General's Life,* 567–68, 578.

58. Ibid., 579–80; Schnabel and Watson, *Korean War,* 274–76; James, *Years of MacArthur,* 3: 492–98.

59. Schnabel and Watson, *Korean War,* 281 82, 287. The JCS often waffled in thier view of how to respond to Chinese intervention. On October 23, Gen. Bradley told the British military chiefs that if the "Chinese Communists come into Korea, we get out." MacArthur's intelligence chief, Gen. Willoughby, covered himself by including in his estimates reports of large numbers of Chinese troops entering Korea. He then discounted the accuracy of these reports.

60. Ibid., 287–90.

61. Pollack, "The Korean War and Sino-American Relations"; Schnabel and Watson, *Korean War,* 292.

62. MacArthur to JCS, November 6, 1950, reprinted in Truman, *Memoirs,* 2: 427–28; *FRUS 1950,* 7: 1051 n.

63. Bradley and Blair, *A Soldier's Life,* 586–88; Truman, *Memoirs,* 2: 428–29; Schnabel and Watson, *Korean War,* 293–96.

64. Schnabel and Watson, *The Korean War,* 288–99; Goulden, *Korea,* 303; Nov. 24 and Dec. 1, 1950, excerpts from Pearson Papers. See also Ronald J. Caridi, *The Korean War and American Politics: The Republican Party as a Case Study* (Philadelphia, 1968), 95.

65. Marshall's message of Nov. 7, 1950, and MacArthur's response of Nov. 8 are printed in Schnabel and Watson, *Korean War,* 296–98.

66. Whiting, *China Crosses the Yalu,* 137, 160–62; Acheson, *Present at the Creation,* 466.

67. MacArthur to Joint Chiefs, Nov. 9, 1950, Schnabel and Watson, *Korean War,* 302.

68. Anglo-French constraint on American policy is discussed in Foot, *The Wrong War,* 91–95.

69. JCS to secretary of defense, Nov. 9, 1950, Schnabel and Watson, *Korean War,* 303–4; ibid., 305–6, 315–16, 326–29; *FRUS 1950,* 7: 1197–98; 1204–8; Bradley and Blair, *A General's Life,* 594.

70. Acheson, *Present at the Creation,* 468.

71. Diary entry of November 15, 1950, Sebald Papers; memorandum of Sebald-MacArthur conversation, Nov. 14, 1950, *FRUS 1950*, 7: 1148–49. Among others, Acheson read this report.

72. Dulles to MacArthur, Nov. 15, 1950, *FRUS 1950*, 6: 1349–52.

73. Conversation of Nov. 17, 1950, ibid., 7: 1175.

74. *FRUS 1950*, 7: 1207–8; MacArthur to Joint Chiefs, Nov. 25, 1950, Schnabel and Watson, *Korean War*, 330–31.

75. Ridgway, *The Korean War*, 37–38.

Chapter 13

1. MacArthur to JCS, Nov. 28, 1950, *FRUS 1950*, 7: 1237–38; summary of MacArthur message to JCS, 1253 n; ibid.; Schnabel and Watson, *Korean War*, 335–37.

2. Diary entry of Nov. 28, Ayers Papers.

3. Memorandum of conversations [at NSC meeting] by Jessup, Nov. 28, 1950, *FRUS 1950*, 7: 1242–49; Bradley and Blair, *A General's Life*, 598–99.

4. Ridgway, *The Korean War*, 62, 73–76; *FRUS 1950*, 7:1253–54, 1259–60, 1276–81, 1320–23; Schnabel and Watson, *Korean War*, 337–46; *New York Times*, Dec. 2, 1950; "MacArthur's Own Story," *U.S. News & World Report* 29 (Dec. 8, 1950): 16–22; Bradley and Blair, *A General's Life*, 601–3, 605; James, *Years of MacArthur*, 3: 541–42.

5. Truman, *Memoirs*, 2: 437; Military Situation in the Far East, 3532–34; Margaret Truman, *Harry S. Truman* (New York, 1985), 443; Ferrell, ed., *Off the Record*, 202, 210.

6. The question of civilian control is discussed in Roger M. Anders, "The Atomic Bomb and the Korean War: Gordon Dean and the Issue of Civilian Control," *Military Affairs* 52 (January 1988), 1–6; *FRUS 1950*, 7: 1041–42; Schnabel and Watson, *Korean War*, 372–73; diary entry of November 30, 1950, Ayers Papers; Bradley and Blair, *A General's Life*, 605–6; Recently Classified Documents, Korea, Nov. 28, 1950, G-3 Files, Records of the JCS, RG 218; Barton J. Bernstein, "New Light on the Korean War," *The International History Review* 3 (April 1981), 257–77; Foot, *The Wrong War*, 116–17. This last source suggests that by November–December 1950 the Joint Chiefs looked with greater favor on the possibility of using the atomic bomb in Korea.

7. *New York Times*, Dec. 1, 1950; Foot, *The Wrong War*, 106–7.

8. Discussion of these topics occurred on December 2–4. See *FRUS 1950*, 7: 1307–8, 1310–13, 1320–22, 1323–34, 1337–38, 1345–47; Acheson, *Present at the Creation*, 473–77.

9. The extensive record of discussion is printed in *FRUS 1950*, 7: see especially 1330, 1335, 1364–66, 1368–69, 1371, 1385, 1395, 1401–2, 1405–6, 1432, 1451, 1456–58, 1464–65, 1476–79; Donovan, *Tumultuous Years*, 318. On British restraint, see Rosemary Foot, "Anglo-American Relations in the Korean Crisis: The British Effort to Avert an Expanded War, December 1950–January 1951," *Diplomatic History* 10 (Winter 1986): 43–57. Mao's remarks about the difficulty of winning Stalin's confidence came in 1962, see Pollack, "The Korean War and Sino-American Relations."

10. Collins, *War in Peacetime*, 229–33; *FRUS 1950*, 7: 1468–72; Bradley and Blair, *A General's Life*, 606–8; Ridgway, *Korean War*, 81–83.

11. Pollack, "The Korean War and Sino-American Relations."

12. *FRUS 1950*, 7: 1524–31; James, *Years of MacArthur*, 3: 543; Kaufman, *The Korean War*, 114–16.

13. Meeting of the president with congressional leaders, Dec. 13, 1950, box 73, Elsey Papers; *FRUS 1950*, 7: 1518–20; Bradley and Blair, *A General's Life*, 609–10; Kaufman, *The Korean War*, 116–17; James, *Years of MacArthur*, 3: 546–47.

14. Memorandum of conversation by Jessup, Dec. 3, 1950, *FRUS 1950*, 7: 1324–34; memorandum of conversation by Rusk, Dec. 19, ibid., 1570–76.

15. JCS to MacArthur, Dec. 29, 1950, ibid., 1625–26; Bradley and Blair, *A General's Life*, 612.

16. Diary entries of Nov. 29–Dec. 29, 1950, Koo Papers; memorandum of discussion with Dulles, Dec. 19, 1950, ibid.

17. Oral History, vol. 7, pt. B, Koo Papers.

18. Records of the Army Staff, G3 091 Korea TS, July 5, 1951, box 38 A, RG 319, cited in Foot, *The Wrong War*, 114–15.

19. MacArthur to Department of the Army, Dec. 30, 1950, *FRUS 1950*, 7: 1630–33; Ridgway to Collins, Dec. 29, 1950, box 22, Ridgway Papers; Frederick J. Hetzel and Harold I. Hitchens, "An Interview with General Matthew B. Ridgway," *Western Pennsylvania Historical Magazine* 65 (Oct. 1982): 279–308.

20. Joint Chiefs of Staff to MacArthur, Jan. 9, 1950, RG 9, MacArthur Papers.

21. MacArthur to Joint Chiefs of Staff, Jan. 10, 1951, RG 6, ibid.

22. Bradley and Blair, *A General's Life*, 616; Acheson, *Present at the Creation*, 515; Truman, *Memoirs*, 2: 492, 493–95; *MSFE*, 1601.

23. *FRUS 1951*, 7: pt. 1, 70–72, 79–81; Schnabel and Watson, *Korean War*, 412–19; James, *Years of MacArthur*, 3: 554–55.

24. Foot, *The Wrong War*, 121–22, argues that Acheson and others in Washington quite seriously considered for a time the "wider war" favored by MacArthur.

25. Collins, *War in Peacetime*, 255; Bradley and Blair, *A General's Life*, 623; Schnabel and Watson, *Korean War*, 432–40.

26. Schnabel and Watson, *Korean War*, 441–65.

27. Pollack, "The Korean War and Sino-American Relations."

28. The great debate is covered extremely well in Kaufman, *The Korean War*, 122–38, and Donovan, *Tumultuous Years*, 321–24.

29. Ibid.

30. Foot, *The Wrong War*, 111–13;

31. James, *Years of MacArthur*, 3: 576.

32. Ibid., 577.

33. Diary entries of Feb. 8 and 17, 1951, Sebald Papers; Donovan, *Tumultuous Years*, 349.

34. MacArthur, *Reminiscences*, 384; "Memorandum on Ending the Korean War," Dec. 14, 1952, Ann Whitman Files, Eisenhower Papers; for 1954 interviews with Jim G. Lucas and Robert Consodine, see *New York Times*, Apr. 9, 1964. D. Clayton James notes that the idea of a radioactive barrier had been discussed by others in previous months. Citing only MacArthur's later pronouncements, William Manchester and Joseph Goulden claim that MacArthur had a well-defined plan for atomic warfare in the spring of 1951 but that his requests were rejected by the Joint Chiefs. There is little evidence to back this up. See Manchester, *American Caesar*, 627; Goulden, *Korea*, 455, 525; James, *Years of MacArthur*, 3: 577–79.

35. MacArthur to Department of the Army, Mar. 10, 1951, CX 57399, DA IN

9243, and JCS response of Mar. 23, 1951, in "Record of Actions Taken by the JCS Relative to the UN Operations in Korea from June 25, 1950 to April 11, 1951," by JSSC, Apr. 27, 1951, box 224, P&O, RG 319.

36. Ridgway, *Korean War*, 145.

37. *New York Times*, Mar. 8 and 16, 1951; *MSFE*, 3540–41; James, *Years of MacArthur*, 3: 582.

38. JCS to MacArthur, Mar. 20, 1951, *FRUS 1951*, 7: pt. 1, 251; MacArthur to JCS, Mar. 21, 1951, ibid., 255–56; draft texts of presidential statement, ibid., 253–54, 263–64; MacArthur to JCS, Mar. 21 (25?), 1951, C58575, DA IN 14526, in "Record of Actions Taken by the JCS Relative to the UN Operations in Korea from June 25, 1950, to Apr. 11, 1951," by JSSC, Apr. 27, 1951, box 225, P&O, RG 319.

39. *FRUS 1951*, 7: pt. 1, 265–66; *New York Times*, Mar. 24, 1951.

40. MacArthur, *Reminiscences*, 387; Whitney, *Rendezvous*, 467; diary entries of Apr. 8 and 11, 1951, Sebald Papers.

41. Diary entry of Mar. 26, 1951, Eichelberger Papers; memorandum of conversation with Lovett, by Acheson, Mar. 24, 1951, box 66, Acheson Papers; outline by Rusk of events of Mar. 24, 1951, *FRUS 1951*, 7: pt. 1, 266–68; Acheson, *Present at the Creation*, 519; Truman, *Memoirs*, 2: 440–4; JCS to MacArthur, Mar. 24, 1951, PSF, Truman Papers; MSFE, 3542; *Washington Post*, Mar. 31, 1951.

42. Entries from the diary of Gordon Dean, Mar. 27–Apr. 5, 1951, cited in Anders, "The Atomic Bomb and the Korean War," For the Dean diary, see Roger M. Anders, ed., *Forging the Atomic Shield: Excerpts from the Office Diary of Gordon E. Dean* (Chapel Hill, N.C., 1987), esp. 127–41.

43. Among other places, the letter to Martin can be found in *FRUS 1951*, 7: pt. 1, 299; Donovan, *Tumultuous Years*, 351–52.

44. MacArthur, *Reminiscences*, 386; diary entry of Apr. 11, 1951, Sebald Papers; James, *Years of MacArthur*, 3: 590–91.

45. Entries of Apr. 5 and 6, 1951, Truman diary, box 278, PSF, Truman Papers.

46. Dean diary entry of Apr. 6, 1951, cited in Anders, ed., *Forging the Atomic Shield*, 137.

47. Ibid; Bradley and Blair, *A General's Life*, 630–31.

48. *FRUS 1951*, 7: pt. 1, 307–9, 316–19, 337–44; 364–66; Schnabel and Watson, *Korean War*, 485–87; Bradley and Blair, *A General's Life*, 630–31; Hastings, *The Korean War*, 240; Anders, "The Atomic Bomb and the Korean War"; Anders, ed., *Forging the Atomic Shield*, 141. Joseph Goulden claims that in March the National Security Agency had intercepted messages from Spanish and Portuguese diplomats in Tokyo in which they reported MacArthur bragging of a plan to "transform the Korean war into a major conflict in which he could dispose of the Chinese Communist question." Goulden attributes this information to Charles Burton Marshall, a diplomat then on the State Department Policy Planning Staff. According to Goulden, these intercepts convinced Truman to sack the general. No documentation of the event has surfaced and several scholars, after speaking with both Goulden and Marshall, discount the story. In any case, MacArthur's views on escalation were hardly secret. Gen. Willoughby was quite close to the Iberian diplomats and may have discussed his own ideas with them. In fact, several American journalists overheard Willoughby "toasting Francisco Franco" at a dinner in Tokyo. See Goulden, *Korea*, xxiii–xxiv, 476–78. On Willoughby, see entry of Feb. 21, 1951, excerpts from Pearson Papers.

49. Edward Almond, Oral History, USMHI; diary entry of Apr. 8, 1951, Sebald Papers; James, *Years of MacArthur*, 3: 599.

50. Entries of Apr. 5–10, 1951, Truman diaries, box 278, PSF, Truman Papers; memorandum of MacArthur dismissal, Apr. 28, 1951, PSF, ibid.; three excellent accounts of the recall decision and chronology appear in James, *Years of Mac-Arthur*, 3: 590–96; Donovan, *Tumultuous Years*, 353–55; Bradley and Blair, *A General's Life*, 631–35. For a further discussion of the role of atomic weapons and of the administration's effort to signal both China and Congress of its intentions, see Roger Dingman "The Uses of Nuclear Weapons in the Korean War," *International Security*, forthcoming.

51. *MSFE*, 3546–47; Donovan, *Tumultuous Years*, 356–58; James, *Years of MacArthur*, 3: 597–99.

52. Diary entry of Apr. 11, 1951, Sebald Papers.

53. Memorandum for diary, Apr. 12, 1951, Ridgway Papers; Ridgway, *Korean War*, 159; Matthew B. Ridgway, *Soldier* (New York, 1956), 223.

54. Diary entry of Apr. 10, 1951, box 278, Truman Papers; Donovan, *Tumultuous Years*, 359–61.

Chapter 14

1. The Remington Rand offer, rumored to total $100,000, was reported on December 24, 1949, by the *Chicago Daily News* and United Press.

2. Donovan, *Tumultuous Years*, 259; Oshinsky, *A Conspiracy So Immense*, 193–96; James, *Years of MacArthur*, 3: 608.

3. Memorandum by Kenneth Hechler for John A. Carroll, Apr. 17, 1951, Elsey Papers.

4. Schedule for Welcoming of General MacArthur, file 74, Elsey Papers.

5. Speech of Apr. 19, 1951, *MSFE*, 3553ff; MacArthur, *Reminiscences*, 400–405; James, *Years of MacArthur*, 3: 616–18.

6. James, *Years of MacArthur*, 3: 621.

7. Doris Fleeson column, May 18, 1951, *Washington Star;* "Battle Hymn of the Waldorf" in MacArthur Recall Clipping File, Records of the Democratic National Committee, Truman Library.

8. *New York Times*, Apr. 21 and 22, 1951.

9. John E. Wiltz, "The MacArthur Hearings of 1951: The Secret Testimony," *Military Affairs* (Dec. 1975), 167–72. This provides an excellent analysis of how the censored material related to that made public in 1951. For the complete record, see Hearings Held by the Committee on Armed Services and the Committee on Foreign Relations: Inquiry Into the Military Situation in the Far East and the Facts Surrounding the Relief of General of the Army Douglas MacArthur from his Assignments in that Area (Declassified Hearings Transcripts), Records of the Senate Foreign Relations Committee, RG 46, Records of the U.S. Senate, National Archives. Hereafter, citations are to the Declassified Hearings Transcripts. The uncensored account has also been commercially microfilmed.

10. "Memorandum on Questions for MacArthur Hearings," n.d., in Full Committee Business for Far Eastern Policy Hearings, U.S. Senate, 1949–61, box 341, Johnson Papers. George Reedy Oral History, Johnson Library; Gerald W. Siegel Oral History, ibid.

11. MacArthur's testimony of May 3–5 is contained in the first 300 pages of

MSFE and the Declassified Hearings Transcripts; *MSFE*, 76; Some General Observations on the Hearings, undated memorandum, U.S. Senate, 1949–61, box 341, Johnson Papers.

12. Wiltz, "The MacArthur Hearings of 1951"; for Acheson's testimony, see *MSFE*, 1837–85, 2338–42, 2827–56; Acheson, *Present at the Creation*, 525–26.

13. See testimony in Wiltz, "The MacArthur Hearing of 1951."

14. Ibid.

15. Ibid.; *MSFE*, 730–44.

16. *MSFE*, 3567–3605; James, *Years of MacArthur*, 3: 631–39.

17. Sulzberger, *A Long Row of Candles*, 672, 685.

18. James, *Years of MacArthur*, 3: 648–51; before the convention, rumors of a MacArthur–Taft "Stop Ike" coalition circulated. Nothing came of it, however.

19. Henry Wallace to Douglas MacArthur, Jan. 29, 1955, and Feb. 11, 1956, and MacArthur to Wallace, Feb. 23, 1956, reel 50, Henry A. Wallace Papers, University of Iowa; C. L. Sulzberger, *The Last of the Giants* (New York, 1970), 757; Kenneth P. O'Donnell, "LBJ and the Kennedy's," *Life* 69 (August 7, 1970), 51; Kenneth P. O'Donnell and David F. Powers, with Joe McCarthy, *"Johnny We Hardly Knew Ye": Memories of John Fitzgerald Kennedy* (New York, 1973), 13–14, 320; James, *Years of MacArthur*, 3: 687.

20. *New York Times*, July 16, 1987. The Chinese also objected to a portrait of Golda Meir, the Israeli premier who had once lived (like MacArthur) in Milwaukee.

BIBLIOGRAPHY

The most comprehensive list of published materials on MacArthur appears in D. Clayton James's *The Years of MacArthur.* See below.

Official Records

All Record Groups (RG) are located in the National Archives or Washington National Records Center (WNRC). Documents obtained through use of the Freedom of Information Act are noted as FOIA.

Allied Council for Japan, RG 43

Bureau of the Budget, RG 51

Department of Defense
    ABC File, RG 165
    Records of the Army Staff, RG 319
    Gen. Omar Bradley File, RG 218
    Adm. William Leahy File, RG 218
    Army Civil Affairs Division, RG 165
    Draper/Voorhees Project Decimal File, RG 335
    Far Eastern Command, RG 332, WNRC
    Joint Chiefs of Staff, RG 218
    Army Plans and Operations Division, RG 319
    Office of the Secretary of the Army and Under Secretary, RG 335
    Department of the Army, RG 336
    Office of the Secretary of Defense, RG 330
    Secretary of War Files (including Henry L. Stimson and John J. McCloy files), RG 107
    Supreme Commander for the Allied Powers, RG 331, WNRC

Economic Cooperation Administration, RG 286, WNRC

Far Eastern Commission, RG 43

National Security Council, Modern Military Branch, National Archives and FOIA

Department of State (DOS)

Decimal Files, RG 59
Tokyo Post Files, RG 84, WNRC
Policy Planning Staff, RG 59 and FOIA
U.S. Mission on Reparations, RG 59
Executive Secretariat, RG 59
Assistant Secretary for Occupied Areas, RG 59
Philippine/Southeast Asia Lot File, RG 59
Chinese Affairs Lot File, RG 59
Office of Public Opinion Studies, RG 59
Research and Analysis Branch, OSS and Bureau of Intelligence and Research, RG 59

Department of the Treasury
Office of the Assistant Secretary for International Affairs, FOIA

State-War-Navy Coordinating Committee, RG 353

U.S. Senate, Records of the Senate Foreign Relations Committee, Hearings Held by the Committee on Armed Services and the Committee on Foreign Relations: Inquiry into the Military Situation in the Far East and the Facts Surrounding the Relief of General of the Army Douglas MacArthur from his Assignments in that Area, RG 46 (uncensored transcript)

British Foreign Office Records, FO 371, Public Record Office, London

Manuscript Collections

University of California, Berkeley
William F. Knowland Papers

Center for Military History
History of the Nonmilitary Aspects of the Occupation of Japan, Unpublished Official Studies by SCAP

Center for Naval History
Admiral Forrest Sherman Papers

Cineworld Transcripts, MacArthur Memorial Bureau of Archives
Faubion Bowers Oral History
Charles Kades Oral History

Clemson University
James F. Byrnes Papers

Columbia University
Wellington Koo Papers
Occupied Japan Oral History Project:
Faubion Bowers
Hugh Borton
Eugene Doomon
Burton Crane
Charles Kades
Kenneth C. Royall
H. Alexander Smith

Detroit Public Library

Joseph Dodge Papers

Duke University
Robert Eichelberger Papers

Dwight D. Eisenhower Presidential Library
Dwight D. Eisenhower Papers

Harvard University
Joseph C. Grew Papers

Herbert Hoover Presidential Library
Herbert Hoover Papers
Hanford MacNider Papers
Walter Robertson Oral History
Robert Wood Papers

Hoover Institution, Stanford University
Charles M. Cooke Papers
Bonner Fellers Papers
George Van Horne Moseley Papers

University of Iowa
Henry A. Wallace Papers (microfilm edition)

Lyndon B. Johnson Presidential Library, University of Texas, Austin
Lyndon B. Johnson Papers
Drew Pearson Papers
George Reedy Oral History
Gerald W. Siegel Oral History

Library of Congress
Joseph and Stewart Alsop Papers
Henry H. Arnold Papers
Claire Chennault Papers
Philip C. Jessup Papers
Harold L. Ickes Papers
William Leahy Papers
John C. O'Laughlin Papers
George Van Horne Moseley Papers
Francis Sayre Papers

MacArthur Memorial Bureau of Archives
Douglas MacArthur Papers
Richard K. Sutherland Papers
Gerald Wilkinson Papers (copy)
Charles A. Willoughby Papers
Faubion Bowers Oral History

George C. Marshall Research Library, Lexington, Virginia
W. Walton Butterworth Papers

Modern Military Records Branch, National Archives
Richard J. Sutherland Papers

University of Maryland
Justin Williams Papers

Princeton University

John Foster Dulles Papers
James Forrestal Papers
George F. Kennan Papers
Karl Lott Rankin Papers
H. Alexander Smith Papers
Whiting Willauer Papers

Franklin D. Roosevelt Presidential Library
Steve Early Papers
Harry Hopkins Papers
Henry Morgenthau Papers and Diaries
Franklin D. Roosevelt Papers

Harry S. Truman Presidential Library
Dean Acheson Papers
Stanley Andrews Papers
Eben Ayers Papers
Matthew Connelly Papers
William H. Draper Oral History
George M. Elsey Papers
John F. Melby Papers
Frank Pace Oral History
Harold Smith Papers
John Sumner Papers
Harry S. Truman Papers
James Webb Papers

United States Military History Institute, Carlisle Barracks
Edward M. Almond Oral History
Matthew B. Ridgway Papers

United States Naval Academy, Nimitz Library, Annapolis
William Sebald Papers

University of Texas, Austin, Humanities Research Center
Morris Ernst Papers

Yale University
Dean Acheson Papers
Henry L. Stimson Papers

Published Government Documents

Department of State. *Foreign Relations of the United States.* Annual volumes for 1941–51, Washington, D.C.

Department of Defense: *The Entry of the Soviet Union into the War Against Japan: Military Plans, 1941–51.* Washington, D.C., 1955.

Supreme Commander for the Allied Powers. *The Political Reorientation of Japan, September 1945–September 1948. Report of the Government Section of SCAP.* Washington, D.C., 1948.

U.S. Congress, House of Representatives, Committee on International Relations. *Selected Executive Session Hearings of the Committee, 1943–50, Vol. 5, Military Assistance Programs, pt. 1, Mutual Defense Assistance Act of 1949.* Washington, D.C., 1976.

U.S. Congress, House of Representatives, Committee on International Relations. *Selected Executive Session Hearings of the Committee, 1943–50, U.S. Policy in the Far East, pt. 1.* Washington, D.C., 1976.

U.S. Senate, Committee on Armed Services and Committee on Foreign Relations, *Hearings to Conduct an Inquiry into the Military Situation in the Far East and the Facts Surrounding the Relief of General of the Army Douglas MacArthur From His Assignments in that Area.* 82d Cong., 1st sess., Washington, D.C., 1951.

U.S. Senate, Committee on Foreign Relations. *Reviews of the World Situation: 1949–50, Hearings Held in Executive Session on the World Situation.* 81st Cong., 1st and 2d sess., Washington, D.C., 1974.

Books

Abell, Tyler, ed. *Drew Pearson Diaries.* New York, 1974.

Acheson, Dean. *Present at the Creation: My Years in the State Department.* New York, 1969.

Allison, John. *Ambassador From the Prairie; or Allison Wonderland.* Boston, 1973.

Ambrose, Stephen. *Eisenhower: Soldier, General of the Army, President Elect, 1890–1952.* New York, 1983.

Anders, Roger M., ed. *Forging the Atomic Shield: Excerpts from the Office Diary of Gordon E. Dean.* Chapel Hill, N.C., 1987.

Anderson, Irvine H. *The Standard Vacuum Oil Company and United States East Asian Policy, 1933–41.* Princeton, N.J., 1975.

Appleman, Roy E. *South to the Naktong, North to the Yalu.* Washington, D.C., 1961.

Arnold, H. H. *Global Mission.* New York, 1949.

Asahi Shimbun. *Pacific Rivals.* New York, 1972.

Ball, William MacMahon. *Japan: Enemy or Ally?* New York, 1949.

Barbey, Daniel E. *MacArthur's Amphibious Navy: Seventh Amphibious Force Operations, 1943–49.* Annapolis, Md., 1969.

Bisson, Thomas A. *Zaibatsu Dissolution in Japan.* Berkeley, Calif., 1954.

Blair, Clay. *The Forgotten War.* New York, 1987.

Blum, John Morton. *From the Morgenthau Diaries: Years of Urgency, 1938–41.* Boston, 1965.

———, ed. *The Price of Vision: The Diary of Henry A. Wallace, 1942–46.* Boston 1973.

Blum, Robert M. *Drawing the Line: The Origins of the American Containment Policy in East Asia.* New York, 1982.

Borden, William S. *The Pacific Alliance: United States Foreign Economic Policy and Japanese Trade Recovery, 1947–55.* Madison, Wis., 1984.

Borg, Dorothy. *The United States and the Far Eastern Crisis of 1933–38.* Cambridge, Mass., 1964.

———, and Heinrichs, Waldo, eds. *The Uncertain Years: Chinese–American Relations, 1947–50.* New York, 1980.

———, and Okamoto, Shumpai. *Pearl Harbor as History.* New York, 1973.

Borton, Hugh. *American Presurrender Planning for Postwar Japan.* New York, 1967.

Bradley, Omar N., and Blair, Clay. *A General's Life.* New York, 1983.

Brereton, Lewis H. *The Brereton Diaries*. New York, 1946.

Buckley, Roger. *Occupation Diplomacy: Britain, the United States and Japan, 1942– 49*. Cambridge, Engl., 1982.

Buhite, Russell D. *Soviet–American Relations, 1945–54*. Norman, Okla., 1981.

Burkman, Thomas, ed. *The Occupation of Japan: The International Context*. Norfolk, Va., 1984.

Butow, R. J. C. *Tojo and the Coming of the War*. Princeton, N.J., 1961.

———. *The John Doe Associates: Backdoor Diplomacy for Peace*. Stanford, Calif., 1974.

Cantril, Hadley, ed. *Public Opinion, 1935–46*. Princeton, N.J., 1951.

Caridi, Ronald J. *The Korean War and American Politics: The Republican Party as a Case Study*. Philadelphia, 1968.

Cary, Otis, ed. *War Wasted Asia: Letters, 1945–46*. Tokyo, 1975.

Catledge, Turner. *My Life and the Times*. New York, 1971.

Cohen, Theodore. *Remaking Japan: The American Occupation as New Deal*, ed. by Herbert Passin. New York, 1987.

Cohen, Warren. *Dean Rusk*. Totawa, N.J., 1980.

———, ed. *New Frontiers in American–East Asian Relations*. New York, 1983.

Collins, J. Lawton. *War in Peacetime: The History and Lessons of Korea*. Boston, 1969.

Coughlin, William J. *Conquered Press: The McArthur Era in Japanese Journalism*. Palo Alto, Calif., 1952.

Cumings, Bruce, ed. *Child of Conflict: The Korean–American Relationship, 1943– 50*. Seattle, 1983.

———. *The Origins of the Korean War: Liberation and the Emergence of Separate Regimes, 1945–47*. Princeton, N.J., 1981.

Dallek, Robert. *Franklin D. Roosevelt and American Foreign Policy, 1932–1945*. New York, 1978.

Dobbs, Charles M. *The Unwanted Symbol: American Foreign Policy, the Cold War and Korea, 1945–50*. Kent, Ohio, 1981.

Doenecke, Justus. *Not to the Swift: The Old Isolationists in the Cold War Era*. Lewisburg, Pa., 1970.

Donovan, Robert J. *Tumultuous Years: The Presidency of Harry S. Truman, 1949– 53*. New York, 1982.

Dorn, Frank. *Walkout With Stilwell in Burma*. New York, 1971.

Dower, John W. *A War Without Mercy*. New York, 1986.

———. *Empire and Aftermath: Yoshida Shigeru and the Japanese Experience, 1878– 1954*. Cambridge, Mass., 1979.

Dugger, Ronnie. *The Politician: The Life and Times of Lyndon Johnson*. New York, 1982.

Eisenhower, Dwight D. *At Ease*. New York, 1967.

Farley, James. *Jim Farley's Story: The Roosevelt Years*. New York, 1948.

Farley, Miriam. *Aspects of Japan's Labor Problems*. New York, 1950.

Feis, Herbert. *Contest Over Japan*. New York, 1967.

Ferrell, Robert, ed. *The Eisenhower Diaries*. New York, 1981.

———, ed. *Off the Record: The Private Papers of Harry S. Truman*. New York, 1980.

Fine, Sidney. *Frank Murphy: The New Deal Years*. Chicago, 1979.

———. *Frank Murphy: The Washington Years*. Ann Arbor, Mich., 1984.

Foot, Rosemary. *The Wrong War: American Policy and the Dimensions of the Korean Conflict, 1950–53*. Ithaca, N.Y., 1985.

Friend, Theodore. *Between Two Empires: The Ordeal of the Philippines, 1929–46*. New Haven, Conn., 1965.

Gaddis, John L. *Strategies of Containment*. New York, 1982.

Gayn, Mark. *Japan Diary*. New York, 1948.

Goulden, Joseph C. *Korea: The Untold Story of the War*. New York, 1982.

Gunther, John. *The Riddle of MacArthur*. New York, 1950.

Hadley, Eleanor. *Anti-Trust in Japan*. Princeton, N.J., 1974.

Harriman, W. Averell, and Abel, Ellie. *Special Envoy to Churchill and Stalin, 1941–1946*. New York, 1975.

Hassett, William D. *Off the Record with FDR*. New Brunswick, N.J., 1958.

Hastings, Max. *The Korean War*. New York, 1987.

Hayes, Samuel P. *The Beginning of American Aid to Southeast Asia: The Griffin Mission of 1950*. Lexington, Mass., 1971.

Heinrichs, Waldo. *American Ambassador: Joseph C. Grew and the Development of the United States Diplomatic Tradition*. Boston, 1966.

Herken, Gregg. *The Winning Weapon: The Atomic Bomb and the Cold War, 1945–50*. New York, 1982.

Hersey, John. *Men on Bataan*. New York, 1943.

Hess, Gary. *The United States' Emergence as a Southeast Asian Power, 1940–50*. New York, 1987.

Higgins, Trumbull. *Korea and the Fall of MacArthur: A Precis in Limited War*. New York, 1960.

Huff, Sid. *My Fifteen Years with General MacArthur*. New York, 1964.

Hunt, Frazier. *The Untold Story of General MacArthur*. New York, 1954.

Ickes, Harold L. *The Secret Diary of Harold Ickes*. 3 vols. New York, 1953–54.

Iriye, Akira. *Power and Culture: The Japanese–American War, 1941–45*. Cambridge, Mass., 1981.

James, D. Clayton. *The Years of MacArthur*. 3 vols. Boston, 1970–85.

Johnson, Chalmers. *Peasant Nationalism and Communist Power: The Emergence of Revolutionary China, 1937–45*. Stanford, 1963.

Kaufman, Burton I. *The Korean War: Challenges in Crisis, Credibility and Command*. New York, 1986.

Kennan, George F. *Memoirs, 1925–50*. Boston, 1967.

Kerkvliet, Benedict John. *The Huk Rebellion: A Study of Peasant Revolt in the Philippines*. Berkeley, 1977.

Khrushchev, Nikita. *Khrushchev Remembers*. Boston, 1970.

Larrabee, Eric. *Commander in Chief: Franklin D. Roosevelt, His Lieutenants, and Their War*. New York, 1987.

Leahy, William D. *I Was There*. New York, 1950.

Leary, William M. *Perilous Missions: Civil Air Transport and CIA Covert Operations in Asia*. University of Alabama Press, 1984.

Leary, William M., ed. *We Shall Return: MacArthur's Commanders and the Defeat of Japan* Lexington, Ky., 1988.

Lebra, Joyce. *Japanese Trained Armies in Southeast Asia*. Hong Kong, 1987.

Lee, Clark, and Henschel, Richard. *Douglas MacArthur*. New York, 1952.

Leutze, James. *A Different Kind of Victory: A Biography of Admiral Thomas C. Hart*. Annapolis, Md., 1981.

Levine, Steven I. *The Anvil of Victory*. New York, 1987.

Lewin, Ronald. *The American Magic*. New York, 1982.

Livingston, Jon, et al., eds. *Postwar Japan: 1945 to the Present*. New York, 1973.

Long, Gavin. *MacArthur As Military Commander*. London, 1969.

Louis, William Roger. *Imperialism at Bay*. New York, 1977.

Lowe, Peter. *The Origins of the Korean War*. New York, 1986.

Luvaas, Jay, ed. *Dear Miss Em*. Westport, Conn., 1972.

MacArthur, Douglas. *Reminiscences* (New York, 1964).

MacDonald, Callum A. *Korea: The War Before Vietnam*. London, 1986.

MacKinnon, Jan, and MacKinnon, Steven. *Agnes Smedley*. Berkeley, Calif., 1987.

Manchester, William. *American Caesar*. Boston, 1978.

Matray, James. *The Reluctant Crusade: American Foreign Policy in Korea, 1941–50*. Honolulu, 1986.

McMahon, Robert J. *Colonialism and Cold War: The United States and the Struggle for Indonesian Independence, 1945–49*. Ithaca, N.Y., 1981.

Messer, Robert. *The End of an Alliance: James F. Byrnes, Roosevelt, Truman and the Origins of the Cold War*. Chapel Hill, N.C., 1982.

Millis, Walter, ed. *The Forrestal Diaries*. New York, 1951.

Minear, Richard H. *Victor's Justice: The Tokyo War Crimes Trials*. Princeton, N.J., 1971.

Montgomery, John D. *Forced to be Free: The Artificial Revolution in Germany and Japan*. Chicago, 1957.

Moore, Joe. *Japanese Workers and the Struggle for Power*. Madison, Wisc., 1983.

Morton, Louis. *The Fall of the Philippines*. Washington, D.C., 1953.

Morton, Louis. *Strategy and Command: The First Two Years*. Washington, D.C., 1962.

Nagai, Yonosuke, and Iriye, Akira. *The Origins of the Cold War in Asia*. New York, 1977.

O'Donnell, Kenneth P., and Powers, David F., with Joe McCarthy. *"Johnny We Hardly Knew Ye": Memories of John Fitzgerald Kennedy*. New York, 1973.

Oshinsky, David M. *A Conspiracy So Immense—The World of Joe McCarthy*. New York, 1983.

Overseas Consultants, Inc. *Report on Industrial Reparations Survey of Japan to the United States of America*. New York, 1948.

Paige, Glenn D. *The Korean Decision*. New York, 1968.

Pelz, Stephen E. *Race to Pearl Harbor*. Cambridge, Mass., 1974.

Perry, John C. *Beneath the Eagle's Wing*. Boston, 1981.

Petillo, Carol M. *Douglas MacArthur: The Philippine Years*. Bloomington, Ind., 1981.

Pilat, Oliver, *Drew Pearson: An Unauthorized Biography*. New York, 1973.

Pogue, Forrest C. *George C. Marshall*. 4 vols. New York, 1964–87.

Poole, Walter S. *The History of the Joint Chiefs of Staff: The Joint Chiefs and National Policy, 1945–53, Vol. IV, 1950–52*. 4 vols., reprinted in 5 vols. Wilmington, Del., 1979.

Prange, Gordon. *At Dawn We Slept*. New York, 1981.

———. *Miracle at Midway*. New York, 1983.

———. *Pearl Harbor: The Verdict of History*. New York, 1986.

———. *Target Tokyo: The Story of the Sorge Spy Ring*. New York, 1984.

Quezon, Manuel L. *The Good Fight*. New York, 1946.

Reeves, Thomas C. *The Life and Times of Joe McCarthy*. New York, 1982.

Rees, David. *Korea: The Limited War*. New York, 1964.

Rhodes, Weldon E. *Flying MacArthur to Victory*. College Station, Tex., 1987.

Ridgway, Matthew B. *Soldier*. New York, 1956.

———. *The Korean War*. New York, 1967.

Roper, Elmo B. *You and Your Leaders: Their Actions and Your Reactions: 1936–56*. New York, 1957.

Rosenman, Samuel I. *Working With Roosevelt*. New York, 1952.

Rotter, Andrew J. *The Path to Vietnam: The Origins of the American Commitment to Southeast Asia*. Ithaca, N.Y., 1987.

Rovere, Richard H., and Schlesinger, Arthur, Jr. *The MacArthur Controversy and American Foreign Policy*. New York, 1965.

Sayre, Francis B. *Glad Adventure*. New York, 1957.

Schaffer, Ronald. *Wings of Judgement: American Bombing in World War II*. New York, 1985.

Schaller, Michael. *The U.S. Crusade in China, 1938–45*. New York, 1979.

———. *The American Occupation of Japan: The Origins of the Cold War in Asia*. New York, 1985.

Schnabel, James E., and Watson, Robert J. *The History of the Joint Chiefs of Staff: The Joint Chiefs and National Policy, 1945–53, Vol. III, The Korean War, pt. 1*. 4 vols., reprinted in 5 vols. Wilmington, Del., 1979.

Schonberger, Howard. *Aftermath of War: Americans and the Remaking of Japan*. Kent, Ohio, 1988.

Sebald, William. *With MacArthur in Japan*. New York, 1965.

Shadegg, Stephen. *Clare Boothe Luce: A Biography*. New York, 1970.

Sherwin, Martin. *A World Destroyed: The Atomic Bomb and the Grand Alliance*. New York, 1977.

Sherry, Michael. *The Rise of American Air Power*. New Haven, 1987.

Sherwood, Robert. *Roosevelt and Hopkins*. 2 vols. New York, 1950.

Simmons, Robert. *The Strained Alliance*. New York, 1975.

Smith, Richard Norton. *Thomas E. Dewey and His Times*. New York, 1982.

Smith, Robert. *MacArthur in Korea: The Naked Emperor*. New York, 1982.

Spanier, John W. *The Truman–MacArthur Controversy and the Korean War*. Cambridge, Mass., 1959.

Spector, Ronald. *Eagle Against the Sun*. New York, 1985.

Steinberg, David J. *Philippine Collaboration in World War II*. Ann Arbor, Mich., 1967.

Stewart, John R. *Notes on the Economic Aspect of the Allied Occupation of Japan*. New York, 1947.

Stimson, Henry, and Bundy, McGeorge. *On Active Service in Peace and War*. New York, 1947.

Stueck, William W. *The Road to Confrontation: American Policy Toward China and Korea, 1947–50*. Chapel Hill, N.C., 1981.

Sulzberger, C. L. *The Last of the Giants*. New York, 1970.

———. *A Long Row of Candles*. Toronto, 1969.

Thomas Burkman, ed. *Proceedings of a Symposium by the MacArthur Memorial*. Norfolk, Va., 1978.

Thorne, Christopher. *Allies of a Kind: The United States, Great Britain and the War Against Japan, 1941–45*. New York, 1978.

———. *The Issue of War*. New York, 1985.

Truman, Margaret. *Harry S. Truman*. New York, 1973.

Truman, Harry S. *Memoirs.* 2 vols. Garden City, N.Y., 1955–56.

Tucker, Nancy. *Patterns in the Dust: Chinese–American Relations and the Recognition Controversy, 1949–50.* New York, 1983.

Tugwell, Rexford G. *The Democratic Roosevelt.* Garden City, N.Y., 1957.

Vandenberg, Arthur H., Jr., ed. *The Private Papers of Senator Vandenberg.* Boston, 1952.

Watson, Mark S. *Chief of Staff: Prewar Plan and Preparations.* Washington, D.C., 1950.

White, Theodore, *In Search of History.* New York, 1978.

Whiting, Allen S. *China Crosses the Yalu: The Decision to Enter the Korean War.* Stanford, Calif., 1960.

Whitney, Courtney. *MacArthur: His Rendezvous With History.* New York, 1955.

Wildes, Harry Emerson. *Typhoon in Tokyo: The Occupation and its Aftermath.* New York, 1954.

Williams, Peter and Wallace, David. *Unit 731: The Japanese Army's Secret of Secrets.* London, 1988.

Willoughby, Charles A., and Chamberlain, John. *MacArthur, 1941–51.* New York, 1954.

Wolf, Robert, ed. *Americans as Proconsuls: The United States Military Government in Germany and Japan.* Carbondale, Ill., 1984.

Woodard, William P. *The Allied Occupation of Japan, 1945–52, and Japanese Religions.* Leiden, Netherlands, 1972.

Yergin, Daniel. *Shattered Peace: The Origins of the Cold War and the National Security State.* Boston, 1977.

Articles

Anders, Roger M. "The Atomic Bomb and the Korean War: Gordon Dean and the Issue of Civilian Control." *Military Affairs* 52 (Jan. 1988): 1–6.

Bernstein, Barton J. "New Light on the Korean War." *The International History Review* 3 (Apr. 1981): 256–77.

———. "The Week We Went to War: American Intervention in the Korean Civil War." *Foreign Service Journal* (January 1977): 6–9, 33–35, *ibid.,* (Feb. 1977): 8–11, 33–35.

Edgerton, Ronald K. "General Douglas MacArthur and the American Military Impact in the Philippines." *Philippine Studies* 25 (1977): 420–40.

Edwards, Corwin. "The Dissolution of Japanese Combines." *Pacific Affairs* 19 (Sept. 1946): 227–40.

Foot, Rosemary. "Anglo-American Relations in the Korean Crisis: The British Effort to Avert an Expanded War, December 1950–January 1951." *Diplomatic History* 10 (Winter 1986): 43–57.

Leary, William M., Jr., "Aircraft and Anti-Communists: CAT in Action, 1949–52." *China Quarterly* (October–December 1972): 654–69.

Levine, Steven I. "A New Look at American Mediation in the Chinese Civil War: The Marshall Mission in Manchuria." *Diplomatic History* 3 (Fall 1979): 349–75.

McCarten, John. "General MacArthur: Fact and Legend." *The American Mercury* 58 (Jan. 1944): 7–18.

Morton, Louis. "War Plan ORANGE: Evolution of a Strategy." *World Politics* 2 (1959), 221–50.

O'Donnell, Kenneth P. "LBJ and the Kennedys." *Life* 69 (Aug. 7, 1970): 51.

Petillo, Carol M. "Douglas MacArthur and Manuel Quezon: A Note on an Imperial Bond." *Pacific Historical Review* 48 (Feb. 1979): 107–17.

Schonberger, Howard. "American Labor's Cold War in Occupied Japan." *Diplomatic History* 3 (Summer 1979): 249–72.

———. "The General and the Presidency: Douglas MacArthur and the Election of 1948." *Wisconsin Magazine of History* 57 (Spring 1974): 201–19.

———. "The Japan Lobby in American Politics" *Pacific Historical Review* 46 (Aug. 1977): 327–59.

Thorne, Christopher. "MacArthur, Australia and the Bristish, 1942–43: The Secret Journal of MacArthur's British Liaison Officer, Part 1." *Australian Outlook* 29 (April 1975): 53–67. "Part 2." *Australian Outlook* 29 (August 1975): 197–210.

Tucker, Nancy B. "American Policy Toward Sino-Japanese Trade in the Postwar Years: Politics and Prosperity." *Diplomatic History* 8 (Summer 1984): 183–208.

Wiltz, John E. "The MacArthur Hearings of 1951: The Secret Testimony." *Military Affairs* (Dec. 1975): 167–72.

———. "Truman and MacArthur: The Wake Island Meeting." *Military Affairs* 42 (Dec. 1978): 168–75.

Wittner, Lawrence S. "MacArthur and the Missionaries: God and Man in Occupied Japan." *Pacific Historical Review* 40 (Feb. 1971): 77–97.

Unpublished Ph.D. Dissertation

Mattern, Carolyn J. "The Man on the Dark Horse: The Presidential Campaigns for General Douglas MacArthur, 1944 & 1948." University of Wisconsin, 1976.

INDEX

Quality Printing and Binding by:
Orange Graphics
P.O. Box 791
Orange, VA 22960 U.S.A.